Polymer Micelles

Special Issue Editors

Shin-ichi Yusa
Pratap Bahadur
Hideki Matsuoka
Takahiro Sato

MDPI • Basel • Beijing • Wuhan • Barcelona • Belgrade

MDPI

Special Issue Editors
Shin-ichi Yusa
University of Hyogo
Japan

Pratap Bahadur
Veer Narmad South Gujarat University
India

Hideki Matsuoka
Kyoto University
Japan

Takahiro Sato
Osaka University
Japan

Editorial Office
MDPI AG
St. Alban-Anlage 66
Basel, Switzerland

This edition is a reprint of the Special Issue published online in the open access journal *Polymers* (ISSN 2073-4360) from 2017–2018 (available at: http://www.mdpi.com/journal/polymers/special_issues/polymer_micelles).

For citation purposes, cite each article independently as indicated on the article page online and as indicated below:

Lastname, F.M.; Lastname, F.M. Article title. *Journal Name* **Year**, *Article number, page range.*

First Edition 2018

ISBN 978-3-03842-807-7 (Pbk)
ISBN 978-3-03842-808-4 (PDF)

Cover photo courtesy of Shin-ichi Yusa

Table of Contents

About the Special Issue Editors

Shin-ichi Yusa is a native of Japan and received B.S. (1993) and M.S. (1995) degrees in polymer chemistry from Osaka University under the direction of Prof. Mikiharu Kamachi and Prof. Yotaro Morishima. He received a Ph.D. from Osaka University (2000) for a thesis entitled: "Synthesis of Cholesterol Bearing Polymers and Their Self-Assembling Properties." He joined Himeji Institute of Technology as an assistant professor in 1997. He became associate professor of University of Hyogo (2008). His research interests are in controlled radical polymerization and characterization of water-soluble polymers. He is currently an Associate Editor of Polymers (MDPI Publishing) and on the Editorial Advisory Board of Langmuir (ACS). He is also on the editorial board of E-Journal of Chemistry (Hindawi Publishing), Open Journal of Polymer Chemistry (Scientific Research Publishing), and Open Chemistry (DE GRUYTER).

Pratap Bahadur (born 1951) PhD, DSc has been associated with Veer Narmad South Gujarat University Surat India since 16 June 1980 (Associate Professor 1980–1988 and as Professor 1988–until superannuation in 2014) and continues as an Emeritus Fellow of University Grants Commission. He has supervised 50 PhD students/published over 250 research articles and has widely travelled all across the globe as a visiting scientist/visiting professor in several countries in Europe, Asia and Americas. His research areas of interest are surface/colloid/polymer science.

Hideki Matsuoka received B.S. (1981) and M.S. (1983) degrees in polymer chemistry from Kyoto University under the direction of Prof. Norio Ise. He received a Ph.D. from Kyoto University (1987) for a thesis entitled: "Small-angle X-ray Scattering Studies on Ordered Structure in Polyelectrolyte Solutions". He became an assistant professor of Department of Polymer Chemistry, Kyoto University in 1983, and was promoted to Associate Professor in 1996. His research interest is the surface and interface chemistry of polymers, especially on amphiphilic polymers, polymer micelles, polymer monolayers and ionic polymers. He was the vice-president of the Division of Colloid and Interface Chemistry, the Chemical Society of Japan for 2013–2017, and is currently the council member of International Association of Colloid and Interface Scientists (IACIS).

Takahiro Sato is a Vice Dean of Graduate School of Science, Osaka University and also a Vice President of the Society of Polymer Science, Japan. He received his PhD from Osaka University in 1985 for his work on "Double-Stranded Helix of a Bacterial Polysaccharide Xanthan in Dilute Solution." In 1985–1986, he was a guest scientist at Polymers Division, National Bureau of Standards (now National Institute of Standards & Technology), USA, working with Dr. Charles C. Han. He became an assistant professor at Osaka University in 1987, and was promoted to a full professor at the same university in 2002. The main research interests of his group include dilute and concentrated solution properties of stiff polymers and helical polymers, self-association behavior of amphiphilic polymers, polyelectrolytes, and hydrogen-bond-forming polymers.

Preface to "Polymer Micelles"

We are pleased to present this Special Issue entitled "Polymer Micelles". Block and graft copolymers show microdomain formation in solid state and self-assemble to core–shell nanoaggregates in selective solvents. Classically, amphiphilic polymers form polymer micelles and vesicular structures in aqueous solution, primarily due to the hydrophobic interaction in analogy to surfactants. Practically, polymer micelles have been applied as nanocarriers in drug delivery systems, solubilizers, associative thickeners, and so on. In recent years, various other interactions such as electrostatic, hydrogen bonds, coordination bonds, and so on have been found to play a role in polymer self-assembly. Additionally, unimolecular micelles may form by intramolecular association within a single polymer chain. Micelles from non-surface active polymers, stimuli-responsive block polymers and from mixtures of oppositely charged copolymers have been of interest in recent years. Due to the advances in polymerization techniques leading to tailormade copolymers from a variety of monomers, characterization/solution behavior using a variety of modern instrumental techniques, theoretical approaches, and emerging areas of applications, polymer self-assembly has gained a great deal of interest in recent years and we need to constantly update the information and knowledge on polymer micelles.

This Special Issue, "Polymer Micelles" covers the synthesis, characterization, solution properties, association behavior, theory, and application of polymer micelles, as well as polymer aggregates. The aim of this Special Issue is to expand our knowledge of polymer micelles by accumulating the latest basic and applicable information. The Special Issue as well as the book contains 14 contributions in the form of two reviews and 12 original articles. Please enjoy reading this collection.

We would like to thank all the authors for their contribution to this issue and the editors of Polymers for their kind help and co-operation.

Shin-ichi Yusa, Pratap Bahadur, Hideki Matsuoka and Takahiro Sato
Special Issue Editors

![polymers logo] *polymers*

MDPI

Article

Preparation of Water-Soluble Polyion Complex (PIC) Micelles Covered with Amphoteric Random Copolymer Shells with Pendant Sulfonate and Quaternary Amino Groups

Rina Nakahata and Shin-ichi Yusa *

Department of Applied Chemistry, University of Hyogo, 2167 Shosha, Himeji, Hyogo 671-2280, Japan; nkht9999@gmail.com
* Correspondence: yusa@eng.u-hyogo.ac.jp; Tel.: +81-79-267-4954

Received: 28 January 2018; Accepted: 17 February 2018; Published: 19 February 2018

Abstract: An amphoteric random copolymer (P(SA)$_{91}$) composed of anionic sodium 2-acrylamido-2-methylpropanesulfonate (AMPS, S) and cationic 3-acrylamidopropyl trimethylammonium chloride (APTAC, A) was prepared via reversible addition-fragmentation chain transfer (RAFT) radical polymerization. The subscripts in the abbreviations indicate the degree of polymerization (DP). Furthermore, AMPS and APTAC were polymerized using a P(SA)$_{91}$ macro-chain transfer agent to prepare an anionic diblock copolymer (P(SA)$_{91}$S$_{67}$) and a cationic diblock copolymer (P(SA)$_{91}$A$_{88}$), respectively. The DP was estimated from quantitative ^{13}C NMR measurements. A stoichiometrically charge neutralized mixture of the aqueous P(SA)$_{91}$S$_{67}$ and P(SA)$_{91}$A$_{88}$ formed water-soluble polyion complex (PIC) micelles comprising PIC cores and amphoteric random copolymer shells. The PIC micelles were in a dynamic equilibrium state between PIC micelles and charge neutralized small aggregates composed of a P(SA)$_{91}$S$_{67}$/P(SA)$_{91}$A$_{88}$ pair. Interactions between PIC micelles and fetal bovine serum (FBS) in phosphate buffered saline (PBS) were evaluated by changing the hydrodynamic radius (R_h) and light scattering intensity (LSI). Increases in R_h and LSI were not observed for the mixture of PIC micelles and FBS in PBS for one day. This observation suggests that there is no interaction between PIC micelles and proteins, because the PIC micelle surfaces were covered with amphoteric random copolymer shells. However, with increasing time, the diblock copolymer chains that were dissociated from PIC micelles interacted with proteins.

Keywords: amphoteric random copolymer; polyelectrolyte; polyion complex; block copolymer; polymer micelle; electrostatic interaction; protein antifouling

---

1. Introduction

A mixture of oppositely charged polyelectrolytes in water forms a water-insoluble polyion complex (PIC) due to attractive electrostatic interactions between the polymer chains [1]. Many researchers study polymer aggregates formed by electrostatic interactions. When oppositely charged diblock copolymers containing nonionic water-soluble poly(ethylene glycol) (PEG) and polyelectrolyte blocks are mixed in water, the polymers spontaneously form water-soluble PIC micelles covered with hydrophilic PEG shells [2,3]. PIC micelles can encapsulate charged compounds such as metal ions, proteins, and nucleic acid in the PIC core via electrostatic interactions [4–11]. Recently, water-soluble PIC micelles were prepared with hydrophilic poly(2-methacryloyloxyethyl phosphorylcholine) (PMPC) coronas without net charge [12–14]. PMPC has pendant phosphorylcholine groups with the same chemical structure as the hydrophilic part of phospholipids comprising cell membranes. The phosphorylcholine group comprises an anionic phosphonium anion and cationic quaternary amino groups. The charges in PMPC are neutralized within a single polymer chain. PMPC is a betaine polymer with good biocompatibility and antithrombogenicity [15–17].

Protein fouling on the surfaces of medical devices due to hydrophobic, electrostatic, and hydrogen bonding interactions causes deterioration of the functionality. Therefore, much attention has been given to the surface modification of medical devices using polymer coatings with antifouling properties. In general, protein antifouling polymers can bind water molecules strongly and are electrically neutral. When an antifouling polymer is coated on a substrate, proteins have minimal contact with the polymer on the substrate due to the presence of water molecules between the proteins and the polymers [18]. In particular, zwitterionic polymers can suppress protein adsorption, because they contain many bound water molecules [19]. Among these, betaine polymers effectively suppress protein adsorption [20–22]. Nanoparticles that are surface-modified with betaine polymers have an increased circulation time in the body compared to bare nanoparticles [23]. Shih et al. reported that amphoteric random copolymers with pendant anionic sulfonate and cationic quaternary amino groups inhibit protein adsorption for stoichiometrically charge neutralized compositions [24]. It is expected that amphoteric random copolymers can be applied to the surface modification of medical devices.

In the present study, we prepared an amphoteric random copolymer ($P(SA)_{91}$) macro chain transfer agent (CTA) composed of the same amounts of anionic sodium 2-acrylamido-2-methylpropanesulfonate (AMPS, S) and cationic 3-acrylamidopropyl trimethylammonium chloride (APTAC, A) via reversible addition-fragmentation chain transfer (RAFT) radical polymerization [25]. AMPS and APTAC were polymerized using $P(SA)_{91}$ macro-CTA to prepare an anionic diblock copolymer ($P(SA)_{91}S_{67}$) and a cationic diblock copolymer ($P(SA)_{91}A_{88}$), respectively (Figure 1). The subscripts in the abbreviations indicate the degree of polymerization (DP). Water-soluble PIC micelles were prepared by stoichiometrically charge neutralized mixing of the oppositely charged diblock copolymers, $P(SA)_{91}S_{67}$ and $P(SA)_{91}A_{88}$, in water. To the best of our knowledge, this is the first report on such micelles. The PIC micelles inhibit interactions with proteins, because their surface is covered with amphoteric random $P(SA)_{91}$ copolymer shells.

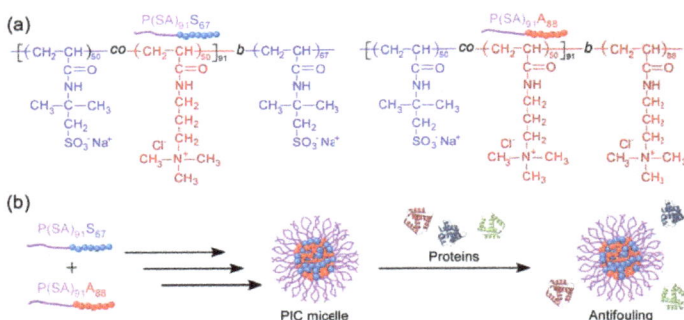

Figure 1. (**a**) Chemical structures of $P(SA)_{91}S_{67}$ and $P(SA)_{91}A_{88}$. (**b**) Schematic representation of water-soluble polyion complex (PIC) micelle formed from a mixture of $P(SA)_{91}S_{67}$ and $P(SA)_{91}A_{88}$; the PIC micelle shows protein antifouling properties because of its amphoteric random copolymer shell.

2. Materials and Methods

2.1. Materials

2-Acrylamido-2-methylpropanesulfonic acid (AMPS, 98%) from Tokyo Chemical Industry (Tokyo, Japan) and 4,4′-azobis(4-cyanopentanoic acid) (V-501, 98%) from Wako Pure Chemical

(Osaka, Japan) were used as received without further purification, and 3-acrylamidopropyl trimethylammonium chloride (APTAC, 75 wt % in water) from Tokyo Chemical Industry was passed through an Aldrich (St Louis, MO, USA) disposable inhibitor removal column. Methanol was dried with molecular sieves 3A and purified by distillation. Phosphate buffered saline (PBS) was prepared by dissolving one PBS tablet (Aldrich) in predetermined amounts of water, while 4-cyanopentanoic acid dithiobenzoate (CPD) was synthesized according to a reported method [26]. Bovine serum albumin (BSA, pH 5.0–5.6 solution) from Wako Pure Chemical and fetal bovine serum (FBS) from GE Healthcare Life Sciences HyClone were used without further purification. Water was purified using an ion-exchange column. Other reagents were used as received.

2.2. Preparation of P(SA)$_{91}$

AMPS (10.0 g, 48.3 mmol) was neutralized with 6 M aqueous NaOH (10.9 mL) to adjust the pH to 6.0. APTAC (9.98 g, 48.3 mmol), V-501 (135 mg, 0.482 mmol), and CPD (270 mg, 0.965 mmol) were dissolved in a mixed solvent of MeOH (5 mL) and water (32 mL), which was added to the aqueous AMPS. The solution was degassed by purging with Ar for 30 min. Polymerization was carried out at 70 °C for 24 h. After the reaction, the total conversion of AMPS and APTAC estimated from ^{1}H NMR was 90.9%. The reaction mixture was dialyzed against 1.5 M aqueous NaCl for one day, and then pure water for one day using a dialysis membrane with a molecular weight cutoff of 14 kDa (EIDIA Co. Ltd, Tokyo, Japan). The polymer (P(SA)$_{91}$) was recovered by a freeze-drying technique (15.3 g, 71.8%). The theoretical degree of polymerization (DP(theory)) and number-average molecular weight (M_n(theory)) were 91 and 1.91×10^4 g/mol, respectively, estimated from the conversion and molar ratio of monomer to CPD (Formulars 1 and 2). The number-average molecular weight (M_n(GPC)) and molecular weight distribution (M_w/M_n) were 1.54×10^4 g/mol and 1.27, respectively, estimated from gel-permeation chromatography (GPC). The APTAC content was 50 mol%, estimated from quantitative ^{13}C NMR measurements. The synthesis route of P(SA)$_{91}$ is shown in Figure S1.

2.3. Preparation of P(SA)$_{91}$S$_{67}$

AMPS (771 mg, 3.72 mmol) was neutralized with 6 M aqueous NaOH (6.2 mL) to adjust the pH to 6.0. P(SA)$_{91}$ (703 mg, 3.68×10^{-2} mmol, M_n(theory) = 1.91×10^4 g/mol, M_w/M_n = 1.27) and V−501 (4.15 mg, 1.48×10^{-2} mmol) were added to the aqueous solution. The solution was degassed by purging with Ar for 30 min. Polymerization was carried out at 70 °C for 24 h (conversion = 90.9%). The reaction mixture was dialyzed against 1.5 M aqueous NaCl for one day, and then pure water for one day. The polymer (P(SA)$_{91}$S$_{67}$) was recovered by a freeze-drying technique (1.23 g, 67.6%). The degree of polymerization (DP(NMR)) of the PAMPS block was 67, estimated from quantitative ^{13}C NMR measurements. M_n(GPC) and M_w/M_n were 2.38×10^4 g/mol and 1.04, respectively, estimated from GPC. The synthesis route of P(SA)$_{91}$S$_{67}$ is shown in Figure S1.

2.4. Preparation of P(SA)$_{91}$A$_{88}$

APTAC (765 mg, 3.69 mmol), P(SA)$_{91}$ (703 mg, 3.68×10^{-2} mmol, M_n(theory) = 1.91×10^4 g/mol, M_w/M_n = 1.27), and V-501 (4.14 mg, 1.48×10^{-2} mmol) were dissolved in water (7.1 mL). The solution was degassed by purging with Ar for 30 min. Polymerization was carried out at 70 °C for 24 h (conversion = 87.7%). The reaction mixture was dialyzed against 1.5 M aqueous NaCl for one day, and then pure water for one day. The polymer (P(SA)$_{91}$A$_{88}$) was recovered by a freeze-drying technique (1.10 g, 74.9%). The DP(NMR) of the PAPTAC block was 88, estimated from quantitative ^{13}C NMR measurements. M_n(GPC) and M_w/M_n were 1.87×10^4 g/mol and 1.14, respectively, estimated from GPC. The synthesis route of P(SA)$_{91}$A$_{88}$ is shown in Figure S1.

2.5. Preparation of PIC micelles

P(SA)$_{91}$S$_{67}$ and P(SA)$_{91}$A$_{88}$ were dissolved separately in 0.1 M aqueous NaCl. The aqueous P(SA)$_{91}$A$_{88}$ was added to the P(SA)$_{91}$S$_{67}$ solution over a period of 5 min with stirring to prepare the

PIC micelles. The mixing ratio of the polymers was represented by the molar fraction of cationic charge (f^+ = [APTAC]/([AMPS] + [APTAC])). The PIC micelles were prepared at f^+ = 0.5 unless otherwise noted, which represents complete charge neutralization.

2.6. Measurements

The ^{1}H NMR and inverse-gated decoupling ^{13}C NMR spectra were obtained in D_2O using a Bruker (Yokohma, Japan) DRX-500 spectrometer. GPC measurements were performed using a Jasco (Tokyo, Japan)UV-2075 detector equipped with a Shodex (Tokyo, Japan) OHpak SB-804 HQ column working at 40 °C with a flow rate of 0.6 mL/min. An acetic acid (0.5 M) solution containing sodium sulfate (0.3 M) was used as the eluent. Sample solutions were filtered with a 0.2 μm pore size membrane filter. M_n and M_w/M_n for the polymers were calibrated using standard poly(2-vinylpyridine) (P2VP) samples. Dynamic light scattering (DLS) measurements were obtained using a Malvern (Worcestershire, UK) Zetasizer Nano ZS with a He-Ne laser (4 mW at 633 nm) at 25 °C. The hydrodynamic radius (R_h) was calculated using the Stokes-Einstein equation, $R_h = k_BT/(6\pi\eta D)$, where k_B is the Boltzmann constant, T is the absolute temperature, and η is the solvent viscosity. The DLS data was analyzed using Malvern Zetasizer software version 7.11. The angular dependence of DLS and static light scattering (SLS) was measured using an Otsuka Electronics Photal (Osaka, Japan) DLS-7000HL light scattering spectrometer equipped with a multi-τ digital time correlator (ALV-5000E), at 25 °C. A He-Ne laser (10 mW at 633 nm) was used as the light source [27–29]. The weight-average molecular weight (M_w), z-average radius of gyration (R_g), and second virial coefficient (A_2) were estimated from SLS measurements [30]. Sample solutions for light scattering measurements were filtered using a membrane filter with 0.2 μm pores. The known Rayleigh ratio of toluene was used to calibrate the instrument. Plots of the refractive index increments against polymer concentration (dn/dC_p) at 633 nm were obtained with an Otsuka Electronics Photal DRM-3000 differential refractometer at 25 °C. The zeta potential was measured using a Malvern Zetasizer Nano-ZS equipped with a He-Ne laser light source (4 mW, at 632.8 nm) at 25 °C. The zeta potential (ζ) was calculated from the electrophoretic mobility (μ) using the Smoluchowski relationship, $\zeta = \eta\mu/\varepsilon$ ($\kappa a \gg 1$), where η is the viscosity of the solvent, ε is the dielectric constant of the solvent, and κ and a are the Debye-Hückel parameter and particle radius, respectively [31]. TEM was performed with a JEOL (Tokyo, Japan) JEM-2100 microscope at an accelerating voltage of 200 kV. A sample for TEM was prepared by placing one drop of the aqueous solution on a copper grid coated with a thin film of Formvar. Excess water was blotted using filter paper. The sample was stained with sodium phosphotungstate and dried under vacuum for one day.

3. Results and Discussion

3.1. Preparation of P(SA)$_{91}$S$_{67}$ and P(SA)$_{91}$A$_{88}$

The monomer conversions (p) of AMPS and APTAC could not be determined individually from ^{1}H NMR measurements, because the peaks of vinyl groups in the monomers completely overlapped around 5.5–6.5 ppm. The sum of p for the AMPS and APTAC monomers was estimated from the integral intensity ratio of the vinyl groups compared to the sum of AMPS and APTAC pendant methylene protons at 3.0–3.5 ppm after polymerization. The monomer reactivity ratios for AMPS and APTAC are assumed to be one, because the polymerizable functional groups had the same chemical structure, i.e., acrylamide-type. When random copolymerization was performed to prepare P(SA)$_{91}$ with equal concentrations of AMPS ([M_{AMPS}]$_0$) and APTAC ([M_{APTAC}]$_0$), DP(theory) and M_n(theory) were calculated from the following formulas:

$$DP(\text{theory}) = \frac{2[M_{AMPS}]_0}{[CTA]_0} \times \frac{p}{100} \tag{1}$$

$$M_n(\text{theory}) = DP(\text{theory}) \times MW_{MAV} + MW_{CTA} \tag{2}$$

where $[CTA]_0$ is the initial concentration of CTA, MW_{MAV} is the average molecular weight of AMPS and APTAC, and MW_{CTA} is the molecular weight of CTA. When $[M_{AMPS}]_0 = [M_{APTAC}]_0$, the total monomer concentration becomes $2[M_{AMPS}]_0$. The values of DP(theory) and M_n(theory) are listed in Table 1.

Table 1. Degree of polymerization (DP), number-average molecular weight (M_n), and molecular weight distribution (M_n/M_n).

Sample	DP (theory) [a]	M_n(theory) [b] $\times 10^4$ (g/mol)	DP (NMR) [c]	M_n(NMR) $\times 10^4$ (g/mol)	M_n(GPC) $\times 10^4$ (g/mol)	M_w/M_n
P(SA)$_{91}$	91	1.91	_[d]	_[d]	1.54	1.27
P(SA)$_{91}$S$_{67}$	61	3.30	67	3.27	2.38	1.04
P(SA)$_{91}$A$_{88}$	90	3.73	88	3.70	1.87	1.15

[a] Theoretical degree of polymerization estimated from Formula 1. [b] Theoretical number-average molecular weight estimated from Formula 2. [c] Estimated from quantitative inverse-gated decoupling ^{13}C NMR spectra in D$_2$O. [d] The values could not be determined from ^{1}H NMR, because the terminal phenyl protons overlapped with the pendant amino protons.

The composition of P(SA)$_{91}$ could not be determined from ^{1}H NMR, because the peaks of the AMPS and APTAC units overlapped (Figure S2). To determine the compositions of P(SA)$_{91}$, quantitative inverse gated decoupling ^{13}C NMR measurements were performed in D$_2$O (Figure 2a). The contents of AMPS and APTAC in the amphoteric random copolymer were determined from the integral intensity ratio of the peaks at 57.6 and 64.2 ppm, attributed to the pendant methylene carbons in AMPS and APTAC, respectively. The compositions of AMPS and APTAC in P(SA)$_{91}$ were both 50 mol %. The DP(NMR) values for the PAMPS and PAPTAC blocks in P(SA)$_{91}$S$_{67}$ and P(SA)$_{91}$A$_{88}$ were found to be 67 and 88, respectively, using quantitative ^{13}C NMR measurements (Figure 2b,c). The M_n(NMR) values for P(SA)$_{91}$S$_{67}$ and P(SA)$_{91}$A$_{88}$ were close to the M_n(theory) values. However, the M_n(GPC) values for P(SA)$_{91}$S$_{67}$ and P(SA)$_{91}$A$_{88}$ deviated from the theoretical values, which indicates that there are interactions between the sample polymers and that the GPC column or the GPC standard sample (P2VP) was unsuitable for determining M_n(GPC) (Figure S3). The M_w/M_n values for P(SA)$_{91}$S$_{67}$ and P(SA)$_{91}$A$_{88}$ were relatively small (1.04–1.15), which suggests that the obtained polymers have well-controlled structures.

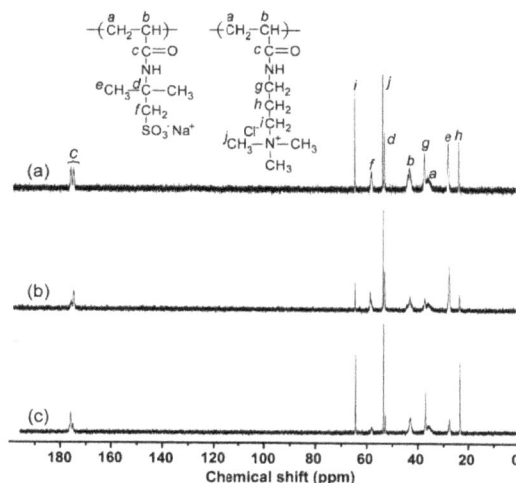

Figure 2. Inverse gated decoupling ^{13}C NMR spectra of (a) P(SA)$_{91}$, (b) P(SA)$_{91}$S$_{67}$, and (c) P(SA)$_{91}$A$_{88}$ in D$_2$O.

3.2. Preparation and Characterization of PIC Micelles

The R_h values for P(SA)$_{91}$S$_{67}$ and P(SA)$_{91}$A$_{88}$ in 0.1 M aqueous NaCl were 5.7 and 6.0 nm, respectively (Figure 3). These had unimodal distributions. The polydispersity indices (PDI) for P(SA)$_{91}$S$_{67}$ and P(SA)$_{91}$A$_{88}$ were 0.147–0.169. The R_h for the stoichiometrically charge neutralized mixture of P(SA)$_{91}$S$_{67}$ and P(SA)$_{91}$A$_{88}$ increased to 29.0 nm, which suggests the formation of PIC micelles. Assuming that the polymer main chain has a planar zig-zag structure, the expanded chain lengths of P(SA)$_{91}$S$_{67}$ and P(SA)$_{91}$A$_{88}$ are 39.5 and 44.8 nm, respectively. These are longer than the R_h of the PIC micelles (= 29.0 nm). This suggests that a PIC micelle has a simple spherical core-shell structure composed of a core formed from the anionic PAMPS and cationic PAPTAC blocks and an amphoteric P(SA)$_{91}$ shell. The PDI for PIC micelles was narrower than for P(SA)$_{91}$S$_{67}$ and P(SA)$_{91}$A$_{88}$, which indicates that the PIC micelles have a uniform size.

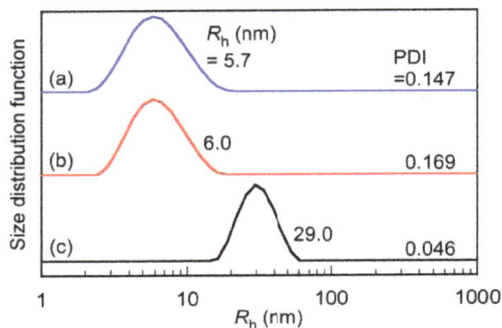

Figure 3. Hydrodynamic radius (R_h) distributions with polydispersity indices (PDI) for (a) P(SA)$_{91}$S$_{67}$; (b) P(SA)$_{91}$A$_{88}$; and (c) PIC micelles in 0.1 M aqueous NaCl.

The relationship between the relaxation rate (Γ) and light scattering angle (θ) was measured for PIC micelles in 0.1 M aqueous NaCl (Figure S4). The scattering vector (q) was calculated using $q = (4\pi n / \lambda)\sin(\theta/2)$, where n is the refractive index of the solvent and λ is the wavelength of the light source (= 632.8 nm). The Γ-q^2 plot was a straight line passing through the origin. Therefore, R_h was determined at $\theta = 90°$, because the diffusion coefficient (D) was independent of θ.

The R_h, LSI, and zeta potential for the mixture of P(SA)$_{91}$S$_{67}$ and P(SA)$_{91}$A$_{88}$ at various f^+ were measured in 0.1 M aqueous NaCl (Figure 4). The total polymer concentration in the solution was maintained at 1.0 g/L. We performed the LSI and zeta-potential experiments at $f^+ = 0$ and 1 were performed at $C_p = 10.0$ g/L, because LSIs were too low to measure R_h and zeta-potential. The compensated LSIs at $f^+ = 0$ and 1 are plotted in Figure 4a. At $f^+ = 0$, the values indicated for aqueous P(SA)$_{91}$S$_{67}$, and the zeta potential was –11.4 mV due to the negative charge of the PAMPS block. At $f^+ = 1$, the zeta potential for P(SA)$_{91}$A$_{88}$ was 8.40 mV due to the positive charge of the PAPTAC block. The zeta potential of the stoichiometrically charge neutralized mixture of P(SA)$_{91}$S$_{67}$ and P(SA)$_{91}$A$_{88}$ at $f^+ = 0.5$ was close to 0 mV. At $f^+ = 0.5$, the R_h and LSI had the highest values for PIC micelles. We will discuss the structure of PIC micelles in more detail, together with the results of the DLS, SLS, and TEM measurements, later in this paper.

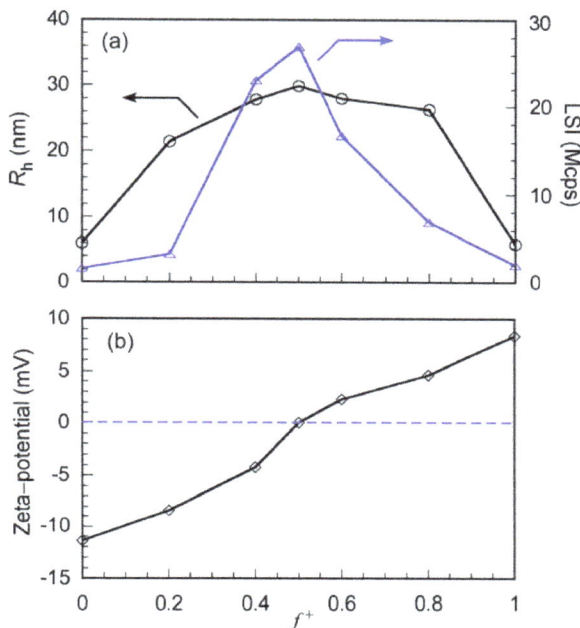

Figure 4. (a) Hydrodynamic radius (R_h, ○) and light scattering intensity (LSI, △) of PIC micelles as a function of f^+ (= [APTAC]/([APTAC] + [AMPS])) in 0.1 M aqueous NaCl; (b) Zeta potential of PIC micelles as a function of f^+ in 0.1 M aqueous NaCl.

The R_h and LSI of PC micelles were plotted against C_p (Figure S5). R_h remained almost constant at about 30 nm and was independent of C_p in the range $0.02 \leq C_p \leq 1.0$ g/L. LSI increased linearly with C_p. These observations suggest that the shape and aggregation number (N_{agg}), which is the number of polymer chains that form one PIC micelle, may be constant and independent of C_p in this region (0.02–1.0 g/L). At $C_p \leq 0.02$ g/L, the LSI for PIC micelles was too low to obtain R_h. We studied the time dependence on R_h and LSI for PIC micelles. The R_h and LSI values remained constant until at least 41 h (Figure S6).

Table 2. Dynamic light scattering (DLS) and static light scattering (SLS) data for P(SA)$_{91}$S$_{67}$, P(SA)$_{91}$A$_{88}$, and PIC micelles in 0.1 M NaCl.

Samples	$M_w \times 10^4$ (g/mol)	N_{agg}	R_g (nm)	R_h (nm)	R_g/R_h	$A_2 \times 10^{-4}$ (cm^3·mol/g^2)	d [a] (g/cm^3)	dn/dC_p (mL/g)
P(SA)$_{91}$S$_{67}$	5.41	1	7.3	5.7	1.28	3.73	0.116	0.118
P(SA)$_{91}$A$_{88}$	4.98	1	6.7	6.0	1.12	3.77	0.0914	0.138
PIC micelles	752	218	26.5	29.0	0.91	0.0127	0.122	0.128

[a] Estimated from the values of M_w and R_h using Formula 3.

SLS measurements for P(SA)$_{91}$S$_{67}$, P(SA)$_{91}$A$_{88}$, and PIC micelles were performed in 0.1 M aqueous NaCl (Table 2). M_w was estimated by extrapolating C_p and θ to zero in Zimm plots (Figure S7). R_g was estimated from the slope of θ at $C_p \rightarrow 0$. A_2 was estimated from the slope of C_p at θ $\rightarrow 0$. The N_{agg} of the PIC micelles was 218, as calculated from the M_w ratio of the PIC micelles and the unimer states of P(SA)$_{91}$S$_{67}$ and P(SA)$_{91}$A$_{88}$. The M_w values of the unimer states of the block copolymers estimated from SLS were close to the weight-average molecular weight calculated from M_n(NMR) and M_w/M_n (Table 1). R_g/R_h is useful for determining the shape of molecular aggregates. The theoretical value of R_g/R_h for a homogeneous hard sphere is 0.778, while that of a random coil is about 1, and this increases substantially for a structure with a lower density and polydispersity, e.g., $R_g/R_h = 1.5$–1.7 for flexible linear chains, and $R_g/R_h \geq 2$ for a rod [32]. The PIC micelles may be spherical, because $R_g/R_h = 0.91$, which is close to 1. The R_g/R_h ratios for $P(SA)_{91}S_{67}$ and $P(SA)_{91}A_{88}$ were above 1, which suggests that the block copolymer chains were relatively expanded due to electrostatic repulsions in the polyelectrolyte blocks in 0.1 M aqueous NaCl. In general, A_2 relates to the interactions between a polymer chain and solvent. A large value of A_2 indicates a good solvent; however, a small or negative value of A_2 indicates a poor solvent [33,34]. For the PIC micelles, A_2 was 1.27×10^{-6} cm^3·mol/g, which was less than the corresponding values for $P(SA)_{91}S_{67}$ and $P(SA)_{91}A_{88}$ $(3.73 \times 10^{-4}$ and 3.77×10^{-4} cm^3·mol/g). This indicates that the solubility of PIC micelles in 0.1 M NaCl decreased compared with those of the unimer states of the block copolymers because of the insoluble PIC core in the PIC micelles. The density (d) for $P(SA)_{91}S_{67}$, $P(SA)_{91}A_{88}$, and PIC micelles was calculated using the following formula:

$$d = \frac{M_w}{N_A \times V} \tag{2}$$

where V is the volume of the block copolymers or PIC micelles calculated from $4/3\pi R_h{}^3$. The d values for $P(SA)_{91}S_{67}$ and $P(SA)_{91}A_{88}$ were 0.116 and 0.0914 g/cm^3, respectively, while d was 0.122 g/cm^3 for the PIC micelles, which was slightly larger than for the diblock copolymers in the unimer states. This suggests that the polymer chains in the PIC micelles were more densely packed than those in the unimer states, because the PIC micelle core was formed by strong electrostatic interactions.

TEM was performed on PIC micelles in 0.1 M aqueous NaCl (Figure 5). The average radius of a PIC micelle was 20.3 nm, estimated from TEM, which was smaller than R_h (= 29.0 nm) estimated from DLS, because the TEM sample was in the dry state. Recently, we have reported the solution properties of amphoteric random copolymers in aqueous solutions [25]. We found out that there are no interpolymer interactions of amphoteric random copolymers in 0.1 M NaCl aqueous solutions. Also, we confirmed that there are on interactions between $P(SA)_{91}$ and anionic and cationic homopolymers using DLS measurements. From these findings, there are no interactions between amphoteric random copolymer shells in PIC micelles. Moreover, there are no interactions between amphoteric random copolymer shells and the PIC core. The PIC micelle system has the same chemical structure in the core and corona, as both domains have the same composition [35].

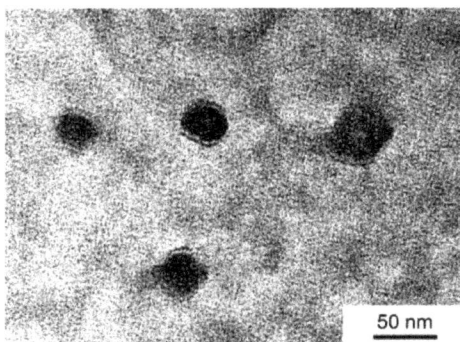

Figure 5. TEM image of PIC micelles with $f^+ = 0.5$ at $C_p = 1.0$ g/L in 0.1 M aqueous NaCl.

The critical micelle concentration (cmc) for PIC micelles in 0.1 M aqueous NaCl was determined from the relationship between the LSI ratio (I/I_0) and C_p (Figure 6). I and I_0 are the LSIs of the solution and solvent, respectively. Although R_h for the PIC micelles cannot be measured for $C_p \leq 0.02$ g/L,

I/I_0 can be measured below 0.02 g/L. The crossing point of the linear portions in the low and high C_p regions was 0.002 g/L, as the cmc. Below the cmc, PIC micelles may dissociate into charge neutralized small aggregates composed of a $P(SA)_{91}S_{67}/P(SA)_{91}A_{88}$ pair [36]. PIC micelles thus have a cmc, suggesting that they are in a dynamic equilibrium state.

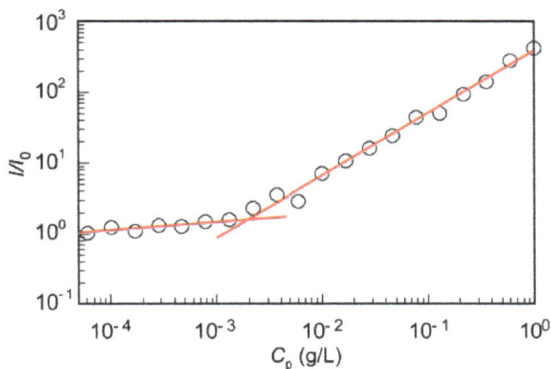

Figure 6. Light scattering intensity ratio (I/I_0) for the aqueous PIC micelles containing 0.1 M NaCl as a function of polymer concentration (C_p). I and I_0 are the light scattering intensities of the solution and solvent, respectively.

PIC micelles formed due to electrostatic interactions of the anionic and cationic blocks. When a low molecular weight salt such as NaCl is added to aqueous PIC micelles, their shape may change because of a screening effect [37,38]. R_h and LSI for the aqueous PIC micelles were plotted as a function of the NaCl concentration ([NaCl]) (Figure 7). At [NaCl] $\leq$ 0.7 M, R_h gradually decreased with increasing [NaCl], and R_h was 25.8 nm at [NaCl] = 0.7 M. R_h decreased rapidly at 0.7 M < [NaCl] $\leq$ 0.8 M, and R_h was about 12 nm at [NaCl] = 0.8 M. The LSI decreased almost monotonously with increasing [NaCl] until [NaCl] = 0.8 M. At [NaCl] > 0.8 M, the LSI was almost constant and independent of [NaCl]. This observation indicates that the PIC micelles partially dissociated due to the screening effect of NaCl. R_h was almost constant at 9–12 nm at [NaCl] > 0.8 M. The R_h values for the unimer states of $P(SA)_{91}S_{67}$ and $P(SA)_{91}A_{88}$ were 5.7–6.0 nm (Figure 3). The R_h values for PIC micelles at [NaCl] > 0.8 M were larger than those for the unimer states. These observations suggest that PIC micelles cannot dissociate into their unimer states at [NaCl] > 0.8 M, presumably because of salting out effects.

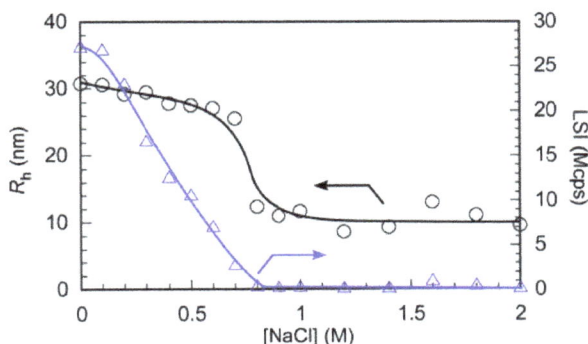

Figure 7. Hydrodynamic radius (R_h, ○) and light scattering intensity (LSI, △) of PIC micelles with f^+ = 0.5 at C_p = 1.0 g/L as a function of NaCl concentration ([NaCl]).

3.3. Interactions between PIC Micelles and Proteins

We evaluated the interactions between PIC micelles (0.1 g/L) and BSA (5.0 g/L) in PBS using DLS (Figure 8). Before mixing, the R_h distributions were unimodal, and the R_h values for PIC micelles and BSA were 29.0 and 4.9 nm, respectively. A mixed solution of PIC micelles and BSA was prepared in PBS, and then DLS was performed on the solution within 1 h. After mixing, the distribution became bimodal, with R_h = 5.1 and 31.5 nm. The R_h value of the large size distribution was close to the R_h (= 29.0 nm) of the PIC micelles. Before mixing, the LSI values for the PIC micelles and BSA were 1.9 and 2.0 Mcps, respectively. The LSI for the mixture of PIC micelles and BSA was 2.1 Mcps, which was close to the values before mixing. This indicates that the PIC micelles do not interact considerably with BSA.

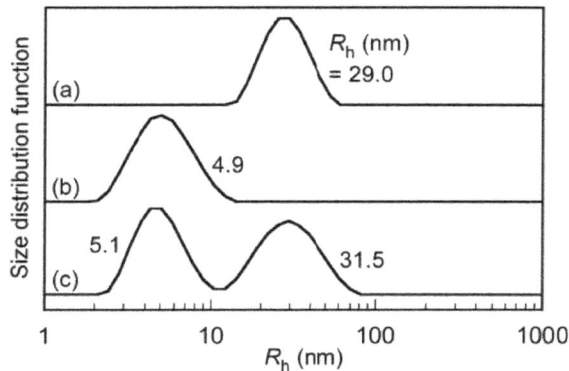

Figure 8. R_h distributions of (**a**) PIC micelles; (**b**) BSA; and (**c**) a mixture of PIC micelles with BSA in PBS at 25 °C.

The time dependences of R_h and LSI for the mixture of PIC micelles and BSA were determined (Figure S8). The R_h distributions and LSI remained almost constant for one day. After two days, the distribution became trimodal. A small third distribution peak with an R_h of about 200 nm was observed after two days. The large aggregates with $R_h \geq 200$ nm may be formed from BSA and unit PIC of a pair of P(SA)$_{91}$S$_{67}$/P(SA)$_{91}$A$_{88}$ dissociated from the PIC micelles. After four days, the area of the third distribution peak increased, and R_h increased to 281 nm. The average R_h and LSI for the mixture were almost constant for three days; however, these increased after four days. These observations suggest that the interaction between PIC micelles and proteins increased with time. The block copolymer chains can dissociate from the PIC micelles because these are in a dynamic equilibrium state. The unit PICs of P(SA)$_{91}$S$_{67}$/P(SA)$_{91}$A$_{88}$ detached from the PIC micelles may interact with proteins due to electrostatic attractions. The total charges in the unit PIC are compensated. However, dangling charge loops may be formed in the complex of cationic PAMPTAC and anionic PAMSP blocks in the unit PIC (Figure S9). In the case of PIC micelles, the surface of the PIC core was completely covered with amphoteric random copolymer shells which have protein antifouling properties. On the other hand, in the case of the unit PIC, the dangling charge loops may be exposed. The dangling charge groups in the unit PIC strongly interacted with proteins to form large aggregates.

The interactions between PIC micelles (0.1 g/L) and FBS (40 g/L) in PBS were evaluated by DLS (Figure 9). While BSA is a negatively charged protein, FBS is a mixture of negatively and positively charged proteins. The R_h distribution for FBS was bimodal with R_h = 5.3 and 31.8 nm. Mixed PBS solutions of PIC micelles and FBS were prepared, and then DLS was performed within 1 h. The R_h values for the mixture of PIC micelles and FBS were similar to those for FBS. The LSI of the mixed solution was 2.1 Mcps, which was similar to the values for the PIC micelles (1.9 Mcps) and FBS

(2.1 Mcps) before mixing. This indicates that there is no interaction between PIC micelles and FBS. The time dependences of R_h distributions for the mixture of PIC micelles and FBS were measured (Figure S10). After two days, the distribution became trimodal. A small third distribution peak with an R_h of about 200 nm was observed after two days, which suggests that the large aggregates were formed from FBS and unit PIC formed.

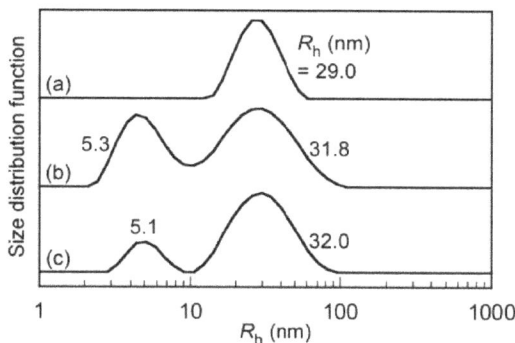

Figure 9. (a) R_h distributions for (**a**) PIC micelles, (**b**) FBS, and (**c**) a mixture of PIC micelles with FBS in PBS buffer at 25 °C.

4. Conclusions

Oppositely charged diblock copolymers of anionic P(SA)$_{91}$S$_{67}$ and cationic P(SA)$_{91}$A$_{88}$ were prepared via RAFT using P(SA)$_{91}$ macro-CTA. Water-soluble PIC micelles were formed from a stoichiometrically charge neutralized mixture of 0.1 M aqueous NaCl solutions of P(SA)$_{91}$S$_{67}$ and P(SA)$_{91}$A$_{88}$. The maximum values of R_h and LSI were observed when f^+ was 0.5, and the zeta potential was close to 0 mV. The PIC micelles were in a dynamic equilibrium state with PIC micelles and charge neutralized small aggregates composed of a P(SA)$_{91}$S$_{67}$/P(SA)$_{91}$A$_{88}$ pair. The particle sizes in the mixture of PIC micelles and proteins in PBS remained almost constant for at least one day, which suggests that there is no interaction between PIC micelles and proteins. The PIC micelles were covered with amphoteric P(SA)$_{91}$ shells that suppressed their interaction with proteins. However, the interactions of the diblock copolymer chains dissociated from PIC micelles and proteins increased with time. If the lengths of the PAMPS block in P(SA)$_{91}$S$_{67}$ and the PAPTSC block in P(SA)$_{91}$A$_{88}$ increased, the dynamic equilibrium may shift to form PIC micelles. Therefore, it is expected that the interaction of PIC micelles with long polyelectrolyte chains and proteins can be suppressed for a long time.

Supplementary Materials: The following are available online at http://www.mdpi.com/2073-4360/10/2/205/s1; Figure S1: Synthesis routes of P(SA)$_{91}$S$_{67}$ and P(SA)$_{91}$A$_{88}$; Figure S2: ^{1}H NMR spectra for (a) P(SA)$_{91}$, (b) P(SA)$_{91}$S$_{67}$, and (c) P(SA)$_{91}$A$_{88}$ in D$_2$O; Figure S3: GPC elution curves for (a) P(SA)$_{91}$S$_{67}$ and (b) P(SA)$_{91}$A$_{88}$ using an acetic acid (0.5 M) solution containing sodium sulfate (0.3 M) as an eluent; Figure S4: Relationship between the relaxation rate (Γ) and the square of the magnitude of the scattering intensity vector (q^2) for PIC micelles at C_p = 1 g/L in 0.1 M aqueous NaCl at 25 °C; Figure S5: Hydrodynamic radius (R_h, ○) and light scattering intensity (LSI, △) of PIC micelles as a function of polymer concentration (C_p) in 0.1 M aqueous NaCl. Figure S6: Relationship between R_h (○) and light scattering intensity (LSI, △) as a function of time for PIC micelle with f^+ = 0.5 at C_p = 1.0 g/L in 0.1 M NaCl aqueous solution; Figure S7: A typical Zimm plot for PIC micelles in 0.1 M aqueous NaCl at 25 °C; Figure S8: (a) Relationship between R_h (○) and light scattering intensity (LSI, △) as a function of time, and (b) R_h distributions for a mixture of PIC micelles/BSA at C_p = 0.1 g/L and [BSA] = 5.0 g/L in PBS at 25 °C; Figure S9: Conceptual illustration of dangling charge groups in the unit PIC of P(SA)$_{91}$S$_{67}$/P(SA)$_{91}$A$_{88}$; Figure S10: R_h distributions for mixture of PIC micelle/FBS at C_p = 0.1 g/L and [FBS] = 40 g/L in PBS at 25 °C.

Acknowledgments: This work was financially supported by a Grant-in-Aid for Scientific Research (25288101 and 16K14008) from the Japan Society for the Promotion of Science (JSPS), JSPS Bilateral Open Partnership

Polymers **2018**, *10*, 205

Joint Research Projects, and the Research Program of "Dynamic Alliance for Open Innovation Bridging Human, Environment and Materials" in "Network Joint Research Center for Materials and Devices."

Author Contributions: Rina Nakahata was a research student who performed the experimental work. Shin-ichi Yusa was responsible for analyzing the experimental data and writing the paper. All authors approved the manuscript.

Conflicts of Interest: The authors declare no conflict of interest.

References

1. Michaels, A.S.; Miekka, R.G. Polycation-polyanion complexes: preparation and properties of poly(vinylbenzyltrimethylammonium) poly(styrensulfonate). *J. Phys. Chem.* **1961**, *65*, 1765–1773. [CrossRef]
2. Wibowo, A.; Osada, K.; Matsuda, H.; Anraku, Y.; Hirose, H.; Kishimura, A.; Kataoka, K. Morphology control in water of polyion complex nanoarchitectures of double-hydrophilic charged block copolymers through composition tuning and thermal treatment. *Macromolecules* **2014**, *47*, 3086–3092. [CrossRef]
3. Yusa, S.; Yokoyama, Y.; Morishima, Y. Synthesis of oppositely charged block copolymers of poly(ethylene glycol) via reversible addition–fragmentation chain transfer (RAFT) radical polymerization and characterization of their polyion complex (PIC) micelles in water. *Macromolecules* **2009**, *42*, 376–383. [CrossRef]
4. Oparin, A.I. Origin and evolution of metabolism. *Comp. Biochem. Physiol.* **1962**, *4*, 371–377. [CrossRef]
5. Oishi, M.; Nagasaki, Y.; Itaka, K.; Nishiyama, N.; Kataoka, K. Lactosylated poly(ethylene glycol)-siRNA conjugate through acid-labile β-thiopropionate linkage to construct pH-sensitive polyion complex micelles achieving enhanced gene silencing in hepatoma cells. *J. Am. Chem. Soc.* **2005**, *127*, 1624–1625. [CrossRef] [PubMed]
6. Van der Gucht, J.; Spruijt, E.; Lemmers, M.; Cohen Stuart, M.A. Polyelectrolyte complexes: Bulk phases and colloidal systems. *J. Colloid Interface Sci.* **2011**, *361*, 407–422. [CrossRef] [PubMed]
7. Müller, M. Sizing, shaping and pharmaceutical applications of polyelectrolyte complex nanoparticles. *Adv. Polym. Sci.* **2014**, *256*, 197–260.
8. Pergushov, D.V.; Mueller, A.H.; Schacher, F.H. Micellar interpolyelectrolyte complexes. *Chem. Soc. Rev.* **2012**, *41*, 6888–6901. [CrossRef] [PubMed]
9. Voets, I.K.; de Keizer, A.; Stuart, M.A.C. Complex coacervate core micelles. *Adv. Colloid Interface Sci.* **2009**, *147*, 300–318. [CrossRef] [PubMed]
10. Steinschulte, A.A.; Gelissen, A.P.H.; Jung, A.; Brugnoni, M.; Caumanns, T.; Lotze, G.; Mayer, J.; Pergushov, D.V.; Plamper, F.A. Facile screening of various micellar morphologies by blending miktoarm stars and diblock copolymers. *ACS Macro Lett.* **2017**, *6*, 711–715. [CrossRef]
11. Dähling, C.; Lotze, G.; Mori, H.; Pergushov, D.V.; Plamper, F.A. Thermoresponsive segments retard the formation of equilibrium micellar interpolyelectrolyte complexes by detouring to various intermediate structures. *J. Phys. Chem. B* **2017**, *121*, 6739–6748. [CrossRef] [PubMed]
12. Nakai, K.; Nishiuchi, M.; Inoue, M.; Ishihara, K.; Sanada, Y.; Sakurai, K.; Yusa, S. Preparation and characterization of polyion complex micelles with phoshobetaine shells. *Langmuir* **2013**, *29*, 9651–9661. [CrossRef] [PubMed]
13. Nakai, K.; Ishihara, K.; Yusa, S. Preparation of giant polyion complex vesicles (G-PICsomes) with polyphosphobetaine shells composed of oppositely charged diblock copolymers. *Chem. Lett.* **2017**, *46*, 824–827. [CrossRef]
14. Nakai, K.; Ishihara, K.; Kappl, M.; Fujii, S.; Nakamura, Y.; Yusa, S. Polyion complex vesicles with solvated phosphobetaine shells formed from oppositely charged diblock copolymers. *Polymers* **2017**, *9*, 49. [CrossRef]
15. Iwasaki, Y.; Ishihara, K. Cell membrane-inspired phospholipid polymers for developing medical devices with excellent biointerfaces. *Sci. Technol. Adv. Mater.* **2012**, *13*, 064101. [CrossRef] [PubMed]
16. Iwasaki, Y.; Ijuin, M.; Mikami, A.; Nakabayashi, N.; Ishihara, K. Behavior of blood cells in contact with water-soluble phospholipid polymer. *J. Biomed. Mater. Res.* **1999**, *6*, 360–367. [CrossRef]
17. Ishihara, K.; Ueda, T.; Nakabayasi, N. Preparation of phospholipid polymers and their properties as hydrogel sheet. *Polym. J.* **1990**, *22*, 355–360. [CrossRef]
18. Ostuni, E.; Chapman, R.G.; Holmlin, R.E.; Takayama, S.; Whitesides, G.M. A survey of structure-property relationships of surfaces that resist the adsorption of protein. *Langmuir* **2001**, *17*, 5605–5620. [CrossRef]

19. Holmlin, R.E.; Chen, X.; Chapman, R.G.; Takayama, S.; Whitesides, G.M. Zwitterionic SAMs that resist nonspecific adsorption of protein from aqueous buffer. *Langmuir* **2001**, *17*, 2841–2850. [CrossRef]

20. Zhao, T.; Chen, K.; Gu, H. Investigations on the interactions of proteins with polyampholyte-coated magnetite nanoparticles. *J. Phys. Chem. B* **2013**, *117*, 14129–14135. [CrossRef] [PubMed]

21. Chang, Y.; Chen, S.; Zhang, Z.; Jiang, S. Highly protein-resistant coatings from well-defined diblock copolymers containing sulfobetaines. *Langmuir* **2006**, *22*, 2222–2226. [CrossRef] [PubMed]

22. Zhang, Z.; Chen, S.; Jiang, S. Dual-functional biomimetic materials: Nonfouling poly(carboxybetaine) with active functional groups for protein immobilization. *Biomacromolecules* **2006**, *7*, 3311–3315. [CrossRef] [PubMed]

23. Muro, E.; Pons, T.; Lequeux, N.; Fragola, A.; Sanson, N.; Lenkei, Z.; Dubertret, B. Small and stable sulfobetaine zwitterionic quantum dots for functional live-cell imaging. *J. Am. Chem. Soc.* **2010**, *132*, 4556–4557. [CrossRef] [PubMed]

24. Shih, Y.; Chang, Y.; Quemener, D.; Yang, H.; Jhong, J.; Ho, F.; Higuchi, A.; Chang, Y. Hemocompatibility of polyampholyte copolymers with well-defined charge bias in human blood. *Langmuir* **2014**, *30*, 6489–6496. [CrossRef] [PubMed]

25. Nakahata, R.; Yusa, S. Solution properties of amphoteric random copolymers bearing pendant sulfonate and quaternary ammonium groups with controlled structures. *Langmuir* **2018**. [CrossRef] [PubMed]

26. Mitsukami, Y.; Donovan, S.M.; Lowe, B.A.; McCormick, L.C. Water-soluble polymers. 81. Direct synthesis of hydrophilic styrenic-based homopolymers and block copolymers in aqueous solution via RAFT. *Macromolecules* **2001**, *34*, 2248–2256. [CrossRef]

27. Koppel, E.D. Analysis of macromolecular polydispersity in intensity correlation spectroscopy: The method of cumulants. *J. Chem. Phys.* **1972**, *57*, 4814–4820. [CrossRef]

28. Huber, K.; Bantle, S.; Lutz, P.; Burchard, W. Hydrodynamic and thermodynamic behavior of short-chain polystyrene in toluene and cyclohexane at 34.5 °C. *Macromolecules* **1985**, *18*, 1461–1467. [CrossRef]

29. Konishi, T.; Yoshizaki, T.; Yamakawa, H. On the "Universal Constants" ρ and Φ. of flexible polymers. *Macromolecules* **1991**, *24*, 5614–5622. [CrossRef]

30. Zimm, H.B. Apparatus and methods for measurement and interpretation of the angular variation of light scattering; Preliminary results on polystyrene solutions. *J. Chem. Phys.* **1948**, *16*, 1099–1116. [CrossRef]

31. Ali, I.S.; Heuts, A.P.J.; van Herk, A.M. Controlled synthesis of polymeric nanocapsules by RAFT-based vesicle templating. *Langmuir* **2010**, *26*, 7848–7858. [CrossRef] [PubMed]

32. Akcasu, A.Z.; Han, C.C. Molecular weight and temperature dependence of polymer dimensions in solution. *Macromolecules* **1979**, *12*, 276–280. [CrossRef]

33. Matsuda, Y.; Kobayashi, M.; Annaka, M.; Ishihara, K.; Takahara, A. Dimensions of a free linear polymer immobilized on silica nanoparticles of a zwitterionic polymer in aqueous solutions with various ionic strength. *Langmuir* **2008**, *28*, 8772–8778. [CrossRef] [PubMed]

34. Cheng, L.; Hou, G.; Miao, J.; Chen, D.; Jiang, M.; Zhu, L. Efficient synthesis of unimolecular polymeric janus nanoparticles and their unique self-assembly behavior in a common solvent. *Macromolecules* **2008**, *41*, 8159–8166. [CrossRef]

35. Savoji, M.T.; Strandman, S.; Zhu, X.X. Switchable vesicles formed by diblock random copolymers with tunable pH-and thermo-responsiveness. *Langmuir* **2013**, *29*, 6823–6832. [CrossRef] [PubMed]

36. Anraku, Y.; Kishimura, A.; Oba, M.; Yamasaki, Y.; Kataoka, K. Spontaneous formation of nanosized unilamellar polyion complex vesicles with tunable size and properties. *J. Am. Chem. Soc.* **2010**, *132*, 1631–1636. [CrossRef] [PubMed]

37. Santis, D.S.; Ladogana, D.R.; Diociaiuti, M.; Masci, G. Pegylated and thermosensitive polyion comples micelles by self-assembly of two oppositely and permanently charged diblock copolymers. *Macromolecules* **2010**, *43*, 1992–2001. [CrossRef]

38. Maggi, F.; Ciccarelli, S.; Diociaiuti, M.; Casciardi, S.; Masci, G. Chitosan nanogels by template chemical cross-linking in polyion complex micelle nanpreactors. *Biomacromolecules* **2011**, *12*, 3499–3507. [CrossRef] [PubMed]

polymers

MDPI

Article

pH-Induced Association and Dissociation of Intermolecular Complexes Formed by Hydrogen Bonding between Diblock Copolymers

Masanobu Mizusaki [1,*], Tatsuya Endo [1], Rina Nakahata [1], Yotaro Morishima [2] and Shin-ichi Yusa [1,*]

[1] Department of Applied Chemistry, Graduate School of Engineering, University of Hyogo, 2167 Shosha, Himeji, Hyogo 671-2280, Japan; ta-endou@toyo-rubber.co.jp (T.E.); nkht9999@gmail.com (R.N.)

[2] Faculty of Engineering, Fukui University of Technology, 6-3-1 Gakuen, Fukui 910-8505, Japan; morisima@fukui-ut.ac.jp

* Correspondence: mizusaki.masanobu@sharp.co.jp (M.M.); yusa@eng.u-hyogo.ac.jp (S.-i.Y.); Tel.: +81-79-267-4954 (S.-i.Y.)

Received: 26 July 2017; Accepted: 15 August 2017; Published: 17 August 2017

Abstract: Poly(sodium styrenesulfonate)–*block*–poly(acrylic acid) (PNaSS–*b*–PAA) and poly(sodium styrenesulfonate)–*block*–poly(N-isopropylacrylamide) (PNaSS–*b*–PNIPAM) were prepared via reversible addition–fragmentation chain transfer (RAFT) radical polymerization using a PNaSS-based macro-chain transfer agent. The molecular weight distributions (M_w/M_n) of PNaSS–*b*–PAA and PNaSS–*b*–PNIPAM were 1.18 and 1.39, respectively, suggesting that these polymers have controlled structures. When aqueous solutions of PNaSS–*b*–PAA and PNaSS–*b*–PNIPAM were mixed under acidic conditions, water-soluble PNaSS–*b*–PAA/PNaSS–*b*–PNIPAM complexes were formed as a result of hydrogen bonding interactions between the pendant carboxylic acids in the PAA block and the pendant amide groups in the PNIPAM block. The complex was characterized by ^{1}H NMR, dynamic light scattering, static light scattering, and transmission electron microscope measurements. The light scattering intensity of the complex depended on the mixing ratio of PNaSS–*b*–PAA and PNaSS–*b*–PNIPAM. When the molar ratio of the N-isopropylacrylamide (NIPAM) and acrylic acid (AA) units was near unity, the light scattering intensity reached a maximum, indicating stoichiometric complex formation. The complex dissociated at a pH higher than 4.0 because the hydrogen bonding interactions disappeared due to deprotonation of the pendant carboxylic acids in the PAA block.

Keywords: block copolymers; RAFT polymerization; complex; hydrogen bonding interactions; pH-responsive

1. Introduction

Non-covalent interactions, such as hydrophobic [1], electrostatic [2,3], van der Waals [4], and hydrogen bonding interactions [5,6], can be a driving force for the complex formation of polymers. In particular, hydrogen bonding interactions are an important driving force for the self-organization of natural polymers such as polysaccharides, proteins, and deoxyribonucleic acid (DNA).

It is known that hydrogen bonding interactions between amide groups or poly(ethylene glycol) (PEG) and carboxylic acid groups promote self-association or complex formation in water [7–9]. Shieh et al. [10] prepared a series of poly(N-isopropylacrylamide-*random*-acrylic acid) (P(NIPAM–*r*–AA)) copolymers and examined their glass transition behavior. They found that the incorporation of acrylic acid units into the PNIPAM polymer enhances the glass transition temperature (T_g) due to intermolecular hydrogen bonding between the pendant isopropyl amide groups and the carboxylic acid groups using ^{1}H NMR and Fourier-transform infrared analyses. As an example of such complex formation,

Bian et al. [11] investigated the interaction between poly(*N,N*-diethylacrylamide) (PDEA), which is an analog of PNIPAM, and poly(acrylic acid) (PAA). The complex is formed between the two polymers through hydrogen bonding interactions with a stoichiometry of $r = 0.6$ (r is the unit molar ratio of PAA/PDEA), and the complex formation depends on pH values. We lately reported [12] the complex formation behavior of poly(sodium styrenesulfonate)–*block*–PEG–*block*–poly(sodium styrenesulfonate) (PNaSS–*b*–PEG–*b*–PNaSS) with poly(methacrylic acid) (PMA). Both PNaSS–*b*–PEG–*b*–PNaSS and PMA were synthesized via reversible addition–fragmentation chain transfer (RAFT) radical polymerization. Below pH 5, water-soluble complexes were formed owing to the hydrogen bonding interactions between the PEG block in PNaSS–*b*–PEG–*b*–PNaSS and the carboxylic acids in PMA. The experimental data indicated that the PNaSS–*b*–PEG–*b*–PNaSS/PMA complex was spherical in shape. At pH greater than 5, the complex dissociated because the hydrogen bonding interaction disappeared due to deprotonation of the pendant carboxylic acids in PMA.

In the present study, we focused on the complex formation behavior owing to the hydrogen bonding interactions as a function of the solution temperature instead of the solution pH. Two species of diblock copolymers (Figure 1a), PNaSS–*block*–poly(acrylic acid) (PNaSS–*b*–PAA) and PNaSS–*block*–PNIPAM (PNaSS–*b*–PNIPAM), were synthesized via RAFT radical polymerization [13]. The PNIPAM blocks of PNaSS-*b*-PNIPAM are also assumed to associate above the lower critical solution temperature (LCST) to form core–corona-type multi-polymer micelles [14]. When PNaSS–*b*–PAA and PNaSS–*b*–PNIPAM were mixed in a 0.1 M NaCl aqueous solution at 20 °C, water-soluble PNaSS–*b*–PAA/PNaSS–*b*–PNIPAM complexes were formed through hydrogen bonding interactions between the PAA and PNIPAM blocks below pH 3.9 (Figure 1b). The complexes were maintained above the LCST of PNaSS–*b*–PNIPAM. The complexes dissociated under basic conditions when the hydrogen bonding interactions disappeared due to deprotonation of the pendant carboxylic acids in the PAA block.

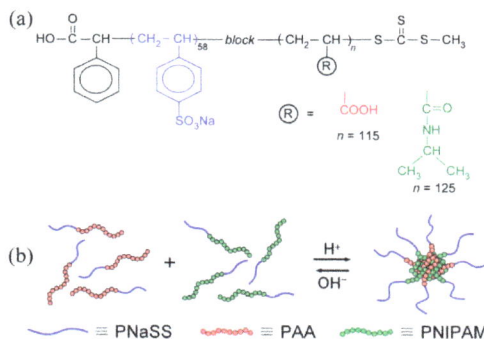

Figure 1. (a) Chemical structures of diblock copolymers used in this study: Poly(sodium styrenesulfonate)$_{58}$–*block*–poly(acrylic acid)$_{125}$ (PNaSS$_{58}$–*b*–PAA$_{125}$) and poly(sodium styrenesulfonate)$_{58}$–*block*–poly(N-isopropylacrylamide)$_{115}$ (PNaSS$_{58}$–*b*–PNIPAM$_{115}$), and (b) schematic representation of polymer chain mixing in the core and corona micelles comprising PNaSS$_{58}$–*b*–PAA$_{125}$ and PNaSS$_{58}$–*b*–PNIPAM$_{115}$.

2. Experimental Section

2.1. Materials

Acrylic acid (AA) from Kanto Chemical (Tokyo, Japan) was dried over 4 Å molecular sieves and distilled under reduced pressure. *N*-isopropylacrylamide (NIPAM) from Aldrich (St. Louis, MO, USA) was purified by recrystallization from a mixed solvent of benzene and *n*-hexane. α-Methyltrithiocarbonate-*S*-phenylaceticacid (MTPA) was synthesized as previously reported [15]. Sodium styrenesulfonate (NaSS) from Tokyo Chemical Industry (Tokyo, Japan) and 4,4′-azobis(4-cyanopentanoic acid) (V-501) from Aldrich were used as received. Methanol was dried using molecular sieves and distilled. Deionized water was used.

2.2. Synthesis of PNaSS Macro-Chain Transfer Agent (PNaSS Macro–CTA)

NaSS (20.6 g, 100 mmol) and MTPA (0.26 g, 1.0 mmol) were dissolved in 180 mL of water, and V-501 (5.6 mg, 0.2 mmol) was added to the aqueous solution. Polymerization was carried out at 70 °C for 3 h under Ar atmosphere. After the polymerization, the mixture was dialyzed against pure water for a week and recovered by a freeze-drying technique (yield 14.8 g, number-average molecular weight (M_n) = 1.22 × 10^4 (gel permeation chromatography (GPC)), molecular weight distribution (M_w/M_n) = 1.19, and degree of polymerization (DP) = 58). The obtained PNaSS could be used as a macro-CTA (PNaSS$_{58}$ macro–CTA).

2.3. Preparation of PNaSS$_{58}$–b–PAA$_{125}$

AA (0.86 g, 12 mmol) was dissolved in 15 mL of water, and PNaSS$_{58}$ macro–CTA (1.05 g, 0.087 mmol) and V-501 (5.0 mg, 0.018 mmol) were added to this solution. The mixture was deoxygenated by purging with Ar gas for 30 min. Block copolymerization was carried out at 70 °C for 2 h. The diblock copolymer was purified by dialysis against pure water for a week and then recovered using a freeze-drying technique. The diblock copolymer, PNaSS$_{58}$–b–PAA$_{125}$, was obtained (yield 1.79 g, M_n = 2.12 × 10^4 (^{1}H NMR), M_w/M_n = 1.18, and DP of the PAA block = 125).

2.4. Preparation of PNaSS$_{58}$–b–PNIPAM$_{115}$

NIPAM (2.26 g, 10.2 mmol) was dissolved in a mixed solvent of water and methanol (12.4 mL, 1/1, *v/v*), and PNaSS$_{58}$ macro–CTA (0.96 g, 0.079 mmol) and V-501 (4.6 mg, 0.016 mmol) were added to this solution. The mixture was deoxygenated by purging with Ar gas for 30 min. Block copolymerization was carried out at 60 °C for 5 h. The diblock copolymer was purified by reprecipitation from a methanol solution into excess ether twice and then recovered using a freeze-drying technique. The diblock copolymer PNaSS$_{58}$–b–PNIPAM$_{115}$ was obtained (yield 1.79 g, M_n = 2.52 × 10^4 (^{1}H NMR), M_w/M_n = 1.39, and DP of the PNIPAM block = 115).

2.5. Preparation of the Water-Soluble Complex

Stock solutions of PNaSS$_{58}$–b–PAA$_{125}$ and PNaSS$_{58}$–b–PNIPAM$_{115}$ were prepared by dissolving each polymer in 0.1 M NaCl aqueous solutions of pH 3 and 10, respectively. To prepare the complex, the PNaSS$_{58}$–b–PNIPAM$_{115}$ aqueous solution was added to the PNaSS$_{58}$–b–PAA$_{125}$ aqueous solution over a period of five min, and the mixture was allowed to stand still for one day. The mixing ratio of the two diblock copolymers was adjusted based on the molar fraction of NIPAM units (f_{NIPAM} = (NIPAM)/((NIPAM) + (AA)), where (NIPAM) and (AA) are the molar concentrations of NIPAM and AA units, respectively). The complex was prepared at f_{NIPAM} = 0.5 unless otherwise noted.

2.6. Measurements

GPC measurements were performed with a Shodex 7.0 µm bead size GF-7F HQ column. A phosphate buffer at pH 9, containing 10 vol % acetonitrile was used as an eluent at a flow rate of 0.6 mL/min at 40 °C. M_n and M_w/M_n were calibrated with standard PNaSS samples of 11 different molecular weights ranging from 1.37 × 10^3 to 2.61 × 10^6. ^{1}H NMR spectra were obtained with a Bruker DRX-500 spectrometer (Buick Rica, MA, USA) operating at 500 MHz. Light scattering measurements were performed using an Otsuka Electronics Photal DLS-7000DL equipped with a digital time correlator (ALV-5000E, Osaka, Japan). Sample solutions were filtered with a 0.2-µm membrane filter. For dynamic light scattering (DLS) measurements, the data obtained were analyzed with ALV software version 3.0 [16,17]. For static light scattering (SLS) measurements, weight-average molecular weight (M_w), and radius of gyration (R_g) were estimated from Zimm plots [18]. Values of dn/dC_P were determined with an Otsuka Electronics Photal DRM-1020 differential refractometer. Transmission electron microscopy (TEM) measurements were carried out using a JEOL JEM-2100 (Tokyo, Japan) at an accelerating voltage of 200 kV. The TEM sample was prepared by placing one droplet of the solution on a copper grid coated with Formvar. The sample was stained by sodium phosphotungstate and dried under reduced pressure. Percent transmittance ($\%T$) measurements were performed using a JASCO V-630 BIO UV–Vis spectrometer (Tokyo, Japan) with a 10 mm path length quartz cell. The temperature was increased from 20 to 80 °C with a heating rate of 1.0 °C·min^{-1} using a JASCO ETC-717 thermostat system.

3. Results and Discussion

We prepared the diblock copolymers, PNaSS$_{58}$–*b*–PAA$_{125}$ and PNaSS$_{58}$–*b*–PNIPAM$_{115}$, via a RAFT technique using PNaSS$_{58}$ macro–CTA. The DP of PNaSS$_{58}$ macro–CTA, which was determined by GPC, was 58 (PNaSS$_{58}$–macro–CTA). The DPs of the PAA and PNIPAM blocks were 125 and 115, respectively. These values were calculated from ^{1}H NMR peak area intensities derived from the PAA or PNIPAM block and an area derived from the PNaSS block. M_n and M_w/M_n of the diblock copolymers were estimated from GPC. The results for the characteristics of the diblock copolymers are listed in Table 1. The M_w/M_n values are relatively small ($M_w/M_n < 1.4$), indicating that the controlled/living polymerizations proceeded successfully [13].

Table 1. Degrees of polymerization (DP) of PNaSS, PAA, and PNIPAM blocks, and number-average molecular weights (M_n), and molecular weight distributions (M_w/M_n) of the diblock copolymers.

Samples	DP of PNaSS [a]	DP of PAA [b]	DP of PNIPAM [b]	M_n(theo) [c] × 10^{-4}	M_n(NMR) [b] × 10^{-4}	M_n(GPC) [a] × 10^{-4}	M_w/M_n [a]
PNaSS$_{58}$–*b*–PAA$_{125}$	58	125		2.07	2.12	3.51	1.18
PNaSS$_{58}$–*b*–PNIPAM$_{115}$	58		115	2.27	2.52	1.21	1.39

[a] Estimated from gel permeation chromatography (GPC) eluted with a phosphate buffer solution at pH 9 containing 10 vol % acetonitrile; [b] Estimated from ^{1}H NMR; [c] Calculated from Equation (4).

When the polymerization is assumed an ideally living process, then the theoretical number-average molecular weight (M_n(theo)) can be estimated as

$$M_n(\text{theo}) = \frac{[M]_0}{[CTA]_0} \frac{x_m}{100} M_m + M_{CTA} \tag{1}$$

where $[M]_0$ is the initial monomer concentration, $[CTA]_0$ is the initial PNaSS$_{58}$ macro–CTA concentration, x_m is the conversion of the monomer, M_m is the molecular weight of the monomer, and M_{CTA} is the molecular weight of PNaSS$_{58}$ macro–CTA. The M_n(NMR) values for PNaSS$_{58}$–*b*–PAA$_{125}$ and PNaSS$_{58}$–*b*–PNIPAM$_{115}$ were calculated from the ^{1}H NMR data. As shown in Table 1, the M_n(NMR) values for PNaSS$_{58}$–*b*–PAA$_{125}$ and PNaSS$_{58}$–*b*–PNIPAM$_{115}$ were in reasonable agreement with the M_n(theo) values. However, the M_n(theo) and M_n(GPC) values for both diblock copolymers were found to be slightly different. This may be because the volume-to-mass ratio for PNaSS is different from those for SS$_{58}$–*b*–PAA$_{125}$ and PNaSS$_{58}$–*b*–PNIPAM$_{115}$ [19,20].

We attempted to monitor complex formation between PNaSS$_{58}$–*b*–PAA$_{125}$ and PNaSS$_{58}$–*b*–PNIPAM$_{115}$ induced by the solution pH using ^{1}H NMR spectra measured in D$_2$O containing 0.1 M NaCl at pH 10 and 3. Figure 2 compares the ^{1}H NMR spectra of mixed solutions at pH 10 and 3. In Figure 2a, the resonance peaks at 1.16 (*f*) and 3.91 ppm (*c*) are attributed to the methyl and methine protons, respectively, in the pendant isopropyl group of the PNIPAM block are observed. Moreover, the resonance bands owing to the pendant phenyl protons in the PNaSS block are also detected at 6.1–7.9 ppm (*a* and *b*). The resonance bands in the 1.28–2.38 ppm region (*d* and *e*) are attributed to the sum of the main chain of the diblock copolymers. At pH 3, the intensities of the resonance peaks *f* and *c* derived from the PNIPAM block remarkably decreased, whereas those of *a* and *b* remained intact, as shown in Figure 2b. Besides this result, the resonance bands in the 1.28–2.38 ppm region (*d* and *e*) became broad. From the significant reduction of the motional freedom for the PNIPAM block at pH 3, one can predict that the complexes are formed from PNaSS$_{58}$–*b*–PAA$_{125}$ and PNaSS$_{58}$–*b*–PNIPAM$_{115}$ owing to the hydrogen bonding interactions between the pendant carboxylic acids in the PAA block and the amide groups in the PNIPAM block. At pH 10, however, the complexes dissociated owing to the disappearance of the hydrogen bonding interactions as a result of deprotonation of the carboxylic acids in the PAA block.

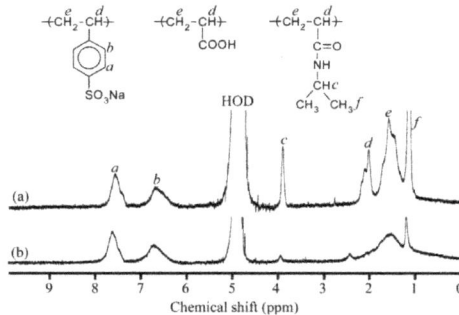

Figure 2. ^{1}H NMR spectra and peak assignment of the mixture of PNaSS$_{58}$–*b*–PAA$_{125}$ and PNaSS$_{58}$–*b*–PNIPAM$_{115}$ in D$_2$O containing 0.1 M NaCl at (**a**) pH 10 and (**b**) pH 3.

Figure 3 shows the light scattering intensities for a mixture of PNaSS$_{58}$–*b*–PAA$_{125}$ and PNaSS$_{58}$–*b*–PNIPAM$_{115}$ in 0.1 M NaCl at pH 3 as a function of f_{NIPAM}. The total polymer concentration was kept constant at 4.4 g/L. An increase in the scattering intensity suggests an increase in the size of the complex. The maximum scattering intensity was observed at $f_{\text{NIPAM}} = 0.5$. This result indicates that a stoichiometric interaction in the mixture of PNaSS$_{58}$–*b*–PAA$_{125}$ and PNaSS$_{58}$–*b*–PNIPAM$_{115}$ led to form a complex with largest aggregation number. The complex with $f_{\text{NIPAM}} = 0.5$ was studied unless otherwise stated.

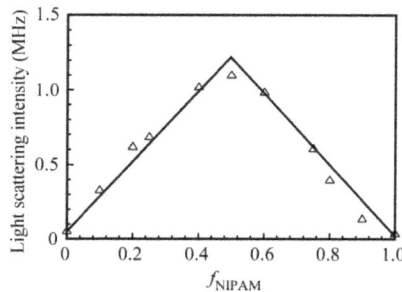

Figure 3. Light scattering intensities of the PNaSS$_{58}$–*b*–PAA$_{125}$/PNaSS$_{58}$–*b*–PNIPAM$_{115}$ complexes as a function of f_{NIPAM} (= (NIPAM)/((NIPAM) + (AA))) in 0.1 M NaCl at pH 3. The total polymer concentration was fixed at 4.4 g/L.

Figure 4 shows the hydrodynamic radius (R_h) distributions for each diblock copolymer and the PNaSS$_{58}$–*b*–PAA$_{125}$/PNaSS$_{58}$–*b*–PNIPAM$_{115}$ complex at pH 3 and 10. The values of R_h were determined by DLS in 0.1 M NaCl and are indicated in the figure. The R_h values for PNaSS$_{58}$–*b*–PAA$_{125}$ and PNaSS$_{58}$–*b*–PNIPAM$_{115}$ were 3.5 and 3.7 nm, respectively, which are reasonable for a unimer state. The R_h values of the complex at pH 3 and 10 were 15.0 and 3.7 nm, respectively. When PNaSS$_{58}$–*b*–PAA$_{125}$ and PNaSS$_{58}$–*b*–PNIPAM$_{115}$ were mixed at pH 3, a water-soluble complex was formed due to hydrogen bonding interactions. On the other hand, at pH 10 the complex dissociated because the hydrogen bonding interactions disappeared as a result of deprotonation of the pendant carboxylic acids.

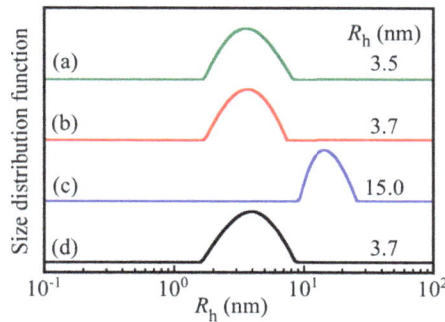

Figure 4. Hydrodynamic radius (R_h) distributions for (**a**) PNaSS$_{58}$–*b*–PAA$_{125}$ and (**b**) PNaSS$_{58}$–*b*–PNIPAM$_{115}$ in 0.1 M NaCl at pH 3, and the PNaSS$_{58}$–*b*–PAA$_{125}$/PNaSS$_{58}$–*b*–PNIPAM$_{115}$ complex at (**c**) pH 3 and (**d**) pH 10.

The relaxation rates (Γ) measured at different scattering angles (θ) are plotted as a function of the square of the magnitude of the scattering vector (q^2) for the mixture of PNaSS$_{58}$–*b*–PAA$_{125}$ and PNaSS$_{58}$–*b*–PNIPAM$_{115}$ in 0.1 M NaCl at pH 3 in Figure 5. A linear plot passing through the origin suggests that the relaxation modes are virtually diffusive [21]. Thus, the R_h values can be estimated at a fixed θ of 90° as the angular dependence is negligible.

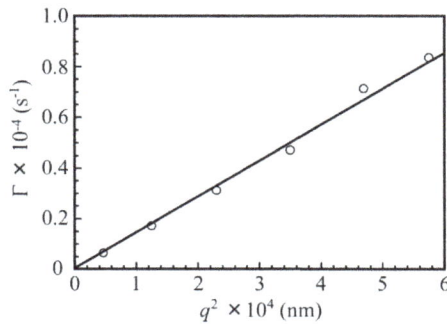

Figure 5. The relaxation rate (Γ) plotted as a function of the square of the magnitude of the scattering vector (q^2) for the PNaSS$_{58}$–*b*–PAA$_{125}$/PNaSS$_{58}$–*b*–PNIPAM$_{115}$ complex at C_p = 4.4 g/L in 0.1 M NaCl at pH 3.

Figure 6 shows the light scattering intensities and R_h values for the mixture of PNaSS$_{58}$–*b*–PAA$_{125}$ and PNaSS$_{58}$–*b*–PNIPAM$_{115}$ in 0.1 M NaCl at pH 3 as a function of the concentration of the mixture. In Figure 6a, the scattering intensity increases sharply with increases in the concentration above a threshold of 0.9 g/L. Figure 6b indicates that the R_h values for the complex are practically constant (R_h is approximately 15 nm) in the concentration range from 0.9 to 4.4 g/L. The results suggest that the formation of the complex between PNaSS$_{58}$–*b*–PAA$_{125}$ and PNaSS$_{58}$–*b*–PNIPAM$_{115}$ starts to occur above a critical concentration of 0.9 g/L. The apparent critical aggregate concentration (CAC) value for the complex, estimated from the plots, is 0.9 g/L.

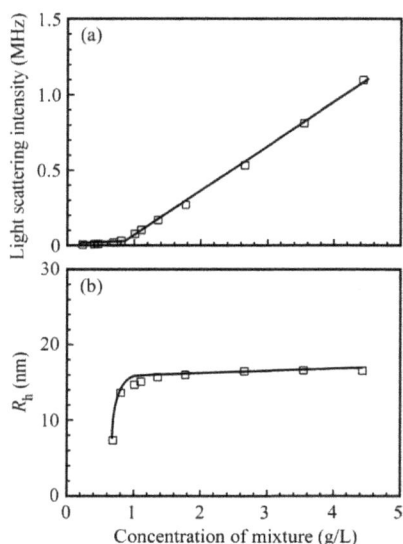

Figure 6. (**a**) Light scattering intensity and (**b**) hydrodynamic radius (R_h) of the PNaSS$_{58}$–*b*–PAA$_{125}$/PNaSS$_{58}$–*b*–PNIPAM$_{115}$ complex as a function of the concentration of the mixture in 0.1 M NaCl at pH 3.

The apparent values of M_w and R_g, determined by SLS measurements, are listed in Table 2 Figure 7 shows a Zimm plot for the PNaSS$_{58}$–*b*–PAA$_{125}$/PNaSS$_{58}$–*b*–PNIPAM$_{115}$ complex, which is formed from a mixture of PNaSS$_{58}$–*b*–PAA$_{125}$ and PNaSS$_{58}$–*b*–PNIPAM$_{115}$ in 0.1 M NaCl at pH 3. The aggregation number (N_{agg}) was defined as the number of polymer chains forming one complex, which can be estimated from the M_w values of the complex and unimer. The result of this calculation gives an N_{agg} of 46 for the complex. The chain numbers of PNaSS$_{58}$–*b*–PAA$_{125}$ and PNaSS$_{58}$–*b*–PNIPAM$_{115}$ for single complex are 22 and 24, respectively, as calculated from $f_{NIPAM} = 0.5$ and the DP values of PAA and PNIPAM.

Table 2. Light scattering data for the PNaSS$_{58}$–*b*–PAA$_{125}$/PNaSS$_{58}$–*b*–PNIPAM$_{115}$ complex.

pH	R_h [a] (nm)	M_w(SLS) [b] $\times 10^{-5}$	R_g [b] (nm)	R_g/R_h	N_{agg} [c]
3	15.0	9.23	12.9	0.86	46
10	3.7	0.20	13.6	3.7	1

[a] Estimated from dynamic light scattering (DLS); [b] Estimated from static light scattering (SLS); [c] Aggregation number calculated from M_w of the aggregate determined at pH 3 and M_w of the corresponding unimer at pH 10 determined from SLS.

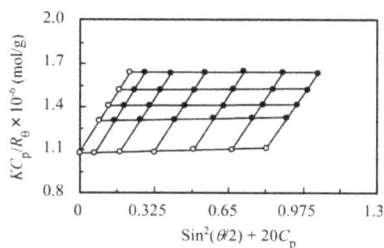

Figure 7. Zimm plot for the PNaSS$_{58}$–*b*–PAA$_{125}$/PNaSS$_{58}$–*b*–PNIPAM$_{115}$ complex with f_{NIPAM} = 0.5 in 0.1 M NaCl at pH 3. Scattering angles (θ) range from 30° to 130° in 20° increments.

The R_g/R_h value indicates the shape of the molecular assemblies. The theoretical R_g/R_h value of a homogeneous hard sphere is 0.778 but this value increases substantially for less dense structures and polydisperse mixtures; for example, R_g/R_h = 1.5–1.7 for flexible linear chains in good solvents, whereas $R_g/R_h \geq 2$ for a rigid rod [22–24]. The R_g/R_h ratio for the complex (Table 2) was 0.86, suggesting that the shape of the complex may be spherical. The R_g/R_h ratio for the unimer was 3.7, which indicates that the unimer was in a relatively expanded conformation with polydispersity.

To confirm the shape and size of the PNaSS$_{58}$–*b*–PAA$_{125}$/PNaSS$_{58}$–*b*–PNIPAM$_{115}$ complex at pH 3, TEM measurements were performed (Figure 8). The complex formed spherical objects with almost uniform contrast, suggesting that it comprises micelles with PAA/PNIPAM cores and PNaSS shells. The average radius estimated from the TEM images for the complex was 13.4 nm, which is similar to the R_h value estimated from DLS.

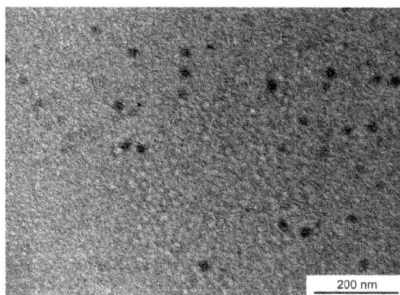

Figure 8. Transmission electron microscopy (TEM) image of the PNaSS$_{58}$–*b*–PAA$_{125}$/PNaSS$_{58}$–*b*–PNIPAM$_{115}$ complex at pH 3.

Figure 9a shows the light scattering intensities for the complex of PNaSS$_{58}$–*b*–PAA$_{125}$ and PNaSS$_{58}$–*b*–PNIPAM$_{115}$ in 0.1 M NaCl as a function of pH. The light scattering intensity increased rapidly as the pH decreased from 4.0 to 3.5, which suggests that the complex was formed below pH 3.5. The light scattering intensity was nearly constant between pH 3.5 and 2.5 suggests that N_{agg} was constant in this pH region because the light scattering intensity is proportional to the molecular mass. Figure 9b shows R_h values for the complex as a function of pH. Above pH 4.0, the R_h values were of the order of 3 nm, suggesting that PNaSS$_{58}$–*b*–PAA$_{125}$ and PNaSS$_{58}$–*b*–PNIPAM$_{115}$ were in a unimer state. As the pH decreased, R_h started to increase at around pH 4.0, reaching a maximum value of 15.0 nm at pH 3.5. As the pH continued to decrease, R_h was nearly constant between pH 3.5 and 2.5, suggesting that not the aggregation number but the compactness of the complex was practically constant.

When the pH value was increased from 3 to 10 and subsequently decreased back to 3, pH-induced R_h changes were found to be completely reversible. Figure 10 shows the pH-induced changes of the

R_h value of the PNaSS$_{58}$–*b*–PAA$_{125}$/PNaSS$_{58}$–*b*–PNIPAM$_{115}$ complex in 0.1 M NaCl cycled between pH 3 and 10 with 20 min intervals. The changes in R_h between the two pH values were completely reproducible, indicating that the association and dissociation of the complex caused by pH changes was reversible over many cycles. This result suggests that the complex may find applications as a pH-responsive controlled association–dissociation system.

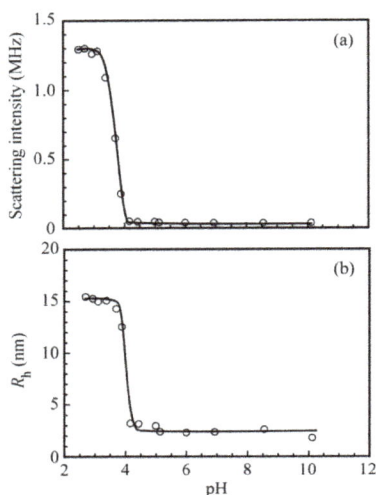

Figure 9. (a) Scattering intensity and (b) hydrodynamic radius (R_h) of the PNaSS$_{58}$–*b*–PAA$_{125}$/PNaSS$_{58}$–*b*–PNIPAM$_{115}$ complex at C_p = 4.4 g/L in 0.1 M NaCl as a function of pH.

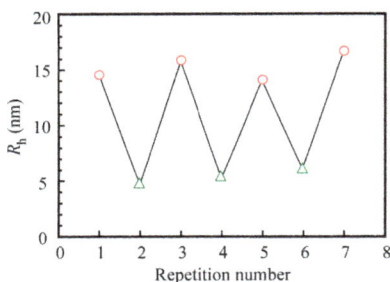

Figure 10. The hydrodynamic radius (R_h) of the PNaSS$_{58}$–*b*–PAA$_{125}$/PNaSS$_{58}$–*b*–PNIPAM$_{115}$ complex at C_p = 4.4 g/L in 0.1 M NaCl at pH 3 (○) and pH 10 (△).

To confirm the formation of the PNaSS$_{58}$–*b*–PAA$_{125}$/PNaSS$_{58}$–*b*–PNIPAM$_{115}$ complex via hydrogen bonding interactions, urea was added to the mixture of PNaSS$_{58}$–*b*–PAA$_{125}$ and PNaSS$_{58}$–*b*–PNIPAM$_{115}$ in 0.1 M NaCl at pH 3. It is known that urea disturbs hydrogen bonding interactions [25]. Figure 11 shows the light scattering intensity for the PNaSS$_{58}$–*b*–PAA$_{125}$/PNaSS$_{58}$–*b*–PNIPAM$_{115}$ complex as a function of the concentration of urea. The light scattering intensity decreased as the concentration of urea increased from 0 to 3.0 mol/L, and the scattering intensity dropped below 0.1 MHz at 3.0 mol/L urea concentration. These observations are indicative of the complete dissociation of the complex caused by an excess amount of urea. Thus, the complex was confirmed to be formed via hydrogen bonding interactions.

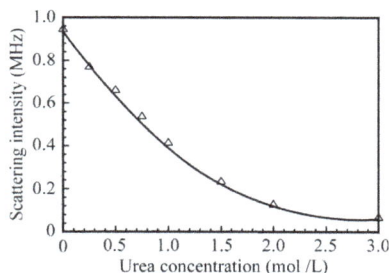

Figure 11. The scattering intensity of the PNaSS$_{58}$–b–PAA$_{125}$/PNaSS$_{58}$–b–PNIPAM$_{115}$ complex at C_p = 4.4 g/L in 0.1 M NaCl as a function of the urea concentration.

The NIPAM block in the PNaSS$_{58}$–b–PNIPAM$_{115}$ diblock copolymer dissolves in water at room temperature, but it separates from aqueous solutions when heated above the lower critical solution temperature (LCST). Figure 12 shows the percent transmittance (%T) values monitored at 600 nm for 0.1 M NaCl aqueous solutions of PNaSS$_{58}$–b–PNIPAM$_{115}$ and a mixture of PNaSS$_{58}$–b–PAA$_{125}$ and PNaSS$_{58}$–b–PNIPAM$_{115}$ at pH 3 and 10, as a function of solution temperature. The diblock copolymer, PNaSS$_{58}$–b–PNIPAM$_{115}$, exhibited a significant %T change at 35–37 °C, which indicates the LCST. The mixture at pH 10 also exhibited a slight %T change at 35–37 °C. On the other hand, the mixture at pH 3 did not show any %T change in the temperature range from 20 to 80 °C. Ordinarily, the mechanism of LCST for PNIPAM can be explained as follows [26,27]. Below LCST, the PNIPAM chains are hydrated because the pendant amide groups form hydrogen bonding with water molecules, whereas above LCST, molecular motions prevail over the hydrogen bonding interactions resulting in the dehydration of the PNIPAM chains, thus leading to phase separation. Therefore, the PNIPAM chains dehydrated causing phase separation. In the case of the PNaSS$_{58}$–b–PAA$_{125}$/PNaSS$_{58}$–b–PNIPAM$_{115}$ complex at pH 3, the pendant amide groups in the PNIPAM block interact with the pendant carboxylic acid in the PAA block irrespective of the temperature. Hence, the pendant amide groups in the PNIPAM block are prevented from forming hydrogen bonds with water molecules. Therefore, the LCST of the complex at pH 3 cannot be observed. However, at pH 10, the mixture dissociated and hydrogen bonding interactions between the pendant amide groups in PNaSS$_{58}$–b–PNIPAM$_{115}$ and water molecules were formed at low temperature. Hence, the LCST can be observed for the mixture of PNaSS$_{58}$–b–PAA$_{125}$ and PNaSS$_{58}$–b–PNIPAM$_{115}$ at pH 10, although the decrease in %T at the LCST was small comparing to the PNaSS$_{58}$–b–PNIPAM$_{115}$ case. This is simply because the concentration of the NIPAM units in the former solution is lower than that in the latter solution.

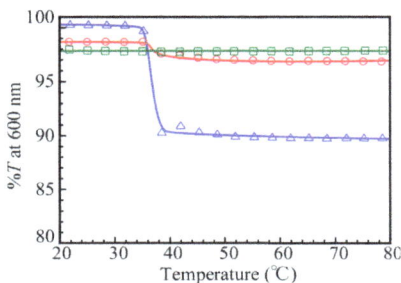

Figure 12. Percent transmittance (%T) at 600 nm for 0.1 M NaCl aqueous solutions of PNaSS$_{58}$–b–PNIPAM$_{115}$ ($\triangle$) and a mixture of PNaSS$_{58}$–b–PAA$_{125}$ and PNaSS$_{58}$–b–PNIPAM$_{115}$ at pH 3 ($\square$) and pH 10 ($\bigcirc$) as a function of solution temperature.

4. Conclusions

The diblock copolymers PNaSS$_{58}$–*b*–PAA$_{125}$ and PNaSS$_{58}$–*b*–PNIPAM$_{115}$ were prepared via RAFT-controlled radical polymerization using a PNaSS$_{58}$ macro–CTA. Both polymerizations proceeded by a controlled mechanism. A mixture of PNaSS$_{58}$–*b*–PAA$_{125}$ and PNaSS$_{58}$–*b*–PNIPAM$_{115}$ formed a water-soluble complex under acidic conditions. The formation of the complex was confirmed by various measurement techniques. The ^{1}H NMR data indicated restricted motion of the PNIPAM block at pH 3 in the complex owing to hydrogen bonding interactions between the pendant carboxylic acids in the PAA block and the pendant amide groups in the PNIPAM block. The DLS and SLS data suggested that the PNaSS$_{58}$–*b*–PAA$_{125}$/PNaSS$_{58}$–*b*–PNIPAM$_{115}$ complex was spherical in shape. When urea was added to the complex aqueous solution, the complex dissociated. This observation indicates that the driving force for the formation of the complex is hydrogen bonding interactions. The %*T* data indicated that the LCST of the complex was not observed at pH 3, owing to the complex formation. However, the complex dissociated to a unimer above pH 4.0.

Acknowledgments: This work was financially supported by a Grant-in-Aid for Scientific Research (17H03071 and 16K14008) from the Japan Society for the Promotion of Science (JSPS), JSPS Bilateral Joint Research Projects, and the Cooperative Research Program of "Network Joint Research Center for Materials and Devices (20174031)".

Author Contributions: Masanobu Mizusaki and Shin-ichi Yusa designed the specific experiments and were responsible for the writing ideas, experimental data analysis; Tatsuya Endo performed the specific experiments, samples preparation; Rina Nakahata TEM observations and data collection; Yotaro Morishima was responsible for the data analysis and manuscript's grammar proofreading.

Conflicts of Interest: The authors declare no conflict of interest.

References

1. Xu, R.; Winnik, M.A.; Riess, G.; Croucher, M.D. Micellization of polystylene-poly(ethylene oxide) block copolymers in water. 5. A test of the star and mean-field models. *Macromolecules* **1992**, *25*, 644–652. [CrossRef]
2. Harada, A.; Kataoka, K. Formation of polyion complex micelles in an aqueous milieu from a pair of oppositely-charged block copolymers with poly(ethylene glycol) segments. *Macromolecules* **1995**, *28*, 5294–5299. [CrossRef]
3. Kabanov, V.A.; Zezin, A.B.; Kasaikin, V.A.; Zakharova, J.A.; Litmanovich, E.A.; Ivleva, E.M. Self-assembly of ionic amphiphiles on polyelectrolyte chains. *Polym. Int.* **2003**, *52*, 1566–1572. [CrossRef]
4. Foreman, M.B.; Coffman, J.P.; Murcia, M.J.; Naumann, C.A.; Cesana, S.; Jordan, R.; Smith, G.S.; Naumann, C.A. Gelation of amphiphilic lipopolymers at the air-water interface: 2D analogue to 3D gelation of colloidal systems with grafted polymer chains. *Langmuir* **2003**, *19*, 326–332. [CrossRef]
5. Jeon, S.H.; Ree, T. Characterization of poly(carboxylic acid)/(poly(ethylene oxide) blends formed through hydrogen bonding by spectroscopic and calorimetric analyses. *J. Polym. Sci. A* **1999**, *26*, 1419–1428. [CrossRef]
6. Lee, W.; Chang, J.; Ju, S. Hydrogen-bond structure at the interfaces between water/poly(methyl methacrylate), water/poly(methacrylic acid), and water/poly(2-aminoethyl-methacrylate). *Langmuir* **2010**, *26*, 12640–12647. [CrossRef] [PubMed]
7. Abe, K.; Koide, M.; Tsuchida, E. Selective complexation of macromolecules. *Macromolecules* **1977**, *10*, 1259–1264. [CrossRef]
8. Velda, J.L.; Liu, Y.; Huglin, M.B. Effect of pH on the swelling behaviour of hydrogels based on *N*-isopropylacrylamide with acidic comonomers. *Macromol. Chem. Phys.* **1998**, *199*, 1127–1134. [CrossRef]
9. Erbil, C.; Akpinar, F.D.; Uyanik, N. Investigation of the thermal aggregation in aqueous poly(*N*-isopropylacrylamide-co-itaconic acid) solutions. *Macromol. Chem. Phys.* **1999**, *200*, 2448–2453. [CrossRef]
10. Shieh, Y.-T.; Lin, P.-Y.; Chen, T.; Kuo, S.-W. Temperature, pH- and CO$_2$-sensitive poly(*N*-isopropylacryl amide-co-acrylic acid) copolymers with high glass transition temperatures. *Polymers* **2016**, *8*, 434–449. [CrossRef]
11. Bian, F.; Liu, M. Complexation between poly(*N,N*-diethylacrylamide) and poly(acrylic acid) in aqueous solution. *Euro. Polym. J.* **2003**, *39*, 1867–1874. [CrossRef]

12. Yokoyama, Y.; Yusa, S. Water-soluble complexes formed from hydrogen bonding interactions between a poly(ethylene glycol)-containing triblock copolymer and poly(methacrylic acid). *Polym. J.* **2013**, *45*, 985–992. [CrossRef]

13. Yusa, S.; Shimada, Y.; Mitsukami, Y.; Yamamoto, T.; Morishima, Y. pH-Responsive micellization of amphiphilic diblock copolymers synthesized via reversible addition-fragmentation chain transfer polymerization. *Macromolecules* **2003**, *36*, 4208–4215. [CrossRef]

14. Yusa, S.; Shimada, Y.; Mitsukami, Y.; Yamamoto, T.; Morishima, Y. Heat-induced association and dissociation behavior of amphiphilic diblock copolymers synthesized via reversible addition-fragmentation chain transfer radical polymerization. *Macromolecules* **2004**, *37*, 7507–7513. [CrossRef]

15. Yusa, S.; Endo, T.; Ito, M. Synthesis of thermo-responsive 4-arm star-shaped porphyrin-centered poly(N,N-diethylacrylamide) via reversible addition-fragmentation chain transfer radical polymerization. *J. Poly. Sci. A* **2009**, *47*, 6827–6838. [CrossRef]

16. Jakeš, J. Testing of the constrained regularization method of inverting Laplace transform on simulated very wide quasielastic light scattering autocorrelation functions. *Czech. J. Phys. B* **1988**, *38*, 1305–1316. [CrossRef]

17. Brown, W.; Nicolai, T.; Hvidt, S.; Stepanek, P. Relaxation time distributions of entangled polymer solutions from dynamic light scattering and dynamic mechanical measurements. *Macromolecules* **1990**, *23*, 357–359. [CrossRef]

18. Zimm, B.H. Apparatus and methods for measurement and interpretation of the angular variation of light scattering; preliminary results on polystyrene solutions. *J. Chem. Phys.* **1948**, *16*, 1099–1116. [CrossRef]

19. Yusa, S.; Fukuda, K.; Yamamoto, T.; Ishihara, K.; Morishima, Y. Synthesis of well-defined amphiphilic block copolymers having phospholipid polymer sequences as a novel biocompatible polymer micelle reagent. *Biomacromolecules* **2005**, *6*, 663–670. [CrossRef] [PubMed]

20. Yusa, S.; Konishi, Y.; Mitsukami, Y.; Yamamoto, T.; Morishima, Y. pH-responsive micellization of amine-containing cationic diblock copolymers prepared by reversible addition-fragmentation chain transfer (RAFT) radical polymerization. *Polym. J.* **2005**, *37*, 480–488. [CrossRef]

21. Xu, R.; Winnik, M.A.; Hallet, F.R.; Riess, G.; Croucher, M.D. Light-scattering study of the association behavior of styrene-ethylene oxide block copolymers in aqueous solution. *Macromolecules* **1991**, *24*, 87–93. [CrossRef]

22. Huber, K.; Bantle, S.; Lutz, P.; Burchard, W. Hydrodynamic and thermodynamic behavior of short-chain polystyrene in toluene and cyclohexane at 34.5 °C. *Macromolecules* **1985**, *18*, 1461–1467. [CrossRef]

23. Akcasu, A.Z.; Han, C.C. Molecular weight and temperature dependence of polymer dimensions in solution. *Macromolecules* **1979**, *12*, 276–280. [CrossRef]

24. Konishi, T.; Yoshizaki, T.; Yamakawa, H. On the "universal constants" ρ and Φ of flexible polymers. *Macromolecules* **1991**, *24*, 5614–5622. [CrossRef]

25. Yin, X.; Stöver, H.D.H. Thermosensitive and pH-sensitive polymers based on maleic anhydride copolymers. *Macromolecules* **2002**, *35*, 10178–10181. [CrossRef]

26. Heskins, M.; Guillet, J.E. Solution properties of poly(N-isopropylacrylamide). *J. Macromol. Sci. A* **1968**, *2*, 1441–1455. [CrossRef]

27. Winnik, F.M. Fluorescence studies of aqueous solutions of poly(N-isopropylacrylamide) below and above their LCST. *Macromolecules* **1990**, *23*, 233–242. [CrossRef]

![polymers logo] *polymers*

MDPI

Review

Micellization of Photo-Responsive Block Copolymers

Oliver Grimm [1,†], Felix Wendler [1,†] and Felix H. Schacher [1,2,*]

[1]   Institute of Organic Chemistry and Macromolecular Chemistry (IOMC), Friedrich-Schiller-University Jena, Humboldtstraße 10, D-07743 Jena, Germany; oliver.grimm@uni-jena.de (O.G.); felix.wendler@uni-jena.de (F.W.)

[2]   Jena Center for Soft Matter (JCSM), Friedrich-Schiller-University Jena, Philosophenweg 7, D-07743 Jena, Germany

*   Correspondence: felix.schacher@uni-jena.de; Tel.: +49-3641-948-250

†   Both authors contributed equally to this work.

Received: 24 July 2017; Accepted: 22 August 2017; Published: 26 August 2017

Abstract: This review focuses on block copolymers featuring different photo-responsive building blocks and self-assembly of such materials in different selective solvents. We have subdivided the specific examples we selected: (1) according to the wavelength at which the irradiation has to be carried out to achieve photo-response; and (2) according to whether irradiation with light of a suitable wavelength leads to reversible or irreversible changes in material properties (e.g., solubility, charge, or polarity). Exemplarily, an irreversible change could be the photo-cleavage of a nitrobenzyl, pyrenyl or coumarinyl ester, whereas the photo-mediated transition between spiropyran and merocyanin form as well as the isomerization of azobenzenes would represent reversible response to light. The examples presented cover applications including drug delivery (controllable release rates), controlled aggregation/disaggregation, sensing, and the preparation of photochromic hybrid materials.

Keywords: block copolymers; photo-responsive; self-assembly; micellization

1. Introduction and Scope

Stimuli-responsive materials are capable of undergoing (reversible) changes with regard to physico-chemical properties such as solubility, polarity, charge, charge density, size or shape. Often, this is exploited in applications, where either response is translated into a certain action or where changes in environmental conditions are reported as or transformed into an optical or chemical readout. Examples are diagnostics, drug delivery, tissue engineering, or even triboelectric materials [1]. Among the different stimuli available, light is attracting more and more attention as it can be controlled both spatially and temporally with utmost precision. Moreover, careful adjustment of the used wavelength allows precisely controlling the amount of energy introduced into the system that is exposed to light. This can be crucial to allow certain processes to occur or to avoid irreversible photo damage to the material. The latter case has to be especially considered if biomedical applications are in focus and tissue integrity or cell viability has to be maintained. Here, the Near infra-red (NIR) window (λ = 650–900 nm) is highly attractive as it allows penetration depths of up to 10 mm through human skin and subcutaneous regions. Absorption by blood constituents is reduced and tissue damage or phototoxic effects are minimized [2,3].

Whereas examples for polymeric materials that respond to changes in pH (e.g., polyelectrolytes) [4] or temperature (e.g., poly(N-isopropylacrylamide)) [5] are manifold, fewer reports deal with photo-responsive polymers and block copolymers. Nevertheless, the last decade has shown that more and more research groups are working on strategies to implement such properties using different approaches; for a general overview, we direct the reader to recent review articles by Zhao, Gohy and Bertrand [6–8].

This review focuses on photo-responsive block copolymers and nanostructured materials generated thereof via self-assembly in selective solvents, whereby most examples report on processes in aqueous media using amphiphilic block copolymers. For bulk applications, the reader is referred to a recent review by Gohy and Bertrand dealing with photo-responsive surfaces, thin films, and gels/hydrogels [8] and an earlier book chapter discussing specifically azo-functionalized block copolymers in the solid state [9]. During this review, we have subdivided the specific examples: (1) according to the wavelength at which the irradiation has to be carried out; and (2) according to whether irradiation with light of a suitable wavelength leads to reversible or irreversible changes in material properties. Thereby, an irreversible change could be the photo-cleavage of a nitrobenzyl ester [10], whereas the photo-mediated transition between spiropyran and merocyanin [11] form would represent reversible response to light. The examples presented cover applications including drug delivery (controllable release rates), controlled aggregation, sensing, and the preparation of photochromic hybrid materials.

2. Block Copolymers in General

Block copolymers represent a class of materials where at least two different monomers are arranged in a sequential manner. As a result, one single material can combine the properties of all involved monomer units. One further peculiarity of block copolymers is that, contrary to statistic, random, or gradient copolymers, the intrinsic immiscibility of unlike polymeric segments leads to microphase separation and thereby a straightforward access to nanostructured materials. This holds true for different (micro) environments, i.e., in the bulk or in solution [12,13].

The synthesis and the even today steadily increasing availability of block copolymers originates in the discovery of living polymerization techniques by Michael Szwarc, more specifically the possibility to produce segmented polymeric materials in a sequential manner and, even more important, without unwanted termination or transfer reactions during the polymerization [14,15]. Whereas ionic (anionic, cationic, and anionic/cationic ring-opening polymerization) techniques are still rather demanding in terms of monomer choice, purification procedures, or the tolerance towards functional groups, another huge step forward in the variety of block copolymer compositions and functionalities occurred after the development of reversible deactivation radical polymerization (RDRP) techniques. Here, the most prominent examples are atom transfer radical polymerization (ATRP) [16–18], nitroxide mediated polymerization (NMP) [19], and reversible addition fragmentation chain transfer polymerization (RAFT) [20,21]. With constant evolution of these techniques, monomers or monomer combinations previously not being accessible could be realized, and a large variety of functional groups were incorporated into polymeric materials and block copolymers of different architecture. Today, polymer and material chemists can rely on a versatile toolbox of reliable polymerization techniques to prepare materials of defined composition, controllable architecture, and narrow dispersity.

Initially, block copolymers were regarded as rather exotic materials, mainly due to (still) increasing synthetic demand, but, during the last 20 years, more and more research groups developed interest in block copolymers. This can be mainly assigned to the straightforward control over both domain size and morphology of block copolymer-based nanostructured materials in the bulk and in solution. Both can be precisely adjusted by molecular parameters such as molar mass, volume fraction of the respective segments, overall architecture, and also segment rigidity (this also includes the presence of crystallizable blocks). Additional elements of control include strategies to influence the self-assembly process itself ("pathway") [22], or to exploit suitable chemical tools to specifically address individual domains or even the block junction [23,24] between two adjacent segments in block copolymer nanostructures [25,26].

3. λ-Dependent Photo-Response in Block Copolymer Nanostructures in Solution

In the following sections, we try to give an overview over relevant photo-responsive chromophores that have been incorporated into block copolymers. We decided to distinguish between examples that refer to reversible or irreversible response upon irradiation with light, as well as with reference to other review articles in this field [6–8]. Furthermore, we have grouped the respective articles according to the wavelength at which the irradiation has to be carried out. Thereby, three wavelength regimes have been defined, Far UV (200–400 nm), Vis (400–700 nm) and Near IR (700–1000 nm), and the most relevant chromophores are depicted in Figure 1.

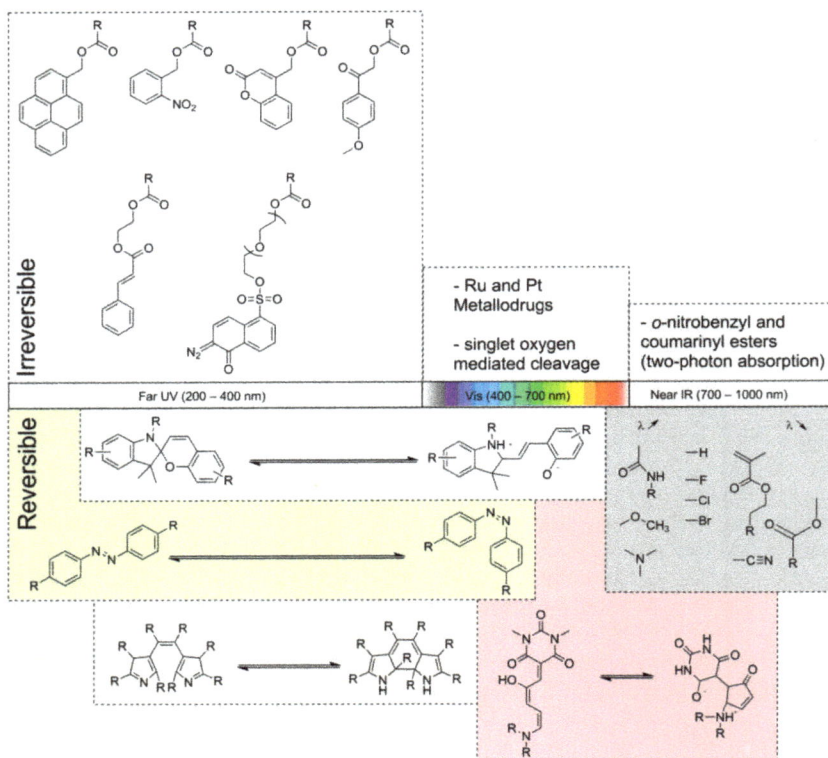

Figure 1. Schematic representation of relevant chromophores that have been incorporated into block copolymers; we distinguish between reversible/irreversible photo-response as well as different excitation regimes: Far UV (200–400 nm), Vis (400–700 nm) and Near IR (700–1000 nm); the grey part depicts different functional groups that are capable of influencing the wavelength at which chromophores respond.

3.1. Far UV (200–400 nm)

3.1.1. Reversible

In general, UV light within the range of 200–400 nm provides enough energy to break bonds, induce cis–trans isomerizations, and enable cycloaddition reactions. The most important molecule classes discussed within this section as chromophores are azobenzenes [27,28], spiropyrans [29], cinnamoyl esters [30], and diarylethenes [31].

The photochromism of heterocyclic 1,2-diarylethenes as a subclass of fulgides has been known since 1905 [32], and the substitution pattern of these photochromic systems was considerably extended in 1988 by Irie and Mohri [33]. The ring closure of single molecule diarylethenes is typically triggered at 325 nm (the back reaction to the cis-form often occurs upon irradiation at 488 nm) [34], but depending on the substitution pattern this wavelength can vary from 313 to 405 nm (correspondingly, the back reaction is triggered from 405 to 546 nm) [33,35]. These photo-responsive diarylethenes can be incorporated into polymers either in the main chain or in the side chain. Some of them could be polymerized, as shown by Stellacci et al. [36] and a redshift of the absorption spectra was observed if compared to the monomer. The modification of a suitable diarylethene with a polymerizable styrene group enabled Nishi et al. to synthesize poly(diarylethene)-*block*-polystyrene diblock copolymers via reversible addition fragmentation chain transfer polymerization (RAFT) [37]. The RAFT endgroup was cleaved off and the thiol was attached to a gold nanoparticle to form photochromic hybrid nanoparticles in toluene (Figure 2). The same technique and also photo-responsive monomer was used by Kobatake [38] to prepare poly(diarylethene)-*block*-PNIPAAm using surface-initiated polymerization from silica nanoparticles. These nanoparticles were investigated both in water and in THF.

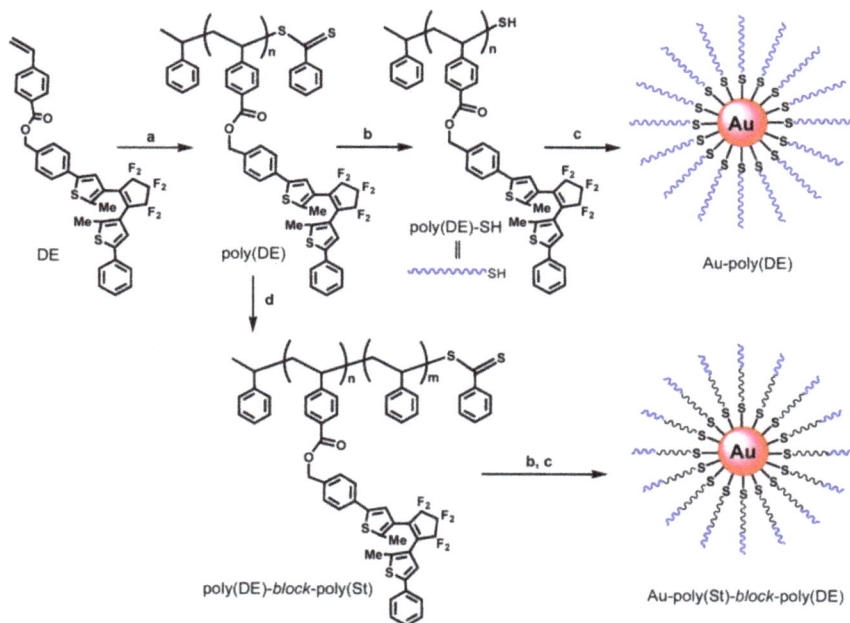

Figure 2. Synthetic scheme for the preparation of Au-poly(diarylethene) and Au-polystyrene-*block*-poly(diarylethene) hybrid nanoparticles. Reprinted with Permission from Ref. [37]. Copyright 2008 American Chemical Society.

The first azo-dye was synthesized by Martius in 1863, and, in the following year, Griess reported the coupling reaction of related diazonium compounds [39]. Nevertheless, it took until 1937 for the reversible cis–trans-isomerization of azobenzene to be proven by Hartley [40]. After irradiation of a sample with blue light in benzene, they observed different dipole moments compared to a non-irradiated sample. This led to a different explanation for this behavior, as up to now the light response of azobenzene was ascribed to some oxidation of the compound [41]. Single molecule azobenzenes have been prepared with a great variety of substituents and exhibit excitation wavelengths starting in the far UV range [27,28] (here, the back isomerization is often mediated by thermal relaxation) to 410 nm [42] (here, the back isomerization occurs upon irradiation at 500 nm). The first polymer containing azobenzene as photo-responsive group was synthesized by Ringsdorf et al. in 1984 [43]. They first prepared the monomer 6-[4-(4-cyano-phenylazo)phenoxyhexyl acrylate and afterwards performed free radical copolymerization with the corresponding benzoate, resulting in a liquid crystalline copolymer. Some years later in 1989, Angeloni et al. compared the spectroscopic properties of main chain and side chain azobenzene polymers with the corresponding monomers and found that the substitution pattern has the largest effect on the excitation maximum [44]. This effect is also shown for block copolymers consisting of a polystyrene (PS), poly(methyl methacrylate) (PMMA), poly(β-acetyl galactose ethyl methacrylate) or poly(ethylene glycol) (PEG) block in combination with various polymeric azobenzenes [45–49]. In that regard, the first block copolymer containing a photo-responsive azobenzene segment was synthesized by Se et al. in 1997 using side chain modification of a polystyrene-*block*-N,N-dimethyl-4-vinylphenethylamine block copolymer prepared by sequential anionic polymerization with *p,p′*-Bis(chloromethyl)azobenzene [50]. In another example, the copolymerisation of methacrylate-based azobenzenes (AzoMA) with N-isopropyl acrylamide (NIPAAm) with various amounts of NIPAAm was shown by Ueki et al. and this significantly affected the thermal response characteristics of the resulting copolymers upon irradiation [51]. The change of the dipole moment of azobenzene was used by Concellón et al. to reversibly load aqueous block copolymer micelles (Figure 3) [52]. Here, RAFT techniques were used to prepare poly(ethylene glycol)-*block*-poly(2,6-diacylaminopyridine) (PEG-b-PDAP), and load these structures with N(1)-[12-(4-(4′-isobutyloxyphenyldiazo)phenoxy)dodecyloxy)] thymine (tAZO$_i$). Afterwards, the loading could be reversibly released upon irradiation with UV-light [53].

The thermally induced color change of spiropyrane-based compounds in solution has been known since the 1920s [54,55]. The origin of this color change is the light induced breakage of a bond between a tertiary carbon and a cyclic heteroatom, thereby switching between a bicyclic spiropyrane form and a zwitterionic merocyanine. Since then, a multitude of substitution patterns have been synthesized [11,56], whereas the first detailed report on the photo-response of this class of molecules was published by Fischer et al. in 1954 [57]. Benzospiropyrane molecules mainly switch to the open form upon irradiation at 365 nm and close the ring typically triggered by irradiation at 560 nm [58–60]. Copolymers containing a spiropyrane moiety were intensively studied by Smets in 1972, including a copolymer with MMA synthesized via free radical polymerization [61]. The first copolymer containing spiropyrane was synthesized by Krongauz et al. in 1981 by modifying benzospiropyrane with an acrylate group, followed by free radical polymerization [62]. The first reported block copolymer was of ABA-type and was prepared by De Los Santos et al. in 1999 [63]. They modified PMMA-b-PU-b-PMMA in the B segment with different spiropyrane molecules. Since then, various examples were found for amphiphilic diblock terpolymers [64,65], mainly consisting of a PEG block, and various poly(benzospiropyrane) segments with different comonomers. In general, their absorption maxima did not differ significantly from the values previously reported for the monomeric compounds [66]. The synthesis of amphiphilic block copolymers leads to many different applications from polymeric liquid crystals to photo-responsive block copolymer micelles [67–71]. Thereby, the formation of micelles can be triggered by addressing the spiropyrane as shown by Guragain et al. [70] or the release rate of encapsulated cargo in case of core-crosslinked micelles can be controlled as shown by Wang et al. [68]. Menon et al. [71] synthesized an amphiphilic block copolymer consisting of poly(spiropyrane methacrylate)-*block*-poly(3-O-4-vinylbenzoyl-D-glucopyranose), which forms 200 nm micelles in aqueous solution that can be loaded in the dark with coumarin, and release their loading upon irradiation with UV light (360 nm). They could further demonstrate that the micelles can be formed again upon irradiation with green light (560 nm). In that respect, the coupling of a thermo-responsive block to the spiropyrane-containing segment enables to control the formation of micelles also by temperature [72]. Very recently, Zhang and coworkers used RAFT polymerization to prepare P(NIPAAm)-b-poly(N-acryloylglycine) diblock copolymers, and subsequently functionalized those with benzospiropyrane ((PNIPAAm$_{94}$-b-P(NAG$_{19}$-co-NAGSP$_{30}$,

Figure 4). These multiple stimuli-responsive block copolymers formed spherical or worm-like micelles in water, depending on the temperature (above or below the LCST of PNIPAAm) or the irradiation with light (switching between merocyanine and spiropyrane form) [67].

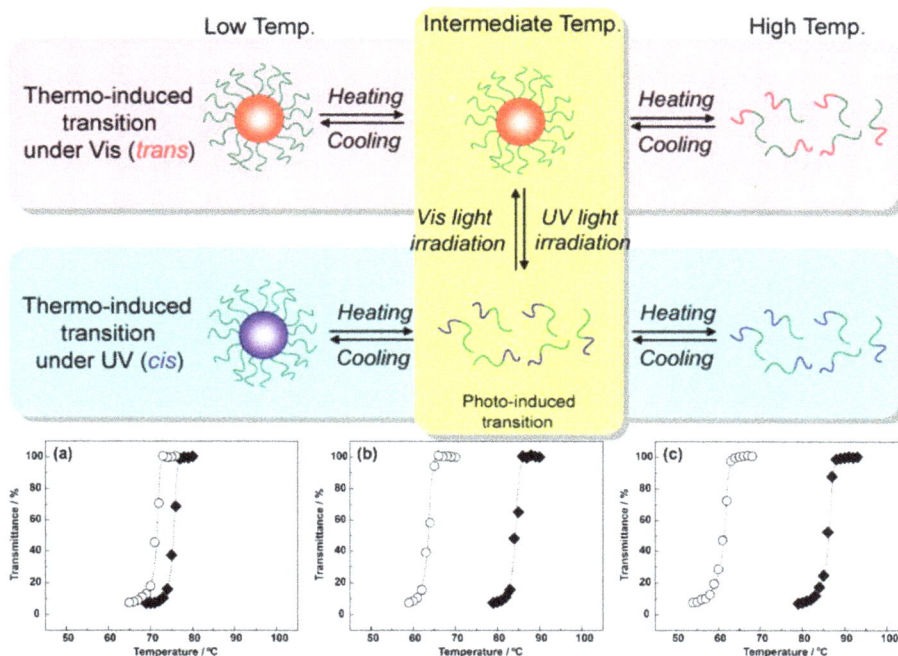

Figure 3. The top row highlights the characteristics of dual stimuli-responsive block copolymer micelles; below, turbidity measurements for: (**a**) p(AzoMA$_{1.9}$-*r*-NIPAAm); (**b**) p(AzoMA$_{5.4}$-*r*-NIPAAm); and (**c**) P(AzoMA$_{8.6}$-*r*-NIPAAm) in [C$_4$mim]PF$_6$ solution are shown upon irradiation with UV light (366 nm, open circles) or in the dark (closed diamonds); the subscript denotes the respective content (mol %) of AzoMA in the random copolymers. Adapted with permission from Ref. [51]. Copyright 2012 American Chemical Society.

Further reversible photo-responsive moieties that can be triggered in the UV region include, e.g., coumarines or anthracenes—Both being capable of undergoing cycloaddition reactions. In 1996, Liu et al. [73] synthesized polystyrene-*block*-poly(2-cinnamoylethyl methacrylate), and prepared nanofibers which were afterwards stabilized via crosslinking. The reversibility of the crosslinking process was first shown by Lendlein and coworkers in 2005, where cinnamic acid was crosslinked upon irradiation at 310 nm via (2 + 2) cycloaddition (de-crosslinking can be triggered by excitation below 260 nm) and they employed such materials in shape memory polymers [74] or reversibly cross-linkable shells in block copolymer micelles [75,76]. This can also be achieved using coumarine moieties in the side chain [77–79] and Zhang et al. reported on amphiphilic block copolymers of poly(ethylene oxide)-*block*-poly((*N*-methacryloxyphthalimide)-*co*-(7-(4-vinyl-benzyloxyl)-4-methylcoumarin)) and the formation of micelles thereof in aqueous media. The core can now be reversibly photo-crosslinked upon irradiation at 365 nm and this enables loading or release of various drugs such as doxorubicin [80]. Anthracene on the other hand can be crosslinked by (4 + 4) cycloadditions and Xie and coworkers used anionic ring-opening polymerization to form poly(L-lactide)-*block*-poly(ethylene glycol) functionalized with anthracene moieties in the side chain—These materials could then be used as light-responsive shape memory block copolymers [81].

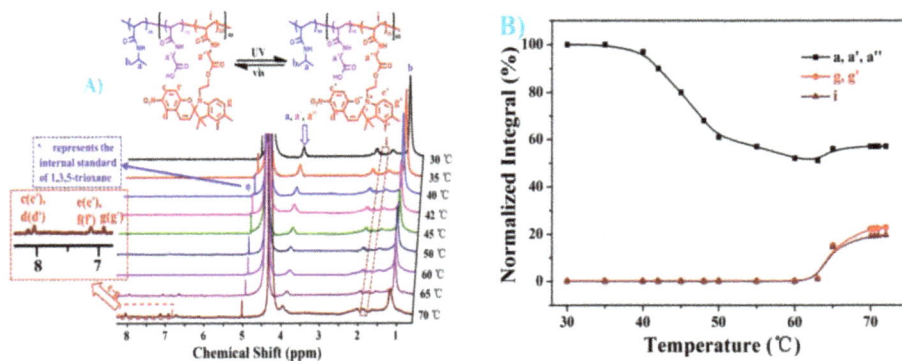

Figure 4. ^{1}H NMR spectra of 1 wt % PNIPAAm$_{94}$-*b*-P(NAG$_{19}$-*co*-NAGSP$_{30}$) micelles dispersed in D$_2$O at different temperatures (**A**); and the temperature-dependent signals of three typical protons of PNIPAAm$_{94}$-*b*-P(NAG$_{19}$-*co*-NAGSP$_{30}$) micelles (**B**). Reproduced from Ref. [67] with permission from The Royal Society of Chemistry.

3.1.2. Irreversible

Most irreversible photoreactions that have been reported showed response to far UV light and include photo cleavage of side groups as well as the introduction of irreversible crosslinks via photocycloadditions or photo rearrangement reactions. The former approach was essentially inspired by photo-labile protecting groups, e.g., pyrenylmethyl esters [82], *o*-nitrobenzyl esters [83], coumarinyl esters [84], and *p*-methoxy-phenacyl esters [85]—All of these have been investigated during the second half of the last century [86]. In this regard, the group of Zhao introduced two fundamental examples both presenting a block copolymer consisting of a hydrophilic PEO block and a hydrophobic and photo-responsive polymethacrylate-based segment [87,88]. The respective cleavage of either the pyrenylmethyl or *o*-nitrobenzyl groups upon UV irradiation (365 nm) shifted the hydrophilic/hydrophobic balance of the amphiphilic block copolymer through the generation of hydrophilic poly(methacrylic acid) (Figure 5). This then led to swelling or even dissociation of the micellar aggregates in aqueous media and a release of encapsulated Nile Red as model drug. Especially *o*-nitrobenzyl chromophores are discussed as promising candidates for biological applications since even stimulation with NIR light is possible (see Section 3.3). Consequently, several block copolymers bearing *o*-nitrobenzyl ester groups in the side chain in combination with blocks of PEO/PEG or POEGMA [89–97], polystyrene [98–100], poly(methyl acrylate) [101], poly(2-ethyl-2-oxazoline) [102], or polydimethylacrylamide [103] have been synthesized and investigated towards their response upon irradiation with far UV light. Exemplarily, Liu and Dong showed the photo-controlled release of the anticancer drug doxorubicin from a biodegradable polypeptide-based poly(*S*-(*o*-nitrobenzyl)-L-cysteine)-*block*-poly(ethylene oxide) block copolymer. The materials formed spherical micelles in aqueous media and exhibited a significant reduction in size after irradiation [104]. In particular, the combination of any photo-responsiveness with other stimuli, e.g., temperature, seems very favorable in that regard because inspired by natural examples a conceivable release mechanism could be initiated more effectively and thus more controlled. In that regard, dual [105–118], triple [119,120], and even quadruple responsive systems have already been presented [121]. For example, Cao et al. prepared a quadruple-responsive (light, temperature, pH and redox) poly(2-nitrobenzyl methacrylate)-*block*-poly(dimethylaminoethyl methacrylate) diblock copolymer where both blocks have been connected through a disulfide linker [122]. UV irradiation led to photo-cleavage of the *o*-nitrobenzyl groups whereas addition of the reducing agent dithiothreitol separated the blocks at the block junction, resulting in disruption of the nanoparticles (spherical structures with diameters of 80 to 140 nm) in both cases.

The poly(dimethylaminoethyl methacrylate) block responded to temperature (shrinkage, diameter decreased to 20–50 nm) and to changes in pH (shrinkage to 40–60 nm at pH 9, swelling at pH 3). Furthermore, the incorporation of *o*-nitrobenzyl esters in the main chain [123,124] or as a block junction synthesized via divergent polymerization [125–129], convergent coupling [129–136], or even as combination of both strategies [137–139] represents another option to prepare photolytically cleavable amphiphilic block copolymers.

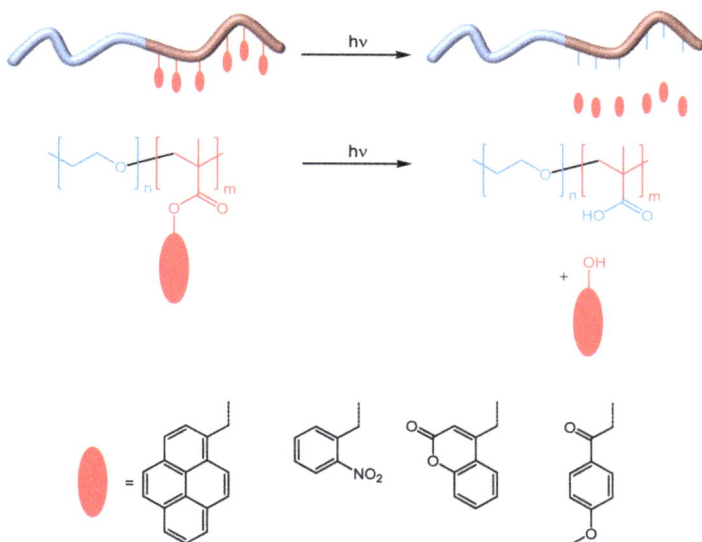

Figure 5. Schematic representation of UV light triggered photo-cleavage of pyrenylmethyl, *o*-nitrobenzyl, coumarinyl and *p*-methoxy-phenacyl esters within block copolymer materials.

In 2009, again Zhao and coworkers demonstrated a similar concept, however they employed coumarinyl moieties instead of pyrenylmethyl and *o*-nitrobenzyl esters as photo-cleavable entities [140]. In analogy to previous studies, a PEO block was combined with a polymethacrylate-based segment containing coumarin functionalities. As another example for the controlled release of therapeutic cargo, Jin et al. reported on the release of the previously covalently bound anticancer drug 5-fluorouracil from a coumarin-functionalized polymer block under irradiation at 254 nm [141]. Quite recently, a block copolymer consisting of a PEO block and a hydrophobic segment containing both phthalimide and coumarin functional groups in the side chain was designed by Zhang et al. [80]. Irradiation with light of 365 and 254 nm wavelength led to both photo-cleavage of the phthalimide esters and simultaneous crosslinking via the coumarin groups. This approach enables the regulation of the amphiphilic imbalance and the crosslinking density of block polymer micelles simultaneously. The afore-mentioned *p*-methoxy-phenacyl esters are also the subject of more recent works by Bertrand et al. [98,142]. Thereby, the absence of a nitro group facilitates controlled polymerization procedures of the respective monomers rendering these compounds favorable if compared to the previously discussed nitrobenzyl esters. Compared to examples involving coumarin groups, photocycloadditions of cinnamic esters can be considered as a different case since most examples have been shown to be reversible. Two early studies reported about (2-cinnamoylethyl methacrylate) based block copolymers and their nanoaggregate formation in organic solvent mixtures, followed by (at least according to these descriptions) irreversible photo-crosslinking experiments [143]. In 2012 the same group demonstrated a dual light-responsive triblock terpolymer consisting of a *o*-nitrobenzyl ester block junction as well as a crosslinkable poly(2-cinnamoylethyl methacrylate) (PCEMA) block [144,145]. Photolysis of THF/water mixtures (80% water, particle sizes: 20–50 nm in diameter) led to both cleavage of the hydrophilic PEO block and a precipitation of the now core-crosslinked nanoparticles. Another example by Yang et al. presented the use of the cinnamic acid (2 + 2) cycloadduct, truxillic acid, as a block junction and demonstrated its use for the photo cleavage of a poly(ethylene glycol)-*block*-poly(acrylate) diblock copolymers [146]. In that context, light induced rearrangement reactions open up another strategy to induce a sudden shift in the hydrophilic/hydrophobic balance of block copolymers. Among others, two different examples, i.e., the Wolff rearrangement of diazonaphtoquinones [147] and the photo-Claisen rearrangement of allylphenyl ethers [148], have been applied to the field of block copolymers. Other irreversible UV light induced photo-responsive block copolymers include the application of photo-decomposable polyurethanes [149,150] and photoacid generators [151]. In the latter case, a PMMA block was combined via sequential RAFT polymerization with a segment containing photoacid-generating sulfonium groups. The self-assembly in the bulk and the lithographic properties in the course of the photochemical reaction were investigated.

3.2. Vis (400–600 nm)

3.2.1. Reversible

The Donor–Acceptor Stenhouse Adduct (DASA) is relatively new to the class of photochromic molecules (Figure 6) [152], whereas the underlying Stenhouse adduct itself is already known since 1850 [153]. Helmy et al. first functionalized a PEG segment with this group and afterwards showed that the resulting materials reversibly respond to irradiation with visible light (570 nm) in toluene, which can also be used for cargo release upon incorporation into micellar systems (Figure 6). DASA was first used in polymeric form by Balamurugan in 2016 by copolymerizing glycidyl methacrylate (GMA) and dimethylacrylamide (DMA) via RAFT polymerization to yield poly(glycidyl methacrylate-*co*-dimethylacrylamide) [P(GMA-*co*-DMA)] [154]. This was then modified with DASA and showed excellent photochromic performance upon irradiation with a crystal clear halogen lamp. It was also shown by Sinawang et al. that it is possible to introduce DASA into the side chain by post polymerization functionalization of poly(styrene-*co*-4-vinylbenzyl chloride) copolymers [155].

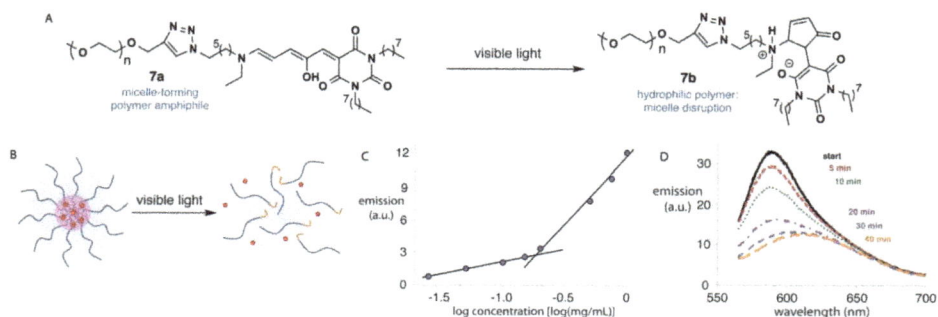

Figure 6. (**A**) Photo-switching of a DASA-functionalized polymeric amphiphile; (**B**) schematic of micelle formation and hydrophobic cargo encapsulation by a photo-responsive amphiphile and micelle disruption and cargo release upon irradiation with visible light; (**C**) fluorescence intensity (emission at 588 nm) vs. log concentration (mg/mL) of the polymeric amphiphile; and (**D**) fluorescence emission spectra of Nile Red in 0.50 mg/mL of the polymeric amphiphile in water at various times of irradiation. Reprinted with permission from Ref. [152]. Copyright 2014 American Chemical Society.

Photo-responsive molecules that can be addressed by irradiation with visible light are of great interest, therefore a redshift by varying the substitution pattern for different chromophores has been discussed in a review by Bleger et al. [156], although only few examples have been incorporated into block copolymers so far. By modification of the linkage from the benzospiropyrane side chain functionality to the block copolymer backbone it was possible for Wang [68] and coworkers to synthesize poly(ethylene glycol)-*block*-poly(spiropyranemethacrylate) (PEG-*b*-PSPMA) diblock copolymers. These amphiphilic materials self-assembled in aqueous media into vesicles, which could be loaded with doxorubicin, gold nanoparticles, or various fluorescence markers. The microstructures of both spiropyrane and merocyanine polymersomes are synergistically stabilized due to hydrophobic interactions, hydrogen bonding, $\pi-\pi$ stacking, and electrostatic (zwitterionic) interactions, with the latter two types being exclusively found for MC polymersomes. Moreover, the reversible photo-triggered SP/MC polymersome transition is accompanied by changes in membrane permeability—Thereby shifting from being non-permeable (450 nm) to selectively permeable (420 nm) towards non-charged, charged, and zwitterionic small molecule species below a critical molar mass (Figure 7) [68].

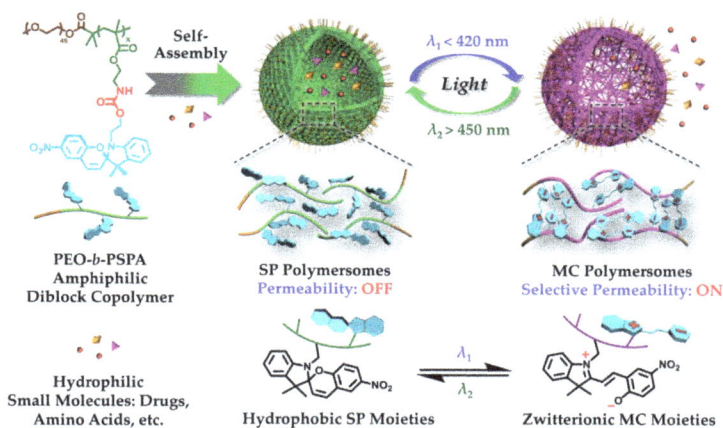

Figure 7. Amphiphilic PEO-*b*-PSPA diblock copolymers self-assemble into polymersomes with hydrophobic bilayers containing carbamate-based hydrogen-bonding motifs; the spiropyran moieties within the polymersome bilayers undergo reversible photo-triggered isomerization between hydrophobic spiropyran (SP, $\lambda_2 > 450$ nm) and zwitterionic merocyanine (MC, $\lambda_1 < 420$ nm) states. Reprinted with permission from Ref. [68]. Copyright 2015 American Chemical Society.

Although not focusing on solution structures, it was further shown by Yu et al. in 2006 that a linearly polarized laser beam (488 nm) can be used to control the self-assembly of nanocylinders from an amphiphilic liquid-crystalline diblock copolymer consisting of flexible poly(ethylene oxide) as hydrophilic block and a poly(methacrylate) containing an azobenzene moiety in the side chain as a hydrophobic liquid-crystalline segment (PEO$_{114}$-*b*-PMAAz$_{60}$) [157]. Upon irradiation at 488 nm, these PEO cylinders could be reoriented in perpendicular direction.

3.2.2. Irreversible

Only few examples have been developed where photo-responsive block copolymers have been activated within the visible light regime. Besides, a classification concerning the actually used chromophores similar to that seen for the above-mentioned systems is not equally straightforward. However, two different general strategies can be identified so far. First, photo-cleavage mechanisms have to be discussed and, in that regard, Sun et al. reported the red-light mediated (520 nm) cleavage from block copolymers consisting of a PEG block and a 6-(4-cyanophenoxy) hexyl methacrylate block that in certain amounts coordinates ruthenium complexes which in turn are potential anticancer metallodrugs [158]. Another example by Zhou and coworkers presented similarly an ABA triblock copolymer with water soluble PEG as A segment and hydrophobic polyurethane having a Pt(IV) prodrug linked to the backbone as the middle block B [159]. Under irradiation with UV or visible light a conversion to Pt(II) occurred which in vivo (demonstrated with BALB/c nude mice) enabled binding to DNA, finally resulting in cell death.

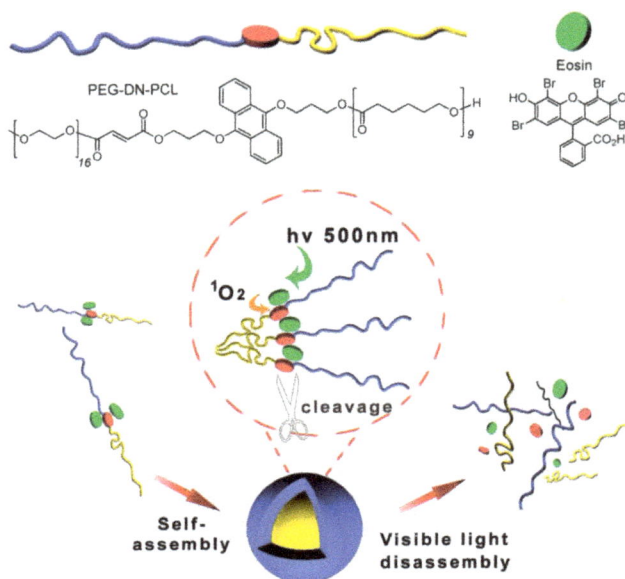

Figure 8. Schematic representation of a poly(ethylene glycol)-*block*-poly(caprolactone) amphiphilic block copolymer featuring a dialkoxyanthracene block junction, its self-assembly and visible light-triggered disassembly via photo-cleavage of the block junction. Reproduced from Ref. [160] with permission from the Royal Society of Chemistry.

The second approach represents in principle also a cleavage reaction, although here the process is mediated by the presence of singlet oxygen. For instance, by using an eosin sensitizer that responds to visible light sources and produces singlet oxygen, dialkoxyanthracenes can be converted to 9,10-anthraquinones by cleavage of the alkoxy moieties. This principle was demonstrated for a poly(ethylene glycol)-*block*-poly(caprolactone) PEG-*b*-PCL amphiphilic block copolymer consisting of a dialkoxyanthracene block junction in 2012 (Figure 8) [160]. Alternatively, porphyrin derivatives can be used as sensitizers for a subsequent cleavage like shown for an ABA-type triblock copolymer with a singlet oxygen sensitive diselenide-containing polyurethane B middle block surrounded by PEG segments [161]. Saravanakuma et al. presented a similar approach for a poly(ethylene glycol)-*block*-poly(caprolactone) diblock copolymer with a cleavable vinyldithioether block junction [162]. In both cases, the visible light mediated release of doxorubicin as a hydrophobic model drug was tested. Quite recently, also amphiphilic diblock copolymers with one block featuring a diselenide linkage in the side-chain were reported [163]. Visible light exposure induced diselenide exchange and, thereby, crosslinking of these drug-loaded nanocarriers which then in turn are capable of undergoing redox-responsive release in close vicinity to a tumor.

3.3. Near IR (700–1000 nm)

Near infra-red (NIR) light-responsive block copolymers are becoming more and more popular since they represent a promising opportunity to overcome issues that are combined especially with UV light irradiation (poor tissue penetration and toxic side effects) [2,3]. Recently, a review article focusing on photo-responsive materials for NIR stimulation has been published by Cho et al. [164]. Besides photo-induced heating strategies, NIR-triggered photoreactions have also been outlined and discussed. The respective examples and some recent work will be discussed in the following section. Here, we decided to focus on irreversible approaches because reversible NIR-responsive block copolymers have not been reported yet. Instead, the general concept using nanoparticles capable of light upconversion addressing photo-responsive block copolymers (mainly after micellization) will shortly be presented as well.

3.3.1. Irreversible

Most examples for irreversible photoreactions in the NIR range are closely connected to the above mentioned studies for far UV light responsive block copolymers since *o*-nitrobenzyl and coumarinyl esters show also photo-cleavage in the NIR window in terms of two-photon absorption occurring [165,166]. This has been already discussed in the early reports by the group of Zhao where they investigated the stimulation via NIR irradiation [88,140]. Cao et al. presented the preparation of polysaccharide-based *N*-succinyl-*N'*-4-(2-nitrobenzyloxy)-succinyl-chitosan amphiphilic block copolymer micelles containing *o*-nitrobenzyl ester groups and further demonstrated the conjugation with a tumor targeting ligand and an encapsulated antitumor drug [167,168]. It is noteworthy that, in addition to the fact that potentially toxic side-products (nitrosobenzaldehyde) are formed, the NIR-triggered reaction of *o*-nitrobenzyl esters usually exhibits rather long reaction times [169]. One proposed solution that was presented by Zhao and coworkers is the application of more efficient NIR two-photon-absorbing chromophores [170]. Accordingly, they synthesized biocompatible poly(ethylene oxide)-*block*-poly(L-glutamic acid) bearing 6-bromo-7-hydroxycoumarin-4-ylmethyl groups and showed the release of two different drug molecules upon irradiation with 794 nm. Another example by Ji et al. showed a block copolymer containing a coumarin functionalized block and another block of poly(hydroxyethylacrylate) which was successfully conjugated with folic acid as a selective cancer target compound [171]. Through hydrophobic interactions, the block copolymer was adsorbed onto hollow silica nanoparticles modified with hydrophobic octadecyl chains. The resulting nanocontainers were pre-loaded with doxorubicin and subsequently controlled NIR light triggered drug release was performed. Very recently, the same group which reported in 2013 on singlet oxygen mediated cleavage of a diselenide-bridged polyurethane middle block surrounded by PEG segments triggered by visible light (mentioned in Section 3.2.2) presented a similar system but containing tellurium as responsive junction point [163]. Here, the tellurium is coordinating cisplatin and a co-loaded FDA approved NIR dye for photodynamic therapy (indocyanine green (ICG)) which is acting as a sensitizer to generate singlet oxygen. Oxidation of tellurium led to both the release of cisplatin and the ICG in turn which in sum increased the anti-tumor efficacy when compared with the treatment of cisplatin alone (Figure 9).

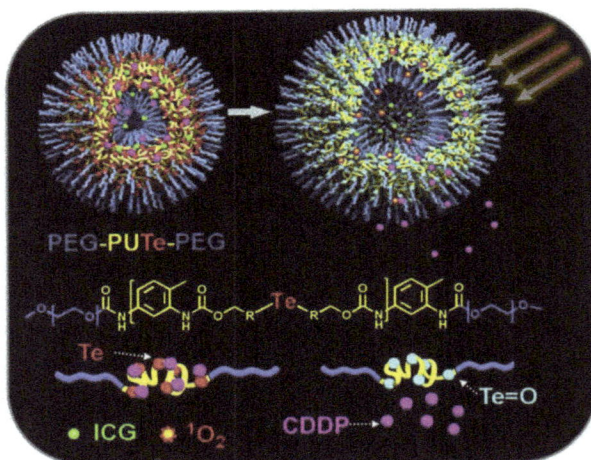

Figure 9. NIR light-triggered release of cisplatin (here abbreviated with CDDP) and indocyanine green (ICG) from an amphiphilic block copolymer micelle; stimulation of ICG leads to the formation of $^1O_2$ and the oxidation of tellurium, thereby drastically weakening the tellurium-cisplatin coordination. Reprinted from Ref. [172] with permission from Elsevier.

3.3.2. Upconversion

Upconverting nanoparticles (UPNP) efficiently absorb NIR light and convert it to lower wavelengths but, even more important, they can be used to assist photochemistry in the UV/Vis range [173]. In that regard, Carling et al. were the first who demonstrated a remotely controlled photoswitching of dithienylethene compounds by using UCNPs of NaYF$_4$:TmYb and NaYF$_4$:ErYb which converted 980 nm NIR light to trigger the UV/Vis responsive process [174]. In 2011, the same authors in cooperation with Zhao and coworkers presented a model system consisting of NaYF$_4$:TmYb upconverting nanoparticles inside poly(ethylene oxide)-*block*-poly(4,5-dimethoxy-2-nitrobenzyl methacrylate) block copolymer micelles (Figure 10) [175]. The desired internal UV light source was used to cleave off the *o*-nitrobenzyl functions leading to dissociation of the micelles and in turn to a release of co-loaded Nile red which was confirmed by fluorescence emission measurements.

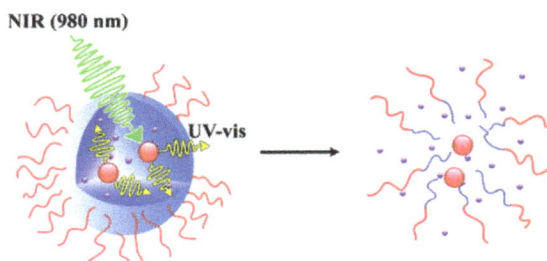

Figure 10. Schematic representation of NIR to UV light conversion via UCNP inside a block copolymer micelle and resulting release of encapsulated guest molecules. Reproduced with permission from Ref. [175]. Copyright 2011 American Chemical Society.

Similarly, a PEO-*b*-P(NIPAM-*co*-NBA) adsorbed onto an UCNP was presented in 2014 and again used for Nile red release experiments [176]. In 2017, a PEG-*b*-PS block copolymer with a block junction having both an *o*-nitrobenzyl moiety and an azobenzene group incorporated was photo-triggered in the presence of UCNPs resulting in disruption of the nanoaggregates [177]. Thus far, no examples for spiropyrane containing block copolymers in combination with UCNP have been demonstrated. For direct modification of these photo-responsive chromophores via UCNPs, the reader is referred to a recent review by Wu and Butt [178]. However, statistical copolymers having spiropyranes incorporated have been already attached to UCNP by Chen and coworkers [179]. These nanoparticles were used to polymerize poly(NIPAAm-*co*-spiropyrane methacrylate) and reversible switching of spiropyrane to the merocyanine form was achieved upon irradiation with light at 980 nm [180]. In another example, lanthanide-based UCNPs co-doped with Yb^{3+} and Tm^{3+} were encapsulated within mesoporous silica and coated with a methacrylate/methacrylamide terpolymer consisting of spiropyrane and PEG grafted groups as well as side-chain conjugated folic acid functions used as receptors for tumor cell targeting [181]. Beforehand, doxorubicin could be loaded into the mesoporous silica layer and the respective release studies were carried out, both in vitro and in vivo.

4. Conclusions and Outlook

Clearly, the field of photo-responsive block copolymers is still evolving, especially when it comes to materials or examples where irradiation has to be carried out at wavelengths distinctly higher than the far UV regime. However, one thing that becomes evident when comparing different approaches reported so far is that it can be difficult to directly compare photo-response of different reports, as often entirely different irradiation setups are used. In other words, quantitative evaluation of photo-response in nanostructured materials alongside with issues such as complete/incomplete reversibility, determination of the amount of unaffected chromophores, or long-time photo-stability assessment is sometimes difficult to judge. This even translates into entirely different classifications of the term "photo-responsive" as well as a broad variety of light sources being used—Sometimes also without determination of the overall light intensity.

Another aspect that is not always considered is whether continuous or repetitive irradiation also leads to photo-damage of the underlying polymer/block copolymer backbone, especially if UV light is used. Nevertheless, many very interesting and promising studies have been reported during the last decade and, although there is a certain variety of applications for such materials, the main research direction with regard to photo-responsive block copolymers in our opinion is towards improved control over spatial and temporal release of encapsulated cargo. Thereby, many examples have been reported where amphiphilic block copolymers containing one segment with covalently or non-covalently attached chromophores self-assembled into micelles within aqueous media and irradiation with light of a suitable wavelength then leads to (burst) release or degradation. In the latter case, irradiation is often accompanied by a sudden increase in solvent quality for the core of such aggregates, leading to swelling and, thus, permeability for guest molecules or even to complete dissolution if the resulting segment then is sufficiently hydrophilic. Nevertheless, most studies still deal with model drugs such as Doxorubicin or even Nile Red and the next step forward, possibly involving methods of upconversion to trigger release, can be anticipated within the next few years.

Acknowledgments: The authors are grateful for financial support from the State of Thuringia (ProExzellenzinitiative "NanoPolar", Felix Wendler and Felix H. Schacher) and the DFG (SCHA1640/9-1, Oliver Grimm and Felix H. Schacher).

Conflicts of Interest: The authors declare no conflict of interest.

References

1. Stuart, M.A.C.; Huck, W.T.S.; Genzer, J.; Müller, M.; Ober, C.; Stamm, M.; Sukhorukov, G.B.; Szleifer, I.; Tsukruk, V.V.; Urban, M.; et al. Emerging applications of stimuli-responsive polymer materials. *Nat. Mater.* **2010**, *9*, 101–113. [CrossRef] [PubMed]
2. Weissleder, R. A clearer vision for in vivo imaging. *Nat. Biotech.* **2001**, *19*, 316–317. [CrossRef] [PubMed]

3. Simpson, C.R.; Matthias, K.; Matthias, E.; Mark, C. Near-infrared optical properties of ex vivo human skin and subcutaneous tissues measured using the Monte Carlo inversion technique. *Phys. Med. Biol.* **1998**, *43*, 2465–2478. [CrossRef] [PubMed]
4. Dobrynin, A.V.; Rubinstein, M. Theory of polyelectrolytes in solutions and at surfaces. *Prog. Polym. Sci.* **2005**, *30*, 1049–1118. [CrossRef]
5. Halperin, A.; Kröger, M.; Winnik, F.M. Poly(*N*-isopropylacrylamide) Phase Diagrams: Fifty Years of Research. *Angew. Chem. Int. Ed.* **2015**, *54*, 15342–15367. [CrossRef] [PubMed]
6. Zhao, Y. Light-Responsive Block Copolymer Micelles. *Macromolecules* **2012**, *45*, 3647–3657. [CrossRef]
7. Gohy, J.-F.; Zhao, Y. Photo-responsive block copolymer micelles: Design and behavior. *Chem. Soc. Rev.* **2013**, *42*, 7117–7129. [CrossRef] [PubMed]
8. Bertrand, O.; Gohy, J.-F. Photo-responsive polymers: Synthesis and applications. *Polym. Chem.* **2017**, *8*, 52–73. [CrossRef]
9. Yu, H.; Ikeda, T. *Smart Light-Responsive Materials*; John Wiley & Sons, Inc.: Hoboken, NJ, USA, 2008; pp. 411–456.
10. Barzynski, H.; Jun, M.; Saenger, D.; Volkert, O. Lithographic printing plates and photoresists comprising a photosensitive polymer. US. Patent 3,849,137, 19 November 1974.
11. Bergmann, E.D.; Weizmann, A.; Fischer, E. Structure and Polarity of Some Polycyclic Spirans. *J. Am. Chem. Soc.* **1950**, *72*, 5009–5012. [CrossRef]
12. Bates, F.S.; Fredrickson, G.H. Block Copolymers—Designer Soft Materials. *Phys. Today* **1999**, *52*, 32–38. [CrossRef]
13. Schacher, F.H.; Rupar, P.A.; Manners, I. Functional Block Copolymers: Nanostructured Materials with Emerging Applications. *Angew. Chem. Int. Ed.* **2012**, *51*, 7898–7921. [CrossRef] [PubMed]
14. Szwarc, M.; Levy, M.; Milkovich, R. Polymerization initiated by electron transfer to monomer. A new method of formation of block polymers1. *J. Am. Chem. Soc.* **1956**, *78*, 2656–2657. [CrossRef]
15. Webster, O.W. Living Polymerization Methods. *Science* **1991**, *251*, 887–893. [CrossRef] [PubMed]
16. Braunecker, W.A.; Matyjaszewski, K. Controlled/living radical polymerization: Features, developments, and perspectives. *Prog. Polym. Sci.* **2007**, *32*, 93–146. [CrossRef]
17. Matyjaszewski, K.; Tsarevsky, N.V. Nanostructured functional materials prepared by atom transfer radical polymerization. *Nat. Chem.* **2009**, *1*, 276–288. [CrossRef] [PubMed]
18. Anastasaki, A.; Nikolaou, V.; Nurumbetov, G.; Wilson, P.; Kempe, K.; Quinn, J.F.; Davis, T.P.; Whittaker, M.R.; Haddleton, D.M. Cu(0)-Mediated Living Radical Polymerization: A Versatile Tool for Materials Synthesis. *Chem. Rev.* **2016**, *116*, 835–877. [CrossRef] [PubMed]
19. Nicolas, J.; Guillaneuf, Y.; Lefay, C.; Bertin, D.; Gigmes, D.; Charleux, B. Nitroxide-mediated polymerization. *Prog. Polym. Sci.* **2013**, *38*, 63–235. [CrossRef]
20. Gregory, A.; Stenzel, M.H. Complex polymer architectures via RAFT polymerization: From fundamental process to extending the scope using click chemistry and nature's building blocks. *Prog. Polym. Sci.* **2012**, *37*, 38–105. [CrossRef]
21. Moad, G.; Rizzardo, E.; Thang, S.H. Living Radical Polymerization by the RAFT Process—A Third Update. *Aust. J. Chem.* **2012**, *65*, 985–1076. [CrossRef]
22. Hayward, R.C.; Pochan, D.J. Tailored Assemblies of Block Copolymers in Solution: It Is All about the Process. *Macromolecules* **2010**, *43*, 3577–3584. [CrossRef]
23. Tonhauser, C.; Golriz, A.A.; Moers, C.; Klein, R.; Butt, H.-J.; Frey, H. Stimuli-Responsive Y-Shaped Polymer Brushes Based on Junction-Point-Reactive Block Copolymers. *Adv. Mater.* **2012**, *24*, 5559–5563. [CrossRef] [PubMed]
24. Rudolph, T.; Barthel, M.J.; Kretschmer, F.; Mansfeld, U.; Hoeppener, S.; Hager, M.D.; Schubert, U.S.; Schacher, F.H. Poly(2-vinyl pyridine)-block-Poly(ethylene oxide) Featuring a Furan Group at the Block Junction—Synthesis and Functionalization. *Macromol. Rapid Commun.* **2014**, *35*, 916–921. [CrossRef] [PubMed]
25. Rudolph, T.; Schacher, F.H. Selective crosslinking or addressing of individual domains within block copolymer nanostructures. *Eur. Polym. J.* **2016**, *80*, 317–331. [CrossRef]
26. Barner-Kowollik, C.; Goldmann, A.S.; Schacher, F.H. Polymer Interfaces: Synthetic Strategies Enabling Functionality, Adaptivity, and Spatial Control. *Macromolecules* **2016**, *49*, 5001–5016. [CrossRef]
27. Natansohn, A.; Rochon, P. Photoinduced Motions in Azo-Containing Polymers. *Chem. Rev.* **2002**, *102*, 4139–4176. [CrossRef] [PubMed]

28. Kumar, G.S.; Neckers, D.C. Photochemistry of azobenzene-containing polymers. *Chem. Rev.* **1989**, *89*, 1915–1925. [CrossRef]
29. Berkovic, G.; Krongauz, V.; Weiss, V. Spiropyrans and Spirooxazines for Memories and Switches. *Chem. Rev.* **2000**, *100*, 1741–1754. [CrossRef] [PubMed]
30. Assaid, I.; Bosc, D.; Hardy, I. Improvements of the Poly(vinyl cinnamate) Photoresponse in Order to Induce High Refractive Index Variations. *J. Phys. Chem. B* **2004**, *108*, 2801–2806. [CrossRef]
31. Irie, M. Diarylethenes for Memories and Switches. *Chem. Rev.* **2000**, *100*, 1685–1716. [CrossRef] [PubMed]
32. Stobbe, H.; Leuner, K. Farblose Alkylfulgide. (8. Abhandlung über Butadiënverbindungen.). *Ber. Dtsch. Chem. Ges.* **1905**, *38*, 3682–3685. [CrossRef]
33. Irie, M.; Mohri, M. Thermally irreversible photochromic systems. Reversible photocyclization of diarylethene derivatives. *J. Org. Chem.* **1988**, *53*, 803–808. [CrossRef]
34. Tanio, N.; Irie, M. Photooptical Switching of Polymer Film Waveguide Containing Photochromic Diarylethenes. *Jpn. J. Appl. Phys.* **1994**, *33*, 1550–1553. [CrossRef]
35. Fukaminato, T.; Sasaki, T.; Kawai, T.; Tamai, N.; Irie, M. Digital photoswitching of fluorescence based on the photochromism of diarylethene derivatives at a single-molecule level. *J. Am. Chem. Soc.* **2004**, *126*, 14843–14849. [CrossRef] [PubMed]
36. Stellacci, F.; Toscano, F.; Gallazzi, M.C.; Zerbi, G. From a photochromic diarylethene monomer to a dopable photochromic polymer: optical properties. *Synth. Met.* **1999**, *102*, 979–980. [CrossRef]
37. Nishi, H.; Kobatake, S. Photochromism and Optical Property of Gold Nanoparticles Covered with Low-Polydispersity Diarylethene Polymers. *Macromolecules* **2008**, *41*, 3995–4002. [CrossRef]
38. Seno, R.; Kobatake, S. Synthesis and characterization of amphiphilic silica nanoparticles covered by block copolymers branching photochromic diarylethene moieties on side chain. *Dyes Pigm.* **2015**, *114*, 166–174. [CrossRef]
39. Griffiths, J., II. Photochemistry of azobenzene and its derivatives. *Chem. Soc. Rev.* **1972**, *1*, 481–493. [CrossRef]
40. Hartley, G.S. The Cis-form of Azobenzene. *Nat. Chem.* **1937**, *140*, 281. [CrossRef]
41. Krollpfeiffer, F.; Mühlhausen, C.; Wolf, G. Zur Kenntnis der Lichtempfindlichkeit von Aryl-β-naphtylamin-azofarbstoffen. *Liebigs Ann. Chem.* **1934**, *508*, 39–51. [CrossRef]
42. Bleger, D.; Schwarz, J.; Brouwer, A.M.; Hecht, S. o-Fluoroazobenzenes as readily synthesized photoswitches offering nearly quantitative two-way isomerization with visible light. *J. Am. Chem. Soc.* **2012**, *134*, 20597–20600. [CrossRef]
43. Ringsdorf, H.; Schmidt, H.-W. Electro-optical effects of azo dye containing liquid crystalline copolymers. *Makromol. Chem.* **1984**, *185*, 1327–1334. [CrossRef]
44. Angeloni, A.S.; Caretti, D.; Carlini, C.; Chiellini, E.; Galli, G.; Altomare, A.; Solaro, R.; Laus, M. Photochromic liquid-crystalline polymers. Main chain and side chain polymers containing azobenzene mesogens. *Liq. Cryst.* **1989**, *4*, 513–527. [CrossRef]
45. Moriya, K.; Seki, T.; Nakagawa, M.; Mao, G.; Ober, C.K. Photochromism of 4-cyanophenylazobenzene in liquid crystalline-coil AB diblock copolymers: the influence of microstructure. *Macromol. Rapid Commun.* **2000**, *21*, 1309–1312. [CrossRef]
46. Frenz, C.; Fuchs, A.; Schmidt, H.-W.; Theissen, U.; Haarer, D. Diblock Copolymers with Azobenzene Side-Groups and Polystyrene Matrix: Synthesis, Characterization and Photoaddressing. *Macromol. Chem. Phys.* **2004**, *205*, 1246–1258. [CrossRef]
47. Wang, G.; Tong, X.; Zhao, Y. Preparation of Azobenzene-Containing Amphiphilic Diblock Copolymers for Light-Responsive Micellar Aggregates. *Macromolecules* **2004**, *37*, 8911–8917. [CrossRef]
48. Hu, J.; Yu, H.; Gan, L.H.; Hu, X. Photo-driven pulsating vesicles from self-assembled lipid-like azopolymers. *Soft Matter* **2011**, *7*, 11345–11350. [CrossRef]
49. Pearson, S.; Vitucci, D.; Khine, Y.Y.; Dag, A.; Lu, H.; Save, M.; Billon, L.; Stenzel, M.H. Light-responsive azobenzene-based glycopolymer micelles for targeted drug delivery to melanoma cells. *Eur. Polym. J.* **2015**, *69*, 616–627. [CrossRef]
50. Se, K.; Kijima, M.; Fujimoto, T. Photochemical isomerization of azobenzene incorporated in poly(*N,N*-dimethyl-4-vinylphenethylamine-*block*-styrene) diblock copolymer by cross linkage. *Polymer* **1997**, *38*, 5755–5760. [CrossRef]
51. Ueki, T.; Nakamura, Y.; Lodge, T.P.; Watanabe, M. Light-Controlled Reversible Micellization of a Diblock Copolymer in an Ionic Liquid. *Macromolecules* **2012**, *45*, 7566–7573. [CrossRef]

52. Concellón, A.; Blasco, E.; Martínez-Felipe, A.; Martínez, J.C.; Šics, I.; Ezquerra, T.A.; Nogales, A.; Piñol, M.; Oriol, L. Light-Responsive Self-Assembled Materials by Supramolecular Post-Functionalization via Hydrogen Bonding of Amphiphilic Block Copolymers. *Macromolecules* **2016**, *49*, 7825–7836. [CrossRef]

53. Concellón, A.; Clavería-Gimeno, R.; Velázquez-Campoy, A.; Abian, O.; Piñol, M.; Oriol, L. Polymeric micelles from block copolymers containing 2,6-diacylaminopyridine units for encapsulation of hydrophobic drugs. *RSC Adv.* **2016**, *6*, 24066–24075. [CrossRef]

54. Dilthey, W.; Berres, C. Die Halochromie acylierter Aminochalkone und verwandter Verbindungen. (Heteropolare Kohlenstoffverbindungen. II. *J. Prakt. Chem.* **1926**, *112*, 299–313. [CrossRef]

55. Löwenbein, A.; Katz, W. Über substituiertespiro-Dibenzopyrane. *Ber. Dtsch. Chem. Ges.* **1926**, *59*, 1377–1383. [CrossRef]

56. Koelsch, C.F. Steric Factors in Thermochromism of Spiropyrans and in Reactivities of Certain Methylene Groups. *J. Org. Chem.* **1951**, *16*, 1362–1370. [CrossRef]

57. Hirshberg, Y.; Fischer, E. Photochromism and reversible multiple internal transitions in some spiroPyrans at low temperatures. Part II. *J. Chem. Soc.* **1954**, 3129–3137. [CrossRef]

58. Raymo, F.M.; Giordani, S. Signal Processing at the Molecular Level. *J. Am. Chem. Soc.* **2001**, *123*, 4651–4652. [CrossRef] [PubMed]

59. Lee, S.K.; Neckers, D.C. Benzospiropyrans as photochromic and/or thermochromic photoinitiators. *Chem. Mater.* **1991**, *3*, 852–858. [CrossRef]

60. Balmond, E.I.; Tautges, B.K.; Faulkner, A.L.; Or, V.W.; Hodur, B.M.; Shaw, J.T.; Louie, A.Y. Comparative Evaluation of Substituent Effect on the Photochromic Properties of Spiropyrans and Spirooxazines. *J. Org. Chem.* **2016**, *81*, 8744–8758. [CrossRef] [PubMed]

61. Smets, G. Photochromic behaviour of polymeric systems and related phenomena. *Pure Appl. Chem.* **1972**, *30*, 1–24. [CrossRef]

62. Krongauz, V.A.; Goldburt, E.S. Crystallization of poly(spiropyran methacrylate) with cooperative spiropyran-merocyanine conversion. *Macromolecules* **1981**, *14*, 1382–1386. [CrossRef]

63. Gonzalez-De Los Santos, E.A.; Lozano-Gonzalez, M.J.; Johnson, A.F. Photoresponsive polyurethane-acrylate block copolymers. I. Photochromic effects in copolymers containing 6′-nitro spiropyranes and 6′-nitro-bis-spiropyranes. *J. Appl. Polym. Sci.* **1999**, *71*, 259–266. [CrossRef]

64. Lee, H.I.; Wu, W.; Oh, J.K.; Mueller, L.; Sherwood, G.; Peteanu, L.; Kowalewski, T.; Matyjaszewski, K. Light-induced reversible formation of polymeric micelles. *Angew. Chem. Int. Ed.* **2007**, *46*, 2453–2457 [CrossRef] [PubMed]

65. Kotharangannagari, V.K.; Sánchez-Ferrer, A.; Ruokolainen, J.; Mezzenga, R. Photoresponsive Reversible Aggregation and Dissolution of Rod–Coil Polypeptide Diblock Copolymers. *Macromolecules* **2011**, *44*, 4569–4573. [CrossRef]

66. Berman, E.; Fox, R.E.; Thomson, F.D. Photochromic Spiropyrans. I. The Effect of Substituents on the Rate of Ring Closure. *J. Am. Chem. Soc.* **1959**, *81*, 5605–5608. [CrossRef]

67. Zhang, Y.; Chen, S.; Pang, M.; Zhang, W. Synthesis and micellization of multi-stimuli responsive block copolymer based on spiropyran. *Polym. Chem.* **2016**. [CrossRef]

68. Wang, X.; Hu, J.; Liu, G.; Tian, J.; Wang, H.; Gong, M.; Liu, S. Reversibly Switching Bilayer Permeability and Release Modules of Photochromic Polymersomes Stabilized by Cooperative Noncovalent Interactions. *J. Am. Chem. Soc.* **2015**, *137*, 15262–15275. [CrossRef] [PubMed]

69. Chen, J.; Zeng, F.; Wu, S.; Chen, Q.; Tong, Z. A core-shell nanoparticle approach to photoreversible fluorescence modulation of a hydrophobic dye in aqueous media. *Chem. Eur. J.* **2008**, *14*, 4851–4860. [CrossRef] [PubMed]

70. Guragain, S.; Bastakoti, B.P.; Ito, M.; Yusa, S.-I.; Nakashima, K. Aqueous polymeric micelles of poly[N-isopropylacrylamide-b-sodium 2-(acrylamido)-2-methylpropanesulfonate] with a spiropyran dimer pendant: Quadruple stimuli-responsiveness. *Soft Matter* **2012**, *8*, 9628. [CrossRef]

71. Menon, S.; Ongungal, R.M.; Das, S. Photocleavable glycopolymer aggregates. *Polym. Chem.* **2013**, *4*, 623–628. [CrossRef]

72. Jin, Q.; Liu, G.; Ji, J. Micelles and reverse micelles with a photo and thermo double-responsive block copolymer. *J. Polym. Sci. A* **2010**, *48*, 2855–2861. [CrossRef]

73. Liu, G.; Qiao, L.; Guo, A. Diblock Copolymer Nanofibers. *Macromolecules* **1996**, *29*, 5508–5510. [CrossRef]

74. Lendlein, A.; Jiang, H.; Junger, O.; Langer, R. Light-induced shape-memory polymers. *Nat. Chem.* **2005**, *434*, 879–882. [CrossRef] [PubMed]

75. Jiang, N.; Cheng, Y.; Wei, J. Coumarin-modified fluorescent microcapsules and their photo-switchable release property. *Colloids Surf. A* **2017**, *522*, 28–37. [CrossRef]

76. Zhang, Z.; Xue, Y.; Zhang, P.; Müller, A.H.E.; Zhang, W. Hollow Polymeric Capsules from POSS-Based Block Copolymer for Photodynamic Therapy. *Macromolecules* **2016**, *49*, 8440–8448. [CrossRef]

77. Jensen, A.I.; Binderup, T.; Kumar, E.P.; Kjaer, A.; Rasmussen, P.H.; Andresen, T.L. Positron emission tomography based analysis of long-circulating cross-linked triblock polymeric micelles in a U87MG mouse xenograft model and comparison of DOTA and CB-TE2A as chelators of copper-64. *Biomacromolecules* **2014**, *15*, 1625–1633. [CrossRef] [PubMed]

78. Wu, L.; Jin, C.; Sun, X. Synthesis, properties, and light-induced shape memory effect of multiblock polyesterurethanes containing biodegradable segments and pendant cinnamamide groups. *Biomacromolecules* **2011**, *12*, 235–241. [CrossRef] [PubMed]

79. Jiang, J.; Qi, B.; Lepage, M.; Zhao, Y. Polymer Micelles Stabilization on Demand through Reversible Photo-Cross-Linking. *Macromolecules* **2007**, *40*, 790–792. [CrossRef]

80. Zhang, X.; Wang, Y.; Li, G.; Liu, Z.; Liu, Z.; Jiang, J. Amphiphilic Imbalance and Stabilization of Block Copolymer Micelles on-Demand through Combinational Photo-Cleavage and Photo-Crosslinking. *Macromol. Rapid Commun.* **2016**, *38*, 1600543. [CrossRef] [PubMed]

81. Xie, H.; He, M.J.; Deng, X.Y.; Du, L.; Fan, C.J.; Yang, K.K.; Wang, Y.Z. Design of Poly(L-lactide)-Poly(ethylene glycol) Copolymer with Light-Induced Shape-Memory Effect Triggered by Pendant Anthracene Groups. *ACS Appl. Mater. Interfaces* **2016**, *8*, 9431–9439. [CrossRef] [PubMed]

82. Iwamura, M.; Ishikawa, T.; Koyama, Y.; Sakuma, K.; Iwamura, H. 1-Pyrenylmethyl esters, photolabile protecting groups for carboxylic acids. *Tetrahedron Lett.* **1987**, *28*, 679–682. [CrossRef]

83. Barltrop, J.A.; Plant, P.J.; Schofield, P. Photosensitive protective groups. *Chem. Commun.* **1966**, 822–823. [CrossRef]

84. Furuta, T.; Torigai, H.; Sugimoto, M.; Iwamura, M. Photochemical Properties of New Photolabile cAMP Derivatives in a Physiological Saline Solution. *J. Org. Chem.* **1995**, *60*, 3953–3956. [CrossRef]

85. Sheehan, J.C.; Umezawa, K. Phenacyl photosensitive blocking groups. *J. Org. Chem.* **1973**, *38*, 3771–3774. [CrossRef]

86. Wuts, P.G.M. *Greene's Protective Groups in Organic Synthesis*; Wuts, P.G.M., Ed.; John Wiley & Sons, Inc.: Hoboken, NJ, USA, 2014; pp. 1203–1262.

87. Jiang, J.; Tong, X.; Zhao, Y. A New Design for Light-Breakable Polymer Micelles. *J. Am. Chem. Soc.* **2005**, *127*, 8290–8291. [CrossRef]

88. Jiang, J.; Tong, X.; Morris, D.; Zhao, Y. Toward Photocontrolled Release Using Light-Dissociable Block Copolymer Micelles. *Macromolecules* **2006**, *39*, 4633–4640. [CrossRef]

89. Lee, J.-E.; Ahn, E.; Bak, J.M.; Jung, S.-H.; Park, J.M.; Kim, B.-S.; Lee, H.-I. Polymeric micelles based on photocleavable linkers tethered with a model drug. *Polymer* **2014**, *55*, 1436–1442. [CrossRef]

90. Liu, X.; He, J.; Niu, Y.; Li, Y.; Hu, D.; Xia, X.; Lu, Y.; Xu, W. Photo-responsive amphiphilic poly(α-hydroxy acids) with pendent o-nitrobenzyl ester constructed via copper-catalyzed azide-alkyne cycloaddition reaction. *Polym. Adv. Technol.* **2015**, *26*, 449–456. [CrossRef]

91. Song, Z.; Kim, H.; Ba, X.; Baumgartner, R.; Lee, J.S.; Tang, H.; Leal, C.; Cheng, J. Polypeptide vesicles with densely packed multilayer membranes. *Soft Matter* **2015**, *11*, 4091–4098. [CrossRef] [PubMed]

92. Zhu, C.; Bettinger, C.J. Photoreconfigurable Physically Cross-Linked Triblock Copolymer Hydrogels: Photodisintegration Kinetics and Structure–Property Relationships. *Macromolecules* **2015**, *48*, 1563–1572. [CrossRef]

93. Wang, X.; Liu, G.; Hu, J.; Zhang, G.; Liu, S. Concurrent Block Copolymer Polymersome Stabilization and Bilayer Permeabilization by Stimuli-Regulated "Traceless" Crosslinking. *Angew. Chem. Int. Ed.* **2014**, *53*, 3138–3142. [CrossRef] [PubMed]

94. Li, Y.; Qian, Y.; Liu, T.; Zhang, G.; Liu, S. Light-Triggered Concomitant Enhancement of Magnetic Resonance Imaging Contrast Performance and Drug Release Rate of Functionalized Amphiphilic Diblock Copolymer Micelles. *Biomacromolecules* **2012**, *13*, 3877–3886. [CrossRef] [PubMed]

95. Xie, Z.; Hu, X.; Chen, X.; Mo, G.; Sun, J.; Jing, X. A Novel Biodegradable and Light-Breakable Diblock Copolymer Micelle for Drug Delivery. *Adv. Eng. Mater.* **2009**, *11*, B7–B11. [CrossRef]

96. Greco, C.T.; Epps, T.H.; Sullivan, M.O. Mechanistic Design of Polymer Nanocarriers to Spatiotemporally Control Gene Silencing. *ACS Biomater. Sci. Eng.* **2016**, *2*, 1582–1594. [CrossRef]
97. Gupta, M.K.; Balikov, D.A.; Lee, Y.; Ko, E.; Yu, C.; Chun, Y.W.; Sawyer, D.B.; Kim, W.S.; Sung, H.-J. Gradient release of cardiac morphogens by photo-responsive polymer micelles for gradient-mediated variation of embryoid body differentiation. *J. Mater. Chem. B* **2017**, *5*, 2019–2033. [CrossRef]
98. Bertrand, O.; Gohy, J.-F.; Fustin, C.-A. Synthesis of diblock copolymers bearing p-methoxyphenacyl side groups. *Polym. Chem.* **2011**, *2*, 2284–2292. [CrossRef]
99. Schumers, J.-M.; Bertrand, O.; Fustin, C.-A.; Gohy, J.-F. Synthesis and self-assembly of diblock copolymers bearing 2-nitrobenzyl photocleavable side groups. *J. Polym. Sci. A* **2012**, *50*, 599–608. [CrossRef]
100. Song, D.-P.; Wang, X.; Lin, Y.; Watkins, J.J. Synthesis and Controlled Self-Assembly of UV-Responsive Gold Nanoparticles in Block Copolymer Templates. *J. Phys. Chem. B* **2014**, *118*, 12788–12795. [CrossRef] [PubMed]
101. Soliman, S.M.A.; Nouvel, C.; Babin, J.; Six, J.-L. o-Nitrobenzyl acrylate is polymerizable by single electron transfer-living radical polymerization. *J. Polym. Sci. A* **2014**, *52*, 2192–2201. [CrossRef]
102. Jana, S.; Saha, A.; Paira, T.K.; Mandal, T.K. Synthesis and Self-Aggregation of Poly(2-ethyl-2-oxazoline)-Based Photocleavable Block Copolymer: Micelle, Compound Micelle, Reverse Micelle, and Dye Encapsulation/Release. *J. Phys. Chem. B* **2016**, *120*, 813–824. [CrossRef]
103. Xu, Z.; Yan, B.; Riordon, J.; Zhao, Y.; Sinton, D.; Moffitt, M.G. Microfluidic Synthesis of Photoresponsive Spool-Like Block Copolymer Nanoparticles: Flow-Directed Formation and Light-Triggered Dissociation. *Chem. Mater.* **2015**, *27*, 8094–8104. [CrossRef]
104. Liu, G.; Dong, C.-M. Photoresponsive Poly(S-(o-nitrobenzyl)-l-cysteine)-b-PEO from a l-Cysteine N-Carboxyanhydride Monomer: Synthesis, Self-Assembly, and Phototriggered Drug Release. *Biomacromolecules* **2012**, *13*, 1573–1583. [CrossRef]
105. Jiang, X.; Lavender, C.A.; Woodcock, J.W.; Zhao, B. Multiple Micellization and Dissociation Transitions of Thermo- and Light-Sensitive Poly(ethylene oxide)-b-poly(ethoxytri(ethylene glycol) acrylate-co-o-nitrobenzyl acrylate) in Water. *Macromolecules* **2008**, *41*, 2632–2643. [CrossRef]
106. Yuan, W.; Guo, W. Ultraviolet light-breakable and tunable thermoresponsive amphiphilic block copolymer: From self-assembly, disassembly to re-self-assembly. *Polym. Chem.* **2014**, *5*, 4259–4267. [CrossRef]
107. Yang, F.; Cao, Z.; Wang, G. Micellar assembly of a photo- and temperature-responsive amphiphilic block copolymer for controlled release. *Polym. Chem.* **2015**, *6*, 7995–8002. [CrossRef]
108. Jiang, X.; Jin, S.; Zhong, Q.; Dadmun, M.D.; Zhao, B. Stimuli-Induced Multiple Sol−Gel−Sol Transitions of Aqueous Solution of a Thermo- and Light-Sensitive Hydrophilic Block Copolymer. *Macromolecules* **2009**, *42*, 8468–8476. [CrossRef]
109. Yao, C.; Wang, X.; Liu, G.; Hu, J.; Liu, S. Distinct Morphological Transitions of Photoreactive and Thermoresponsive Vesicles for Controlled Release and Nanoreactors. *Macromolecules* **2016**. [CrossRef]
110. Shrivastava, S.; Matsuoka, H. Photocleavable amphiphilic diblock copolymer micelles bearing a nitrobenzene block. *Colloid Polym. Sci.* **2016**, *294*, 879–887. [CrossRef]
111. Fang, J.-Y.; Lin, Y.-K.; Wang, S.-W.; Li, Y.-C.; Lee, R.-S. Synthesis and characterization of dual-stimuli-responsive micelles based on poly(N-isopropylacrylamide) and polycarbonate with photocleavable moieties. *React. Funct. Polym.* **2015**, *95*, 46–54. [CrossRef]
112. Sun, T.; Li, P.; Oh, J.K. Dual Location Dual Reduction/Photoresponsive Block Copolymer Micelles: Disassembly and Synergistic Release. *Macromol. Rapid Commun.* **2015**, *36*, 1742–1748. [CrossRef] [PubMed]
113. Wu, W.-C.; Kuo, Y.-S.; Cheng, C.-H. Dual-stimuli responsive polymeric micelles: Preparation, characterization, and controlled drug release. *J. Polym. Res.* **2015**, *22*, 80. [CrossRef]
114. Jin, Q.; Cai, T.; Wang, Y.; Wang, H.; Ji, J. Light-Responsive Polyion Complex Micelles with Switchable Surface Charge for Efficient Protein Delivery. *ACS Macro Lett.* **2014**, *3*, 679–683. [CrossRef]
115. Kalva, N.; Parekh, N.; Ambade, A.V. Controlled micellar disassembly of photo- and pH-cleavable linear-dendritic block copolymers. *Polym. Chem.* **2015**, *6*, 6826–6835. [CrossRef]
116. Wu, Y.; Hu, H.; Hu, J.; Liu, T.; Zhang, G.; Liu, S. Thermo- and Light-Regulated Formation and Disintegration of Double Hydrophilic Block Copolymer Assemblies with Tunable Fluorescence Emissions. *Langmuir* **2013**, *29*, 3711–3720. [CrossRef] [PubMed]
117. Huo, H.; Ma, X.; Dong, Y.; Qu, F. Light/temperature dual-responsive ABC miktoarm star terpolymer micelles for controlled release. *Eur. Polym. J.* **2017**, *87*, 331–343. [CrossRef]

.18. He, L.; Hu, B.; Henn, D.M.; Zhao, B. Influence of cleavage of photosensitive group on thermally induced micellization and gelation of a doubly responsive diblock copolymer in aqueous solutions: A SANS study. *Polymer* **2016**, *105*, 25–34. [CrossRef]

119. Tao, Z.; Peng, K.; Fan, Y.; Liu, Y.; Yang, H. Multi-stimuli responsive supramolecular hydrogels based on Fe^{3+} and diblock copolymer micelle complexation. *Polym. Chem.* **2016**, *7*, 1405–1412. [CrossRef]

120. Kumar, S.; Dory, Y.L.; Lepage, M.; Zhao, Y. Surface-Grafted Stimuli-Responsive Block Copolymer Brushes for the Thermo-, Photo- and pH-Sensitive Release of Dye Molecules. *Macromolecules* **2011**, *44*, 7385–7393. [CrossRef]

121. Wang, X.; Jiang, G.; Li, X.; Tang, B.; Wei, Z.; Mai, C. Synthesis of multi-responsive polymeric nanocarriers for controlled release of bioactive agents. *Polym. Chem.* **2013**, *4*, 4574–4577. [CrossRef]

122. Cao, Z.; Wu, H.; Dong, J.; Wang, G. Quadruple-Stimuli-Sensitive Polymeric Nanocarriers for Controlled Release under Combined Stimulation. *Macromolecules* **2014**, *47*, 8777–8783. [CrossRef]

123. Han, D.; Tong, X.; Zhao, Y. Fast Photodegradable Block Copolymer Micelles for Burst Release. *Macromolecules* **2011**, *44*, 437–439. [CrossRef]

124. Han, D.; Tong, X.; Zhao, Y. Block Copolymer Micelles with a Dual-Stimuli-Responsive Core for Fast or Slow Degradation. *Langmuir* **2012**, *28*, 2327–2331. [CrossRef] [PubMed]

125. Cabane, E.; Malinova, V.; Meier, W. Synthesis of Photocleavable Amphiphilic Block Copolymers: Toward the Design of Photosensitive Nanocarriers. *Macromol. Chem. Phys.* **2010**, *211*, 1847–1856. [CrossRef]

126. Zhao, H.; Sterner, E.S.; Coughlin, E.B.; Theato, P. o-Nitrobenzyl Alcohol Derivatives: Opportunities in Polymer and Materials Science. *Macromolecules* **2012**, *45*, 1723–1736. [CrossRef]

127. Zhao, H.; Gu, W.; Thielke, M.W.; Sterner, E.; Tsai, T.; Russell, T.P.; Coughlin, E.B.; Theato, P. Functionalized Nanoporous Thin Films and Fibers from Photocleavable Block Copolymers Featuring Activated Esters. *Macromolecules* **2013**, *46*, 5195–5201. [CrossRef]

128. Xuan, J.; Han, D.; Xia, H.; Zhao, Y. Dual-Stimuli-Responsive Micelle of an ABC Triblock Copolymer Bearing a Redox-Cleavable Unit and a Photocleavable Unit at Two Block Junctions. *Langmuir* **2014**, *30*, 410–417. [CrossRef] [PubMed]

129. Li, L.; Lv, A.; Deng, X.-X.; Du, F.-S.; Li, Z.-C. Facile synthesis of photo-cleavable polymers via Passerini reaction. *Chem. Commun.* **2013**, *49*, 8549–8551. [CrossRef] [PubMed]

130. Schumers, J.-M.; Gohy, J.-F.; Fustin, C.-A. A versatile strategy for the synthesis of block copolymers bearing a photocleavable junction. *Polym. Chem.* **2010**, *1*, 161–163. [CrossRef]

131. Gungor, E.; Armani, A.M. Photocleavage of Covalently Immobilized Amphiphilic Block Copolymer: From Bilayer to Monolayer. *Macromolecules* **2016**, *49*, 5773–5781. [CrossRef]

132. Lee, R.-S.; Li, Y.-C.; Wang, S.-W. Synthesis and characterization of amphiphilic photocleavable polymers based on dextran and substituted-ε-caprolactone. *Carbohyd. Polym.* **2015**, *117*, 201–210. [CrossRef] [PubMed]

133. Shota, Y.; Hidemi, T.; Shuya, Y.; Seiichi, N.; Kazuo, Y. Synthesis of Amphiphilic Diblock Copolymer Using Heterobifunctional Linkers, Connected by a Photodegradable N-(2-Nitrobenzyl)imide Structure and Available for Two Different Click Chemistries. *Bull. Chem. Soc. Jpn.* **2016**, *89*, 481–489. [CrossRef]

134. Gao, Y.; Qiu, H.; Zhou, H.; Li, X.; Harniman, R.; Winnik, M.A.; Manners, I. Crystallization-Driven Solution Self-Assembly of Block Copolymers with a Photocleavable Junction. *J. Am. Chem. Soc.* **2015**, *137*, 2203–2206. [CrossRef] [PubMed]

135. Coumes, F.; Malfait, A.; Bria, M.; Lyskawa, J.; Woisel, P.; Fournier, D. Catechol/boronic acid chemistry for the creation of block copolymers with a multi-stimuli responsive junction. *Polym. Chem.* **2016**, *7*, 4682–4692. [CrossRef]

136. Katz, J.S.; Zhong, S.; Ricart, B.G.; Pochan, D.J.; Hammer, D.A.; Burdick, J.A. Modular Synthesis of Biodegradable Diblock Copolymers for Designing Functional Polymersomes. *J. Am. Chem. Soc.* **2010**, *132*, 3654–3655. [CrossRef] [PubMed]

137. Gamys, C.G.; Schumers, J.-M.; Vlad, A.; Fustin, C.-A.; Gohy, J.-F. Amine-functionalized nanoporous thin films from a poly(ethylene oxide)-*block*-polystyrene diblock copolymer bearing a photocleavable o-nitrobenzyl carbamate junction. *Soft Matter* **2012**, *8*, 4486–4493. [CrossRef]

138. Zhao, H.; Gu, W.; Sterner, E.; Russell, T.P.; Coughlin, E.B.; Theato, P. Highly Ordered Nanoporous Thin Films from Photocleavable Block Copolymers. *Macromolecules* **2011**, *44*, 6433–6440. [CrossRef]

139. Yang, L.; Lei, M.; Zhao, M.; Yang, H.; Zhang, H.; Li, Y.; Zhang, K.; Lei, Z. Synthesis of the light/pH responsive polymer for immobilization of α-amylase. *Mater. Sci. Eng. C* **2017**, *71*, 75–83. [CrossRef] [PubMed]

140. Babin, J.; Pelletier, M.; Lepage, M.; Allard, J.-F.; Morris, D.; Zhao, Y. A New Two-Photon-Sensitive Block Copolymer Nanocarrier. *Angew. Chem. Int. Ed.* **2009**, *48*, 3329–3332. [CrossRef] [PubMed]
141. Jin, Q.; Mitschang, F.; Agarwal, S. Biocompatible Drug Delivery System for Photo-Triggered Controlled Release of 5-Fluorouracil. *Biomacromolecules* **2011**, *12*, 3684–3691. [CrossRef]
142. Bertrand, O.; Fustin, C.-A.; Gohy, J.-F. Multiresponsive Micellar Systems from Photocleavable Block Copolymers. *ACS Macro Lett.* **2012**, *1*, 949–953. [CrossRef]
143. Guo, A.; Liu, G.; Tao, J. Star Polymers and Nanospheres from Cross-Linkable Diblock Copolymers. *Macromolecules* **1996**, *29*, 2487–2493. [CrossRef]
144. Rabnawaz, M.; Liu, G. Preparation and Application of a Dual Light-Responsive Triblock Terpolymer. *Macromolecules* **2012**, *45*, 5586–5595. [CrossRef]
145. Ding, J.; Liu, G. Hairy, Semi-shaved, and Fully Shaved Hollow Nanospheres from Polyisoprene-block-poly(2-cinnamoylethyl methacrylate). *Chem. Mater.* **1998**, *10*, 537–542. [CrossRef]
146. Yang, H.; Jia, L.; Wang, Z.; Di-Cicco, A.; Lévy, D.; Keller, P. Novel Photolabile Diblock Copolymers Bearing Truxillic Acid Derivative Junctions. *Macromolecules* **2011**, *44*, 159–165. [CrossRef]
147. Chen, C.-J.; Liu, G.-Y.; Shi, Y.-T.; Zhu, C.-S.; Pang, S.-P.; Liu, X.-S.; Ji, J. Biocompatible Micelles Based on Comb-like PEG Derivates: Formation, Characterization, and Photo-responsiveness. *Macromol. Rapid Commun.* **2011**, *32*, 1077–1081. [CrossRef] [PubMed]
148. Yoshida, E.; Kuwayama, S. Micelle formation induced by photo-Claisen rearrangement of poly(4-allyloxystyrene)-block-polystyrene. *Colloid Polym. Sci.* **2009**, *287*, 789–793. [CrossRef]
149. Ishida, Y.; Takeda, Y.; Kameyama, A. Synthesis of block copolymer with photo-decomposable polyurethane and its photo-initiated domino decomposition. *React. Funct. Polym.* **2016**, *107*, 20–27. [CrossRef]
150. Tian, M.; Cheng, R.; Zhang, J.; Liu, Z.; Liu, Z.; Jiang, J. Amphiphilic Polymer Micellar Disruption Based on Main-Chain Photodegradation. *Langmuir* **2016**, *32*, 12–18. [CrossRef] [PubMed]
151. Kim, Y.J.; Kang, H.; Leolukman, M.; Nealey, P.F.; Gopalan, P. Synthesis of Photoacid Generator-Containing Patternable Diblock Copolymers by Reversible Addition−Fragmentation Transfer Polymerization. *Chem. Mater.* **2009**, *21*, 3030–3032. [CrossRef]
152. Helmy, S.; Leibfarth, F.A.; Oh, S.; Poelma, J.E.; Hawker, C.J.; Read de Alaniz, J. Photoswitching using visible light: A new class of organic photochromic molecules. *J. Am. Chem. Soc.* **2014**, *136*, 8169–8172. [CrossRef] [PubMed]
153. Stenhouse, J. Ueber die Oele, die bei der Einwirkung der Schwefelsäure auf verschiedene Vegetabilien entstehen. *Liebigs Ann. Chem.* **1850**, *74*, 278–297. [CrossRef]
154. Balamurugan, A.; Lee, H.-I. A Visible Light Responsive On–Off Polymeric Photoswitch for the Colorimetric Detection of Nerve Agent Mimics in Solution and in the Vapor Phase. *Macromolecules* **2016**, *49*, 2568–2574. [CrossRef]
155. Sinawang, G.; Wu, B.; Wang, J.; Li, S.; He, Y. Polystyrene Based Visible Light Responsive Polymer with Donor-Acceptor Stenhouse Adduct Pendants. *Macromol. Chem. Phys.* **2016**, *217*, 2409–2414. [CrossRef]
156. Bleger, D.; Hecht, S. Visible-Light-Activated Molecular Switches. *Angew. Chem. Int. Ed.* **2015**, *54*, 11338–11349. [CrossRef] [PubMed]
157. Yu, H.; Iyoda, T.; Ikeda, T. Photoinduced alignment of nanocylinders by supramolecular cooperative motions. *J. Am. Chem. Soc.* **2006**, *128*, 11010–11011. [CrossRef] [PubMed]
158. Sun, W.; Parowatkin, M.; Steffen, W.; Butt, H.-J.; Mailänder, V.; Wu, S. Ruthenium-Containing Block Copolymer Assemblies: Red-Light-Responsive Metallopolymers with Tunable Nanostructures for Enhanced Cellular Uptake and Anticancer Phototherapy. *Adv. Healthc. Mater.* **2016**, *5*, 467–473. [CrossRef] [PubMed]
159. Zhou, D.; Guo, J.; Kim, G.B.; Li, J.; Chen, X.; Yang, J.; Huang, Y. Simultaneously Photo-Cleavable and Activatable Prodrug-Backboned Block Copolymer Micelles for Precise Anticancer Drug Delivery. *Adv. Healthc. Mater.* **2016**, *5*, 2493–2499. [CrossRef] [PubMed]
160. Yan, Q.; Hu, J.; Zhou, R.; Ju, Y.; Yin, Y.; Yuan, J. Visible light-responsive micelles formed from dialkoxyanthracene-containing block copolymers. *Chem. Commun.* **2012**, *48*, 1913–1915. [CrossRef] [PubMed]
161. Han, P.; Li, S.; Cao, W.; Li, Y.; Sun, Z.; Wang, Z.; Xu, H. Red light responsive diselenide-containing block copolymer micelles. *J. Mater. Chem. B* **2013**, *1*, 740–743. [CrossRef]
162. Saravanakumar, G.; Lee, J.; Kim, J.; Kim, W.J. Visible light-induced singlet oxygen-mediated intracellular disassembly of polymeric micelles co-loaded with a photosensitizer and an anticancer drug for enhanced photodynamic therapy. *Chem. Commun.* **2015**, *51*, 9995–9998. [CrossRef]

163. Zhai, S.; Hu, X.; Hu, Y.; Wu, B.; Xing, D. Visible light-induced crosslinking and physiological stabilization of diselenide-rich nanoparticles for redox-responsive drug release and combination chemotherapy. *Biomaterials* **2017**, *121*, 41–54. [CrossRef] [PubMed]

164. Cho, H.J.; Chung, M.; Shim, M.S. Engineered photo-responsive materials for near-infrared-triggered drug delivery. *J. Ind. Eng. Chem.* **2015**, *31*, 15–25. [CrossRef]

165. Bort, G.; Gallavardin, T.; Ogden, D.; Dalko, P.I. From One-Photon to Two-Photon Probes: "Caged" Compounds, Actuators, and Photoswitches. *Angew. Chem. Int. Ed.* **2013**, *52*, 4526–4537. [CrossRef] [PubMed]

166. Klán, P.; Šolomek, T.; Bochet, C.G.; Blanc, A.; Givens, R.; Rubina, M.; Popik, V.; Kostikov, A.; Wirz, J. Photoremovable Protecting Groups in Chemistry and Biology: Reaction Mechanisms and Efficacy. *Chem. Rev.* **2013**, *113*, 119–191. [CrossRef] [PubMed]

167. Cao, J.; Huang, S.; Chen, Y.; Li, S.; Li, X.; Deng, D.; Qian, Z.; Tang, L.; Gu, Y. Near-infrared light-triggered micelles for fast controlled drug release in deep tissue. *Biomaterials* **2013**, *34*, 6272–6283. [CrossRef] [PubMed]

168. Cao, J.; Chen, D.; Huang, S.; Deng, D.; Gu, Y.; Tang, L. Multifunctional near-infrared light-triggered biodegradable micelles for chemo- and photo-thermal combination therapy. *Oncotarget* **2016**. [CrossRef] [PubMed]

169. Aujard, I.; Benbrahim, C.; Gouget, M.; Ruel, O.; Baudin, J.-B.; Neveu, P.; Jullien, L. o-Nitrobenzyl Photolabile Protecting Groups with Red-Shifted Absorption: Syntheses and Uncaging Cross-Sections for One- and Two-Photon Excitation. *Chem. Eur. J.* **2006**, *12*, 6865–6879. [CrossRef] [PubMed]

170. Kumar, S.; Allard, J.-F.; Morris, D.; Dory, Y.L.; Lepage, M.; Zhao, Y. Near-infrared light sensitive polypeptide block copolymer micelles for drug delivery. *J. Mater. Chem.* **2012**, *22*, 7252–7257. [CrossRef]

171. Ji, W.; Li, N.; Chen, D.; Qi, X.; Sha, W.; Jiao, Y.; Xu, Q.; Lu, J. Coumarin-containing photo-responsive nanocomposites for NIR light-triggered controlled drug release via a two-photon process. *J. Mater. Chem. B* **2013**, *1*, 5942. [CrossRef]

172. Li, F.; Li, T.; Cao, W.; Wang, L.; Xu, H. Near-infrared light stimuli-responsive synergistic therapy nanoplatforms based on the coordination of tellurium-containing block polymer and cisplatin for cancer treatment. *Biomaterials* **2017**, *133*, 208–218. [CrossRef] [PubMed]

173. Chen, G.; Qiu, H.; Prasad, P.N.; Chen, X. Upconversion Nanoparticles: Design, Nanochemistry, and Applications in Theranostics. *Chem. Rev.* **2014**, *114*, 5161–5214. [CrossRef] [PubMed]

174. Carling, C.-J.; Boyer, J.-C.; Branda, N.R. Remote-Control Photoswitching Using NIR Light. *J. Am. Chem. Soc.* **2009**, *131*, 10838–10839. [CrossRef] [PubMed]

175. Yan, B.; Boyer, J.-C.; Branda, N.R.; Zhao, Y. Near-Infrared Light-Triggered Dissociation of Block Copolymer Micelles Using Upconverting Nanoparticles. *J. Am. Chem. Soc.* **2011**, *133*, 19714–19717. [CrossRef] [PubMed]

176. Zhang, R.; Yao, R.; Ding, B.; Shen, Y.; Shui, S.; Wang, L.; Li, Y.; Yang, X.; Tao, W. Fabrication of Upconverting Hybrid Nanoparticles for Near-Infrared Light Triggered Drug Release. *Adv. Mater. Sci. Eng.* **2014**, *2014*, 9. [CrossRef]

177. Wang, J.; Wu, B.; Li, S.; He, Y. NIR light and enzyme dual stimuli-responsive amphiphilic diblock copolymer assemblies. *J. Polym. Sci. A Polym.* **2017**, *55*, 2450–2457. [CrossRef]

178. Wu, S.; Butt, H.-J. Near-Infrared-Sensitive Materials Based on Upconverting Nanoparticles. *Adv. Mater.* **2016**, *28*, 1208–1226. [CrossRef] [PubMed]

179. Chen, C.; Kang, N.; Xu, T.; Wang, D.; Ren, L.; Guo, X. Core-shell hybrid upconversion nanoparticles carrying stable nitroxide radicals as potential multifunctional nanoprobes for upconversion luminescence and magnetic resonance dual-modality imaging. *Nanoscale* **2015**, *7*, 5249–5261. [CrossRef] [PubMed]

180. Chen, S.; Gao, Y.; Cao, Z.; Wu, B.; Wang, L.; Wang, H.; Dang, Z.; Wang, G. Nanocomposites of Spiropyran-Functionalized Polymers and Upconversion Nanoparticles for Controlled Release Stimulated by Near-Infrared Light and pH. *Macromolecules* **2016**, *49*, 7490–7496. [CrossRef]

181. Xing, Q.; Li, N.; Jiao, Y.; Chen, D.; Xu, J.; Xu, Q.; Lu, J. Near-infrared light-controlled drug release and cancer therapy with polymer-caged upconversion nanoparticles. *RSC Adv.* **2015**, *5*, 5269–5276. [CrossRef]

polymers

MDPI

Article

Surface Active to Non-Surface Active Transition and Micellization Behaviour of Zwitterionic Amphiphilic Diblock Copolymers: Hydrophobicity and Salt Dependency

Sivanantham Murugaboopathy [1,2] and Hideki Matsuoka [1,*]

1   Department of Polymer Chemistry, Kyoto University, Katsura, Kyoto 615-8510, Japan;
    sivanantham@prist.ac.in
2   Centre for Research and Development, Department of Physics, PRIST University, Vallam, Thanjavur 613 403,
    Tamil Nadu, India
*   Correspondence: matsuoka.hideki.3s@kyoto-u.jp; Tel.: +81-75-383-2619; Fax: +81-75-383-2475

Received: 8 August 2017; Accepted: 30 August 2017; Published: 5 September 2017

Abstract: We have synthesized a range of zwitterionic amphiphilic diblock copolymers with the same hydrophilic block (carboxybetaine) but with different hydrophobic blocks (n-butylmethacrylate (n-BMA) or 2-ethylhexylacrylate (EHA)) by the reversible addition–fragmentation chain transfer (RAFT) polymerization method. Herein, we systematically examined the role of hydrophobicity and salt concentration dependency of surface activity and micellization behaviour of block copolymer. Transition from surface active to non-surface active occurred with increasing hydrophobicity of the hydrophobic block of block copolymer (i.e., replacing P(n-BMA) by PEHA). Foam formation of block copolymer slightly decreased with the similar variation of the hydrophobic block of block copolymer. Block copolymer with higher hydrophobicity preferred micelle formation rather than adsorption at the air–water interface. Dynamic light scattering studies showed that block copolymer having P(n-BMA) produced near-monodisperse micelles, whereas block copolymer composed of PEHA produced polydisperse micelles. Zimm plot results revealed that the value of the second virial coefficient (A_2) changed from positive to negative when the hydrophobic block of block copolymer was changed from P(n-BMA) to PEHA. This indicates that the solubility of block copolymer having P(n-BMA) in water may be higher than that of block copolymer having PEHA in water. Unlike ionic amphiphilic block copolymer micelles, the micellar shape of zwitterionic amphiphilic block copolymer micelles is not affected by addition of salt, with a value of packing parameters of block copolymer micelles of less than 0.3.

Keywords: non-surface activity; self-assembly; hydrophobicity; zwitterionic amphiphilic block copolymer; light scattering; polymer micelle

---

1. Introduction

Over the past few decades, amphiphilic block copolymers are gaining attention because of their wide range of applications in our daily life, in either solid or solution form [1–3]. They have been used as thermoplastic elastomers, drug delivery systems, emulsifiers, coating materials and templating materials in nano-lithography [1–4]. Studies on novel amphiphilic block copolymers with novel properties are being conducted.

Amphiphilic block copolymers with various kinds of ionic groups such as anionic, cationic, non-ionic and zwitterionic have been studied by different research groups [5–22]. In particular, ionic amphiphilic block copolymer (IABC) systems have been studied extensively [5–18], whereas zwitterionic amphiphilic block copolymers (ZABCs) have hardly been studied [19–22]. In the IABC systems, several factors such as molecular weight of both hydrophobic and hydrophilic blocks of block copolymers, salt, pH, type of ionizing groups (strong or weak), glass transition temperature of hydrophobic block affect their properties namely, surface activity, foam formation and micellization behaviour [7–15]. Kaewsaiha et al. [10] reported that IABC showed non-surface activity when IABC had comparable hydrophilic and hydrophobic block lengths but surface activity was observed when the hydrophobic block length was three times longer than the hydrophilic block. In that study, hydrophobic adsorption force, which depends on the relative molecular weight of hydrophobic and hydrophilic blocks, played a vital role in deciding the surface activity of IABC. Matsuoka and his co-workers [14,15] have examined the role of molecular weight on surface activity and micellization behaviours of IABC. Molecular weight of the polymers decided the non-surface activity of IABC. The degree of polymerization must be more than 30 for both ionic and hydrophobic blocks to be non-surface active [14]. A longer ionic block length also suppresses non-surface activity [15]. The role of hydrophobicity on non-surface activity behaviours of IABC having the same hydrophilic block (poly(styrene sulphonate)) and different hydrophobic blocks ((poly(n-butyl acrylate), polystyrene, and poly(pentafluorostyrene)) has also been investigated [14]. IABC with higher hydrophobicity had higher non-surface activity. This phenomenon can be explained by stable micelle formation: a micelle situation with the highest hydrophobic IABC is more stable than the adsorped state at the air–water interface. Hence, stable micelle formation was found to be one of the key factors for non-surface activity. Another important factor is the image charge effect [14,15]. The image charge effect occurred because of the presence of polyions in the ionic block and these polyions were electrostatically repelled from the air–water interface by the image charge effect at the air–water interface [23]. In fact, transition from non-surface active to surface active was observed by salt addition [8–10]. Hence, two more vital conditions, i.e., the image charge effect and stable micelle formation, are key factors for the non-surface activity of IABC.

Theodoly et al. [18] recently reported that formation of frozen micelles is the main criteria for non-surface activity. However, we found that non-surface activity was exhibited by non-frozen micelles obtained from poly(hydrogenated isoprene)-b-poly(styrene sulphonate) [11]. In their study, the formation of the non-frozen micelle was confirmed by small-angle neutron scattering measurements that revealed the micellar structural transition from sphere to rod after salt addition.

Recently, we investigated the effect of salt on the surface activity and micellization behaviour of ZABC containing n-butylacrylate (n-BA) and carboxybetaine [22], and found that ZABC underwent transition from surface active to non-surface active by addition of salt. We also found that the surface activity and micellization behaviour of ZABC were opposite those of IABC both in the presence and absence of salt. In this study, we carried out systematic investigation on the role of hydrophobicity of the hydrophobic block on the surface activity and micellization behaviour of ZABC using block copolymer composed of n-butylmethacrylate (n-BMA) or 2-ethylhexylacrylate (EHA) as a hydrophobic block and carboxybetaine as a hydrophilic block. The hydrophobicity of EHA is higher than that of either n-BMA or n-BA (order of hydrophobicity: EHA > n-BMA > n-BA). [24,25]. In addition, the glass transition temperature of n-BMA (20 °C) is higher than that of either n-BA (-54 °C) or EHA (-70 °C) [26]. Therefore, we studied the effect of hydrophobicity, salt, block length and glass transition temperature on the surface activity and micellization behaviour of ZABC. Surface tension, static light scattering (SLS) and dynamic light scattering (DLS) measurements were applied to carefully and systematically investigate the interfacial properties of ZABC at air–water interface as well as their hydrodynamic properties in aqueous media.

2. Experimental Section

2.1. Materials

n-Butylmethacrylate (n-BMA), 2-ethylhexylacrylate (EHA), 4,4′-azocyanovaleric acid (ACVA), 2,2′-azobisisobutyronitrile (AIBN), *N*,*N*-dimethylformamide (DMF),diethyl ether and methanol were products of Wako Pure Chemicals (Osaka, Japan). Carboxybetaine (GLBT) was supplied by Osaka Organic Chemicals Co. Ltd. (Osaka, Japan). Distillation method was used to purify *n*-BMA and EHA. The chain transfer agent (CTA) 4-cyanopentanoic acid-4-dithiobenzoate was synthesized as reported previously [27,28]. Deuterium oxide (D$_2$O), deuterated methanol (CD$_3$OD) anddeuterated chloroform (CDCl$_3$) were product of Cambridge Isotope Laboratory (Tewksbury, MA, USA). Water used for sample solution preparation and dialysis was ultrapure water of resistance 18 MΩ cm by the Milli-Q System (Millipore, Bedford, MA, USA).

2.2. Synthesis

Scheme 1 shows the method of synthesis of both homopolymers (PGLBT) and diblock copolymers (P(*n*-BMA)-*b*-PGLBT) and (PEHA-*b*-PGLBT) with various block lengths via reversible addition–fragmentation polymerization (RAFT). The polymerization conditions used for the synthesis of homopolymers (PGLBT) are shown in Table S1. First, GLBT, CTA and ACVA were mixed well with the mixed solvents containing water and DMF at the ratioof 4:1 in a Schlenk tube using a magnetic stirrer. Then, the mixture was degassed under Ar gas atmosphere for three freeze–pump–thaw cycles and then filled with Ar gas. After degassed, RAFT polymerization was carried out at 70 °C for 2 h. When the polymerization was completed, the product was dialyzed in Milli-Q water for 3 days and the homopolymer PGLBT was extracted by freeze-drying. GPC experiments were conducted to determine the molecular weight and its distribution of PGLBT and results are summarized in Table S2. ^{1}H NMR experiments (see Figure 1) were used to confirm the formation of homopolymers (PGLBT).

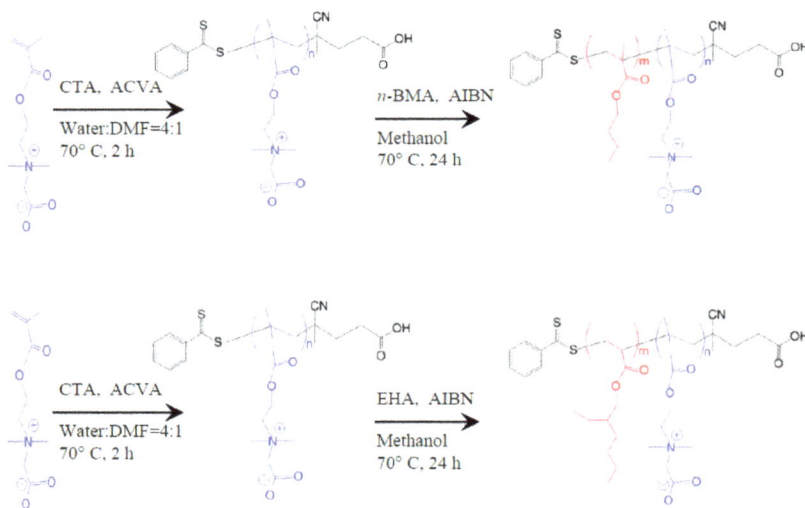

Scheme 1. Synthesis of homopolymer (PGLBT) and diblock copolymers (P(*n*-BMA)-*b*-PGLBT) and (PEHA-*b*-PGLBT).

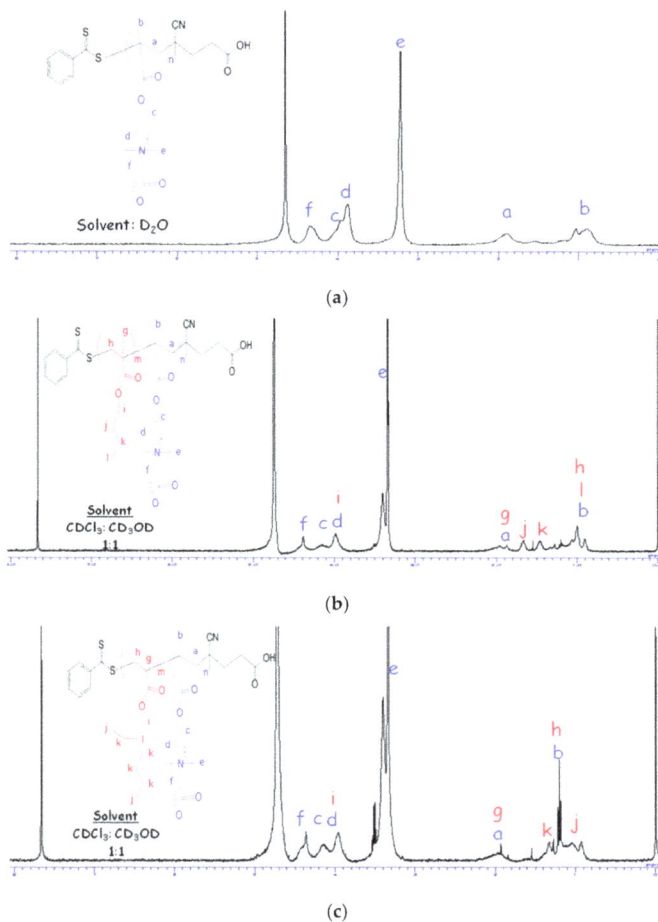

Figure 1. ¹H NMR spectra of: (**a**) homopolymer PGLBT in D₂O; (**b**) diblock copolymer P(*n*-BMA)-*b*-PGLBT in CDCl₃:CD₃OD (1:1); and (**c**) diblock copolymer P(EHA)-*b*-PGLBT in CDCl₃:CD₃OD (1:1).

2.3. ¹H Nuclear Magnetic Resonance (¹H NMR)

¹H NMR spectra of homopolymer and block copolymers were recorded using a JEOL 400WS (JEOL, Tokyo, Japan). To record the ¹H NMR spectra of homopolymers (macro CTA-PGLBT), we used D₂O as a solvent. However, mixed solvents ofCD₃OD and CDCl₃ at a ratio of 1:1 were used for recording the ¹H NMR spectra of diblock copolymers (P(*n*-BMA)-*b*-PGLBT) and (PEHA-*b*-PGLBT).

2.4. Gel Permeation Chromatography (GPC)

GPC measurements were conducted using a JASCO system (Tokyo, Japan) LC-2000 with a UV detector (UV-2075), a refractive index detector (RI-2031) and a Shodex OH pack (SB-804 HQ). The eluent was a mixture of 0.3 M sodium sulphate (Na₂SO₄) and 0.5 M acetic acid (CH₃COOH) adjusted topH3. The sample solution of concentration of 2 mg/mL was used for injection. The number averaged molecular weight (M_n) and the polydispersity index (M_w/M_n) of PGLBT (macro CTA) are shown in Table S2.

2.5. Surface Tension Measurements

Surface tension of polymer solutions was measured by a FACE CBVP-Z Surface Tensiometer (Kyowa Interface Science Co., Ltd., Tokyo, Japan) by the Wilhelmy plate (platinum) method.

2.6. Foam Formation and Foam Height Measurements

Block copolymer solutions of concentration of 1 mg/mL were prepared using Milli-Q water with or without salt (1 M NaCl). These solutions were mechanically shaken for 1 min in identical containers to check the foam forming ability of these polymer solutions. Foam height was also measured as a function of time.

2.7. Light Scattering Measurements

Photal DLS-7000 light scattering setup (Otsuka Electronic, Osaka, Japan) was used for SLS and DLS measurements. This setup was composed of a goniometer, a multi-tau correlator (GC-1000) and a 15 mW He–Ne laser with a wavelength of 632.8 nm. The intensity–intensity autocorrelation function (ICF) $g^{(2)}(q,t)$ was measured at different scattering vectors $q = (4\pi n/\lambda)\sin(\theta/2)$, where n is the refractive index of the solvent, θ is the scattering angle and λ is the wavelength of incident laser beam. Using Siegert relation [29], the field correlation function $g^{(1)}(q,t)$ was obtained from ICF $g^{(2)}(q,t)$. From the field correlation function $g^{(1)}(q,t)$, the decay rate Γ was evaluated by single or double-exponential fitting. From the slope of the plot of Γ vs. q^2, the translational diffusion coefficient (D) was calculated using the relation ($\Gamma = Dq^2$). The Stokes-Einstein equation was used to determine the hydrodynamic radius (R_h) from the value of D.

SLS measurements were carried out by varying the scattering angle (θ) from 30° to 150° at 10° intervals. Zimm plots were used to obtain the weight-average molecular weight (M_w), radius of gyration (R_g) and second virial coefficient (A_2) using the following equation [30,31].

$$\frac{KC_p}{I(q)} = \frac{1}{M_w}\left(1 + \frac{1}{3}\langle R_g\rangle^2 q^2\right) + 2A_2 C_p \tag{1}$$

where K is an optical constant ($4\pi^2 n^2 (dn/dC_p)^2/N_A\lambda^4$), $I(q)$ is the scattered intensity at given q, C_p is the polymer concentration, dn/dC_p is the refractive index increment against C_p, and N_A is Avogadro's number.

2.8. Specific Refractive Index Increment Measurements (dn/dc_p)

Photal differential refractometer DRM-3000S (Otsuka Electronic, Osaka, Japan) was used to determine the specific refractive index increment (dn/dc_p).This instrument had a He-Ne laser (wavelength: 632.8 nm) as a light source.

3. Results and Discussion

3.1. Hydrophobicity and Salt-Dependent Air–Water Interfacial Properties

Previously, we investigated the influence of salt on the surface activity and micellization behaviour of ZABC containing *n*-BA and carboxybetaine [22]. In that study, ZABC having *n*-BA showed transition from surface active to non-surface active after addition of salt. The surface activity behaviour of ZABC was opposite that of IABC both in the presence and absence of salt. Herein, we examined the role of hydrophobicity and salt on the surface activity and micellization behaviour of ZABC having *n*-BMA or EHA as a hydrophobic block and carboxybetaine as a hydrophilic block. As mentioned earlier, the hydrophobicity of EHA is higher than that of *n*-BMA and *n*-BA (order of hydrophobicity: EHA > *n*-BMA > *n*-BA) [24]. Thus, we examined how the hydrophobicity of hydrophobic block of ZABC affects their air–water interfacial properties.

To explore the influence of hydrophobicity and salt on the surface tension of different ZABCs, we measured the surface tension on ZABCs with P(n-BMA) and PEHA and variation of the surface tension is plotted as a function of polymer concentrations, as shown in Figure 2. First, we describe the effect of hydrophobicity on the surface activity of ZABC. ZABC having P(n-BMA) showed moderate surface active behaviour at higher polymer concentrations (predominant for n-BMA$_{42}$-b-GLBT$_{300}$) (Figure 2a–c). This indicates that ZABC with P(n-BMA) in water behaves like an almost surface active polymer which means that ZABC may be adsorped at the air–water interface (higher concentration region). This phenomenon was analogous to that of ZABC with P(n-BA) [22]. ZABC having P(n-BA) was more surface active than ZABC having P(n-BMA) [22]. This could be attributed to the lower hydrophobicity of the hydrophobic block P(n-BA) compared with that of P(n-BMA). Armes et al. [19] reported that the block copolymers having sulphopropylbetaine was slightly surface active. In contrast, almost non-surface active nature was observed when the P(n-BMA) block of ZABC was replaced by PEHA (Figure 2d–f). This implies that ZABC having a higher hydrophobic block (PEHA) acts like a non-surface active polymer which prefers a micelle state rather than adsorped state at the air–water interface since the former is more stable than latter. Earlier, Matsuoka et al. [14] reported the effect of hydrophobicity on non-surface activity behaviours of IABC having the same hydrophilic block (poly-(styrene sulphonate)) and different hydrophobic blocks ((poly(n-butyl acrylate), polystyrene, and poly(pentafluorostyrene)). The more hydrophobic is the hydrophobic block of IABC, the higher is its non-surface activity. Hence, our observations on the non-surface activity of ZABC are consistent with the previous studies on non-surface activity of IABC [14,15]. Hence, a stable micelle formation is one of the key factors of the non-surface active nature also for ZABC in addition to IABC.

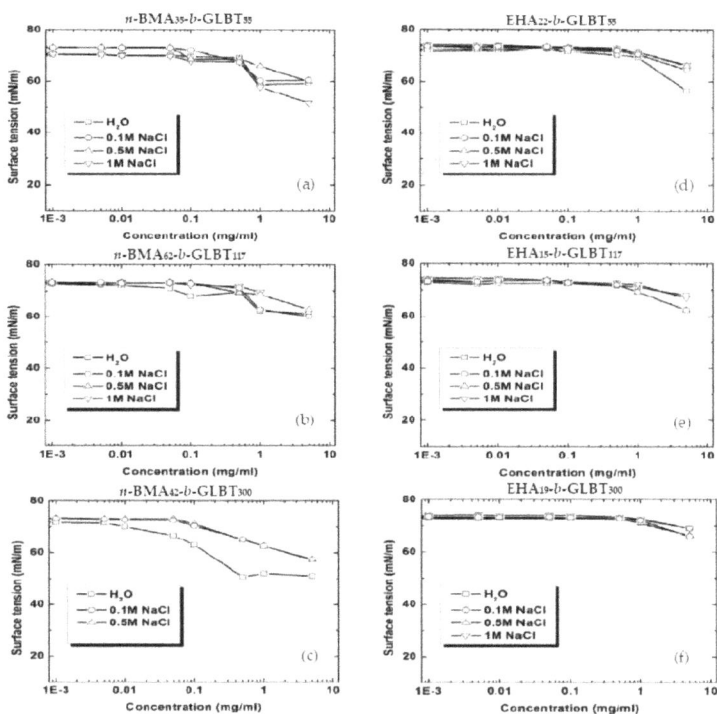

Figure 2. Comparison of the surface tension of ZABC's having P(n-BMA) and PEHA as a function of their concentrations in water and various concentrations of sodium chloride solutions.

Next, we examined the role of salt addition on the surface activity of ZABC. The surface tension of ZABC containing P(n-BMA) (specifically for n-BMA$_{42}$-b-GLBT$_{300}$) with increase in polymer concentration seems to decrease after addition of salt (Figure 2c). This implies that, upon addition of salt, the ZABC with P(n-BMA) (mainly, n-BMA$_{42}$-b-GLBT$_{300}$) undergoes transition from slightly surface active to non-surface active. Recently similar transition was observed for ZABC with P(n-BA) [22]. This transition was attributed to conversion of betaine (zwitterionic) into anionic polymer in the presence of salt which was reported previously [32–36]. This might be related to the change of situation of betaine ions: They are in intra- and intermolecular salts in the absence of added salt, but they change to ions with counterions with added salt ions. On the other hand, salt addition increases the adsorption of IABC at the air–water interface considerably [7–14]. This is due to the screening of image charge effect by added salt ions. Hence, the salt-dependent surface tension behaviour of ZABC showed a trend opposite that of IABC reported previously [14,15,18]. However, salt addition hardly affected the surface activity of ZABC containing PEHA (Figure 2d–f).

The effects of hydrophobicity, salt and block length on the foam forming behaviour of ZABC have also been investigated. Figures 3 and 4 show the hydrophobicity and chain length-dependent foam-forming behaviour of ZABC solutions in the absence and the presence of 1 M sodium chloride, respectively. These figures clearly show that the net hydrophobicity and chain length of ZABC are the dominant factors for the foam forming behaviour of ZABC. The higher are the net hydrophobicity and block length of the block copolymer, the lower is the foam formation. For instance, since ZABC having PEHA is more hydrophobic than those having P(n-BMA), the former showed poorer foam forming behaviour than the latter. The foam forming behaviour of ZABC with P(n-BMA), decreased with the increase in block length except for n-BMA$_{42}$-b-GLBT$_{300}$. The reason for this behaviour could be the smaller net hydrophobicity of n-BMA$_{42}$-b-GLBT$_{300}$ than n-BMA$_{101}$-b-GLBT$_{156}$. We observed a similar trend of decrease in foam formation with the increase in block length for ZABC with P(n-BA) [22]. This might be due to a larger amount of adsorbed polymer with lower molecular weight at the air–water interface than that with higher molecular weight polymers. This might be related to micelle formation in bulk and its stability. Foam formation can be related to CMC of polymers that were determined from SLS measurements. CMC of the polymers such as EHA$_{22}$-b-GLBT$_{55}$, EHA$_{15}$-b-GLBT$_{117}$ and EHA$_{20}$-b-GLBT$_{156}$, respectively, were 0.006, 0.01 and 0.03 mg/mL in water, whereas the CMC of these polymers in 1 M NaCl were 0.0032, 0.0034 and 0.007 mg/mL. Smaller values of CMCs indicate that polymers prefer stable micelle formation even at lower concentration of polymers rather than adsorbed at air–water interface. Some of the polymer solutions, particularly ZABC with P(n-BMA) except n-BMA$_{42}$-b-GLBT$_{300}$, appear bluish, which implies that these solutions have near monodisperse micelles. In the next section, we will explain the evidence for the presence of near-monodisperse micelles in terms of polydispersity indices of micelles. A higher balance/symmetry between hydrophobic and hydrophilic blocks may be the key factor for the bluishness of these micellar solutions. For example, polymer solutions that are formed from the block copolymers n-BMA$_{35}$-b-GLBT$_{55}$, n-BMA$_{62}$-b-GLBT$_{117}$ and n-BMA$_{101}$-b-GLBT$_{156}$ appear bluish (see Figure 3). However, the block copolymers with a lower balance/symmetry such as n-BMA$_{42}$-b-GLBT$_{300}$ and all the ZABC with PEHA form a turbid solution (see Figures 3 and 4). Since zwitterionic block-copolymer (GLBT) does not have upper critical solution temperature (UCST), turbidity of the samples is not due to UCST behaviour of the polymer. In addition, we have kept the temperature constant at room temperature. Thus, Turbidity may possibly due to formation of bigger aggregates. The proof for the formation of aggregates will be explained in terms of polydispersity indices of micelles using DLS results in the upcoming section. Previously, we showed that a nearly equal number of chain lengths of hydrophilic and hydrophobic block is the main criteria for bluishness of solution for ZABC with P(n-BA). The hydrophobicity of P(n-BMA) is higher than that of P(n-BA) [22]. Hence, even the block copolymer of ZABC with P(n-BMA) with a shorter hydrophobic block forms a bluish solution, i.e., stable micelle formation.

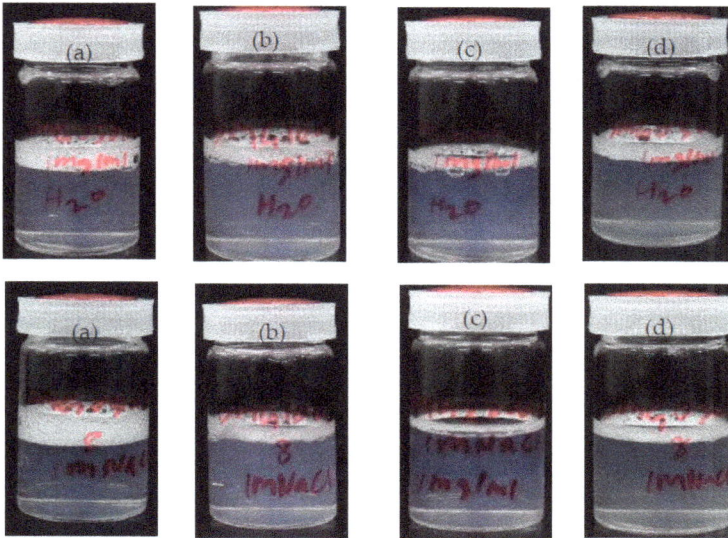

Figure 3. Influence of chain length and salt on foam forming abilities of different ZABCs containing P(n-BMA): (**a**) n-BMA$_{35}$-b-GLBT$_{55}$; (**b**) n-BMA$_{62}$-b-GLBT$_{117}$; (**c**) n-BMA$_{101}$-b-GLBT$_{156}$; and (**d**) n-BMA$_{42}$-b-GLBT$_{300}$ of concentration (1 mg/mL) in water (**top row**) and 1 MNaCl (**bottom row**).

Figure 4. Effect of chain length and salt on foam forming abilities of various ZABCs having PEHA: (**a**) EHA$_{22}$-b-GLBT$_{55}$; (**b**) EHA$_{15}$-b-GLBT$_{117}$; (**c**) EHA$_{20}$-b-GLBT$_{156}$; and (**d**) EHA$_{19}$-b-GLBT$_{300}$ of concentration (1 mg/mL) in water (**top row**) and 1 M NaCl (**bottom row**).

Time-dependent foam heights of the ZABC with P(n-BMA) and PEHA are shown in Figure 5. Foam height and stability of ZABC containing P(n-BMA) is higher than those of ZABC containing

PEHA. For the shortest block length polymers, foam height of former is four times greater than that of latter. ZABC with a shorter chain length have a higher foam height and stability than the ZABC with longer chain lengths. These studies showed that the surface active properties of ZABC depend on the net hydrophobicity of ZABC which is similar to those of IABC [14,15,18].

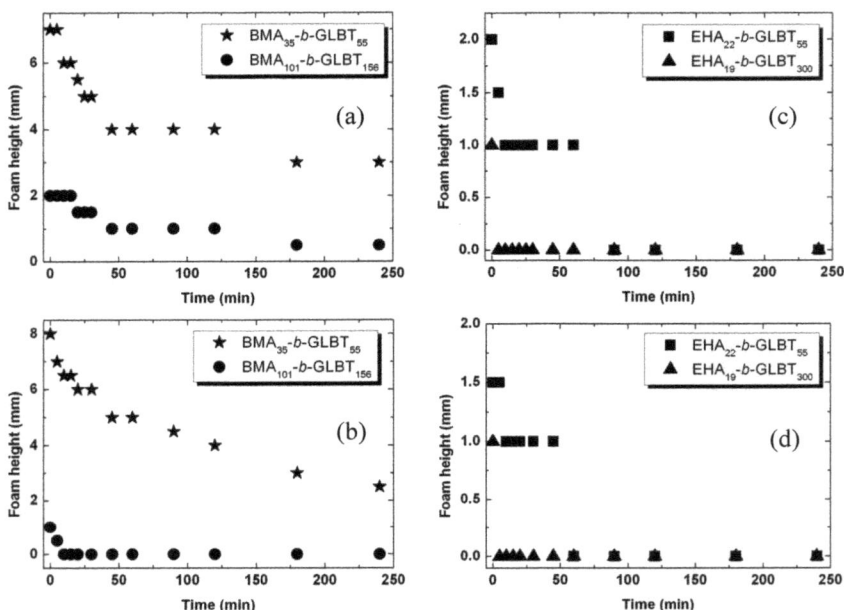

Figure 5. Time-dependent foam height of different ZABC solutions with P(*n*-BMA) and PEHA (concentration = 1 mg/mL): (**a**,**c**) in water; and (**b**,**d**) in 1 M NaCl.

The present study confirmed that micelle formation and its stability are key factors for non-surface activity also for ZABC in addition to the charged state of ionic block, although an opposite trend to IABC was found for the salt effect on ZABC.

3.2. Effect of Salt on CMC

Critical micelle concentrations (CMCs) of ZABC were determined by SLS to elucidate the micellization behaviour of these ZABC. It is difficult to determine CMC of ZABC containing P(*n*-BMA) by SLS unlike PEHA. This may be due to their very low CMC values. Another possible reason may be that the glass transition temperature of P(*n*-BMA) is around room temperature (20 °C) [26]. In contrast, since ZABC with PEHA have a longer hydrophilic chain, CMC was expected to be higher value and hence it was easier to determine their CMC by SLS. Figure 6 shows the salt-dependent CMC of ZABCs. Initially, CMC decreased significantly with the increase in salt concentration up to 0.1 M and then slightly increased with further increase in salt concentration. CMC of low molecular weight ionic surfactants decreased with the increase in salt concentration. Hence, there seems to be a contradiction between the well-known Corrin–Harkins law [37] and the present findings. On the other hand, CMC of non-surface active IABC was found to increase with the increase in salt concentrations [10,16]. Hence, "negative Corrin–Harkins behaviour" is characteristic of non-surface active polymers. ZABCs are slightly surface active without added salt, which can be adsorbed at the water surface and are hard to form micelles, which results in higher CMC. By addition of salt, the situation of zwitterionic group changes as mentioned above and shows a slightly negative charge, which results in appearance of non-surface activity. Hence, the polymer is non-surface active in the presence of salt, CMC increased with increasing added salt concentration. This negative Corrin–Harakins behaviour can be explained as follows: with increasing added salt, the image charge effect at the air–water interface is shielded. Hence, the polymer can easily be adsorbed at the water surface, which makes micelle formation in bulk solution difficult.

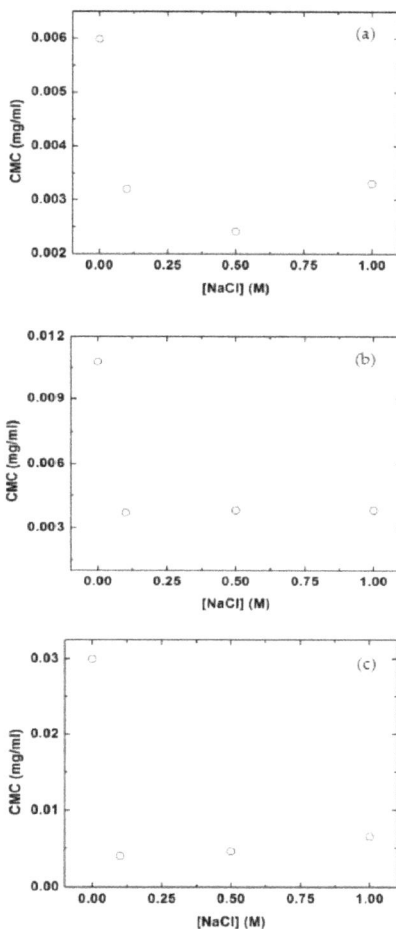

Figure 6. Variation of critical micelle concentration of: (**a**) EHA$_{22}$-*b*-GLBT$_{55}$; (**b**) EHA$_{15}$-*b*-GLBT$_{117}$; and (**c**) EHA$_{20}$-*b*-GLBT$_{156}$ as a function of sodium chloride solutions.

3.3. Influence of Salt on Hydrodynamic Radius

The hydrodynamic radii of the ZABC micelles having P(*n*-BMA) and PEHA were evaluated by the DLS technique. The roles of hydrophobicity and salt concentrations on the hydrodynamic behaviour of ZABC micelles were examined. Figure 7 shows the polymer concentration dependence of hydrodynamic radius of ZABC micelles having P(*n*-BMA) and PEHA at various salt concentrations. Irrespective of polymer concentration, the hydrodynamic radius of the block copolymer micelle was almost constant. This observation is consistent with our previous observations, although an occasional small increase has been reported [22]. An interesting observation here is the transition from near-monodisperse micelles to polydisperse micelles when the hydrophobic block of ZABC was varied from P(*n*-BMA) to PEHA (Table 1). Polydispersity indices of the micelles can be linked with the turbidity of polymer solutions. ZABC micelle solutions having lower polydispersity indices seem to be transparent (bluish) (Figure 3 and Table 1) while ZABC micelle solutions having higher polydispersity indices were turbid (Figure 4 and Table 1). This observation is probably due to the difference in their hydrophobicity and this will be discussed in the next section. Armes and co-workers [19] observed polydisperse micelles by direct dissolution of the sulphopropylbetaine copolymers in water but near-monodisperse micelles were formed when the preliminary dissolution was carried out in a non-selective solvent (2,2,2-trifluoroethanol). In addition, we observed that at a given polymer concentration the hydrodynamic radius of the ZABC micelle increased with the increase in salt concentration (see inset of Figure 7). This anti-polyelectrolyte effect could be responsible for stretching of betaine block chain by salt addition. Analogous phenomena have been observed for block copolymer containing sulphopropylbetaine [19] and carboxybetaine with P(*n*-BA) [22] after addition of salt.

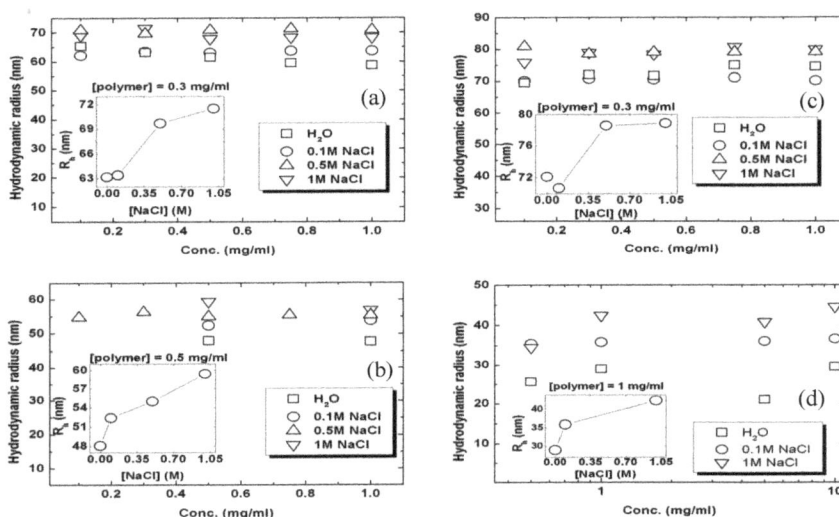

Figure 7. Polymer concentrations dependence of hydrodynamic radius: (**a**) *n*-BMA$_{62}$-*b*-GLBT$_{117}$; (**b**) *n*-BMA$_{101}$-*b*-GLBT$_{156}$; (**c**) *n*-BMA$_{42}$-*b*-GLBT$_{300}$; and (**d**) EHA$_{20}$-*b*-GLBT$_{156}$. Inset shows variation of hydrodynamic radius with different concentrations of sodium chloride.

Table 1. Effect of hydrophobicity and salt on dissymmetry ratio and polydispersity indices of the micelles.

Diblock Copolymer	Z_d [a]		PDI [b]	
	Water	1 MNaCl	Water	1 MNaCl
n-BMA$_{35}$-*b*-GLBT$_{55}$	1.6	1.5	0.16	0.13
n-BMA$_{62}$-*b*-GLBT$_{117}$	2.1	2	0.13	0.13
n-BMA$_{101}$-*b*-GLBT$_{156}$	1.5	1.5	0.11	0.08
n-BMA$_{42}$-*b*-GLBT$_{300}$	2.7	2.6	0.2	0.19
EHA$_{22}$-*b*-GLBT$_{55}$	3.5	3.2	0.29	0.28
EHA$_{15}$-*b*-GLBT$_{117}$	4.4	4.5	0.29	0.29
EHA$_{20}$-*b*-GLBT$_{156}$	3.9	3.9	0.3	0.32
EHA$_{19}$-*b*-GLBT$_{300}$	3.5	3.7	0.19	0.21

[a] Dissymmetry ratio $Z_d = I_{45}/I_{135}$; I_{45} and I_{135} are scattering intensities at angles 45° and 135° respectively.
[b] PDI: Polydispersity indices of the micelles.

The dissymmetry ratio is defined as a ratio of light scattering intensities at 45° and 135°. The dissymmetry ratio varied with respect to relative block length of hydrophobic and hydrophilic blocks of ZABC. In the present work, ZABC containing P(*n*-BMA) is more symmetric (almost comparable hydrophobic and hydrophilic block lengths) as compared to ZABC having PEHA (hydrophilic block is longer than hydrophobic block). ZABC with P(*n*-BMA) except *n*-BMA$_{42}$-*b*-GLBT$_{300}$ showed a lower dissymmetry ratio value which is less than 2.1, but ZABC composed of PEHA had a higher dissymmetry ratio of around 3.5 (Table 1). However, the dissymmetry ratio was almost constant even after addition of salt. In addition, the dissymmetry ratio can also be correlated with the turbidity of polymer solutions. For instance, ZABC containing (P(*n*-BMA)) having a lower dissymmetry ratio appeared transparent except *n*-BMA$_{42}$-*b*-GLBT$_{300}$ (Figure 3 and Table 1), whereas ZABC having a higher dissymmetry ratio (PEHA) appeared turbid (Figure 4 and Table 1). DLS studies revealed that the micellization properties of ZABC consisting of P(*n*-BMA) seems to be different from those of ZABC having PEHA.

3.4. Hydrophobicity, Salt and Block Length-Dependent Aggregation Number and Second Virial Coefficient of Micelles

We analysed the SLS data using Zimm plots (Figure S1) as described previously [22]. Zimm plots are summarized in Table 2. First, the effects of the hydrophobic and hydrophilic block lengths on the aggregation numbers of ZABC were examined. The aggregation number (N_{agg}) was inversely proportional to the chain length of polymers. For example, in the case of ZABC with P(*n*-BMA), when the block length of P(*n*-BMA) was changed from 35 to 101 units and that of PGLBT was increased from 55 to 156 units, the aggregation number (N_{agg}) decreased from 555 to 117, by five times. We found recently that, for ZABC with P(*n*-BA), N_{agg} decreased by five times for similar variation of soluble block length [22]. The present findings for ZABC having P(*n*-BMA) were analogous to our recent results for ZABC with P(*n*-BA) [22]. However, the N_{agg} of ZABC having PEHA was reduced 10 times (from 891 to 91) when the chain length of the PGLBT block varied from 55 to 156 units and that of PEHA was almost constant. The difference in decrease of N_{agg} with change in hydrophobic block (from P(*n*-BMA) to PEHA) could be due to higher hydrophobic nature of PEHA chain. On the other hand, Khougaz et al. [38] found that the aggregation number of IABC was affected by the length of the hydrophobic block more than the hydrophilic block. In addition, Armes et al. [19] found that for the block copolymer having sulphopropylbetaine, the value of N_{agg} increased more with the increase in the length of the hydrophobic block rather than with the decrease in the hydrophilic block.

Table 2. Micellar properties of ZABC micelles.

Diblock Copolymer	N_{agg} [a]		$A_2 \times 10^5$ (cm^3·mol·g^{-2}) [b]		R_g (nm) [c]		β [d]	
	Water	1 MNaCl	Water	1 MNaCl	Water	1 MNaCl	Water	1 MNaCl
n-BMA$_{35}$-*b*-GLBT$_{55}$	555	146	4.7	0.25	65.9	58.1	0.031	0.02
n-BMA$_{101}$-*b*-GLBT$_{156}$	117	32	2.7	1.0	58.8	55.8	0.032	0.01
EHA$_{22}$-*b*-GLBT$_{55}$	891	152	−3.5	−17.6	203.3	172.8	0.1	0.03
EHA$_{20}$-*b*-GLBT$_{156}$	91	13	−3.1	−88.1	218.7	139.8	0.032	0.01

[a] N_{agg}: Aggregation number. [b] A_2: Second virial coefficient. [c] R_g: Radius of gyration. [d] β: Packing parameter.

Next, we noticed that N_{agg} decreased with addition of salt. In the P(*n*-BMA)-*b*-PGLBT systems, salt addition decreased N_{agg} from 555 to 146 in *n*-BMA$_{35}$-*b*-GLBT$_{55}$ and from 117 to 32 in BMA$_{101}$-*b*-GLBT$_{156}$. We have observed similar phenomena for ZABC composed of PEHA where N_{agg} was about 6 times lower (Table 2). Our observations are consistent with earlier studies on block copolymers having sulphopropylbetaine [19] and carboxybetaine [22]. In contrast, Khougaz et al. [38] showed that the aggregation numbers of IABC initially increased with increasing salt concentration and then reached saturation. From the present study, it is clear that the increase in chain length of polymer affects the value of N_{agg} slightly higher margin as compared to salt.

Studies on the effect of hydrophobicity and salt on the second virial coefficient (A_2) showed that the value of A_2 changed from positive to negative when the hydrophobic block of ZABC is changed from P(n-BMA) to PEHA (Table 2). This indicates that the solubility of ZABC having P(n-BMA) in water could be higher than that of ZABC having PEHA in water. Similar to N_{agg}, the value of A_2 also decreased with increase in block length of ZABC. For example, the values of A_2 decreased with the increase in both soluble and insoluble blocks of ZABC containing P(n-BMA) and values were 4.7×10^{-5} and 2.7×10^{-5} cm$^3 \cdot$mol$\cdot$g^{-2} for n-BMA$_{35}$-b-GLBT$_{55}$ and n-BMA$_{101}$-b-GLBT$_{156}$, respectively, in water. In addition, the values of A_2 decreased with the increase in salt concentration. In the presence of 1 M NaCl, the values of A_2 of n-BMA$_{35}$-b-GLBT$_{55}$ and n-BMA$_{101}$-b-GLBT$_{156}$ decreased to 0.25×10^{-5} and 1×10^{-5} cm$^3 \cdot$mol$\cdot$g^{-2}, respectively. Analogous phenomenon has been observed for ZABC having PEHA (see Table 2). The decrease in polymer-solvent interaction is responsible for the decrease in value of A_2 with increase in block length and NaCl concentration. These results are consistent with the decrease in CMC with increase in salt concentration. Khougaz and co-workers [38] reported that in IABC, the values of A_2 for fixed hydrophobic block length PS(23) were increased from -3.5×10^{-4} to -0.36×10^{-4} cm$^3 \cdot$mol$\cdot$g^{-2} with the increase in chain length of hydrophilic block (PANa) from 44 to 300. This behaviour implies that IABC with a longer PANa chain interacts with the solvent in a more favourable manner than IABC with a shorter PANa chain. The solubility of the block copolymer was expected to increase with the increase in the soluble block length. However, the values of the radius of gyration (R_g) were hardly affected by the increase in the chain length and salt concentration (Table 2).

To determine the value of core radius (R_c) of micelles, we substituted the known values of M_w and N_{agg} (from Zimm plots) in the equation $R_c = \sqrt[3]{\frac{3N_{agg}NM_w}{4\pi\rho N_A}}$, where N is the block length of the hydrophobic chain, M_w is the molecular weight of the hydrophobic monomer, ρ is the density of the bulk polymer (for P(n-BMA) and PEHA approximately 1.0 mg/mL and 0.9 mg/mL respectively) and N_A is the Avogadro's number. We estimated the packing parameter (β) of micelle using the equation $\beta = \frac{V_H}{L_c A_0}$, where V_H is the volume occupied by the hydrophobic chain, L_c is the counter length of the hydrophobic chain ($\approx$core radius, R_c) and A_0 is the surface area of the hydrophilic chain. The morphology of polymeric aggregates was identified from the values of β [39,40]. From the values of V_H ($=4\pi R_c^3/3N_{agg}$), L_c ($=R_c$) and $A_0 = \left(4\pi((R_h + R_c)/2)^2\right)/N_{agg}$, packing parameters were calculated and summarized in Table 2. Irrespective of the polymer, the values of β for the ZABC micelles were less than 0.3 and this implies that micelles formed from the block copolymers are spherical even after addition of salt. The surface area of the hydrophilic chain increased after addition of salt and hence the packing parameter tended to decrease below 0.3. For instance, the surface area of hydrophilic chain of n-BMA$_{35}$-b-GLBT$_{55}$ increased from 74 to 267 nm^2 when the concentration of NaCl increased from 0 M NaCl to 1 M NaCl. This phenomenon enables ZABC micelles to retain their spherical shape after salt addition. We observed similar behaviour for ZABC micelles having P(n-BA) [22]. The results obtained from the Zimm plots are consistent with the hydrodynamic behaviour of ZABC micelles. However, our preceding study on the micellization behaviours of IABC with strong acid groups, poly(hydrogenated isoprene)-b-poly(styrene sulphonate), showed transition from sphere to rod after addition of salt [11]. The surface area of the hydrophilic chain and corona thickness of IABC micelles decreased after addition of salt and hence the packing parameter might be above 0.3. Thus, the micelles formed from IABC may undergo transition from sphere to rod.

Figure 8 illustrates the influence of hydrophobicity and salt on the surface activity and micellization behaviour of ZABC. Increase in the hydrophobicity of ZABC by changing the hydrophobic block (from P(n-BMA) to PEHA), caused transition from surface active to non-surface active. Similarly, addition of salt caused transition of the ZABC composed of P(n-BMA) (particularly for n-BMA$_{42}$-b-GLBT$_{300}$) from surface active to non-surface active polymers. Further salt addition increased the hydrodynamic radius of the micelle and decreased the aggregation number of the micelle.

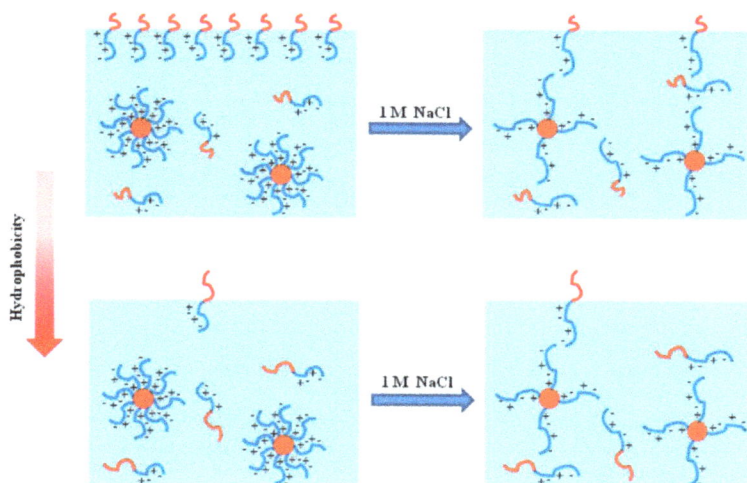

Figure 8. Schematic illustration of influence of hydrophobicity and salt on the air–water interfacial and micellization properties of ZABCs having P(*n*-BMA) and PEHA. Blue and red curves respectively indicate zwitterionic and hydrophobic polymers. Hydrophobic core of the micelles are represented by red circles or balls.

4. Conclusions

Various ZABCs with P(*n*-BMA) or PEHA as a hydrophobic block and carboxybetaine as a hydrophilic block were successfully synthesized by the RAFT polymerization method. The two parameters, hydrophobicity and salt, were varied separately to examine their role on the surface activity and micellization behaviour of ZABC. Surface tension measurements and foam formation observations revealed that the hydrophobicity of hydrophobic block present in ZABC is the predominant factor for the surface active and foam forming behaviour of ZABC. When the hydrophobicity of the hydrophobic block of ZABC was increased by introducing PEHA instead of P(*n*-BMA), ZABC showed transition from surface active to non-surface active. Similar observations were observed for IABC [14]. A possible reason for this behaviour could be that ZABC with higher hydrophobicity can form micelles, which is more stable than the adsorbed state at the air–water interface. Salt addition caused transition of ZABC with P(*n*-BMA) from surface active to non-surface active. This may be due to zwitterionic to anionic transition of the betaine block. The value of CMC of ZABC slightly increased after addition of salt, which is typical for non-surface active polymers. The increase in the value of hydrodynamic radius with increase in salt concentration at a given polymer concentration might be due to the anti-polyelectrolyte effect, i.e., increase in chain length of hydrophilic corona. This observation is supported by the increase in surface area of hydrophilic betaine corona after the addition of salt. However, the aggregation number and second virial coefficient of the micelles tended to decrease with the increase in chain length, and with the addition of salt. DLS and Zimm plot results revealed that ZABC containing P(*n*-BMA) could form more monodisperse micelles than those containing PEHA. When the hydrophobic block P(*n*-BMA) was replaced by PEHA in ZABC having a fixed zwitterionic hydrophilic block, the value of A_2 changed from positive to negative. This is due to the decrease in the solubility of the polymer. Unlike IABC micelles, the micellar shape of ZABC micelles was not affected by the addition of salt, which was confirmed from the packing parameter values of block copolymer micelles (less than 0.3). The present study revealed that ZABC becomes non-surface active when the betaine block is changed to an ionic state by addition of salt or when very stable micelles are formed in bulk solutions with high enough hydrophobicity.

Supplementary Materials: Supplementary Materials are available online at www.mdpi.com/2073-4360/9/9/412/s1.

Acknowledgments: Sivanantham Murugaboopathy gratefully acknowledges Global Centre of Excellence (GCOE) for providing post-doctoral fellowship. We thank Yoshiyuki Saruwatari, Osaka Organic Chemical Industry Ltc., for kindly supplying the GLBT monomer. This work was supported by a grant-in-aid for Scientific Research on Innovative Areas "Molecular Soft-Interface Science" (20106006) from the Ministry of Education, Culture, Sports, Science and Technology of Japan, to which our sincere gratitude is due.

Author Contributions: Sivanantham Murugaboopathy synthesized the polymers, carried out the experiments, analysed the data, and was involved in the interpretation of the results and the writing of the paper. Hideki Matsuoka supervised the research and contributed in the data analyses, interpretation of results and the writing of the paper.

Conflicts of Interest: The authors declare no conflict of interest.

References

1. Bates, F.S.; Fredrickson, G.H. Block copolymers—Designer soft materials. *Phys. Today* **1999**, *52*, 32–38. [CrossRef]
2. Hamley, I.W. *The Physics of Block Copolymers*; Oxford University Press: Oxford, UK, 1998; ISBN 9780198502180
3. Bahadur, P. Block copolymers—Their microdomain formation (in solid state) and surfactant behaviour (in solution). *Curr. Sci.* **2001**, *80*, 1002–1007.
4. Kataoka, K.; Harada, A.; Nagasaki, Y. Block copolymer micelles for drug delivery: Design, characterization and biological significance. *Adv. Drug Deliv. Rev.* **2001**, *47*, 113–131. [CrossRef]
5. Wittmer, J.; Joanny, J.F. Charged diblock copolymers at interfaces. *Macromolecules* **1993**, *26*, 2691–2697. [CrossRef]
6. Netz, R.R.; Andelman, D. Neutral and charged polymers at interfaces. *Phys. Rep.* **2003**, *380*, 1–95. [CrossRef]
7. Matsuoka, H.; Matsutani, M.; Mouri, E.; Matsumoto, K. Polymer micelle formation without Gibbs monolayer formation—Synthesis and characteristics of amphiphilic diblock copolymer having sulfonic acid groups. *Macromolecules* **2003**, *36*, 5321–5330. [CrossRef]
8. Matsumoto, K.; Ishizuka, T.; Harada, T.; Matsuoka, H. Association behavior of fluorine-containing and non-fluorine-containing methacrylate-based amphiphilic diblock copolymer in aqueous media. *Langmuir* **2004**, *20*, 7270–7282. [CrossRef] [PubMed]
9. Matsuoka, H.; Maeda, S.; Kaewsaiha, P.; Matsumoto, K. Micellization of non-surface-active diblock copolymers in water, Special characteristics of poly(styrene)-block-poly(styrenesulfonate). *Langmuir* **2004**, *20*, 7412–7421. [CrossRef] [PubMed]
10. Kaewsaiha, P.; Matsumoto, K.; Matsuoka, H. Non-surface activity and micellization of ionic amphiphilic diblock copolymers in water, Hydrophobic chain length dependence and salt effect on surface activity and the critical micelle concentration. *Langmuir* **2005**, *21*, 9938–9945. [CrossRef] [PubMed]
11. Kaewsaiha, P.; Matsumoto, K.; Matsuoka, H. Sphere-to-rod transition of non-surface-active amphiphilic diblock copolymer micelles: A small-angle neutron scattering study. *Langmuir* **2007**, *23*, 9162–9169. [CrossRef] [PubMed]
12. Nayak, R.R.; Yamada, T.; Matsuoka, H. Non-surface activity of cationic amphiphilic diblock copolymers. *IOP Conf. Ser. Mater. Sci. Eng.* **2011**, *24*, 012024. [CrossRef]
13. Ghosh, A.; Yusa, S.; Matsuoka, H.; Saruwatari, Y. Non-surface activity and micellization behavior of cationic amphiphilic block copolymer synthesized by reversible addition-fragmentation chain transfer process. *Langmuir* **2011**, *27*, 9237–9244. [CrossRef] [PubMed]
14. Matsuoka, H.; Chen, H.; Matsumoto, K. Molecular weight dependence of non-surface activity for ionic amphiphilic diblock copolymers. *Soft Matter* **2012**, *35*, 9140–9146. [CrossRef]
15. Ghosh, A.; Yusa, S.; Matsuoka, H.; Saruwatari, Y. Chain length dependence of non-surface activity and micellization behaviour of cationic amphiphilic diblock copolymers. *Langmuir* **2014**, *30*, 3319–3328. [CrossRef] [PubMed]
16. Matsuoka, H.; Onishi, T.; Ghosh, A. pH-responsive non-surface-active/surface-active transition of weakly ionic amphiphilic diblock copolymers. *Colloid Polym. Sci.* **2014**, *292*, 797–806. [CrossRef]
17. Amiel, C.; Sikka, M.; Schneider, J.W.; Tsao, Y.H.; Tirrell, M.; Mays, J.W. Adsorption of hydrophilic-hydrophobic block copolymers on silica from aqueous solutions. *Macromolecules* **1995**, *28*, 3125–3134. [CrossRef]

18. Theodoly, O.; Jacquin, M.; Muller, P.; Chhun, S. Adsorption kinetics of amphiphilic diblock copolymers: From kinetically frozen colloids to macrosurfactants. *Langmuir* **2009**, *25*, 781–793. [CrossRef] [PubMed]

19. Tuzar, Z.; Pospisil, H.; Plestil, J.; Lowe, A.B.; Baines, F.L.; Billingham, N.C.; Armes, S.P. Micelles of hydrophilic-hydrophobic poly(sulfobetaine)-based block copolymers. *Macromolecules* **1997**, *30*, 2509–2512. [CrossRef]

20. Lowe, A.B.; Billingham, N.C.; Armes, S.P. Synthesis and properties of low-polydispersity poly(sulfopropylbetaine)s and their block copolymers. *Macromolecules* **1999**, *32*, 2141–2148. [CrossRef]

21. Yusa, S.; Fukuda, K.; Yamamoto, T.; Ishihara, K.; Morishima, Y. Synthesis of well-defined amphiphilic block copolymers having phospholipid polymer sequences as a novel biocompatible polymer micelle reagent. *Biomacromolecules* **2005**, *6*, 663–670. [CrossRef] [PubMed]

22. Sivanantham, M.; Matsuoka, H. Salt-dependent surface activity and micellization behaviour of zwitterionic amphiphilic diblock copolymers having carboxybetaine. *Colloid Polym. Sci.* **2015**, *293*, 1317–1328.

23. Onsager, L.; Samaras, N.N.T. The surface tension of Debye-Hückel electrolytes. *J. Chem. Phys.* **1934**, *2*, 528–536. [CrossRef]

24. Leonard, E.C. *Vinyl and Diene Monomers, High Polymers Series*; Wiley-Interscience: New York, NY, USA, 1970; Volume XXIV.

25. Vijayendran, B.R. Polymer polarity and surfactant adsorption. *J. Appl. Polym. Sci.* **1979**, *23*, 733–742. [CrossRef]

26. Dumitriu, S.; Popa, V. *Polymeric Biomaterials*; CRC Press: Boca Raton, FL, USA, 2002; ISBN 9781420094725.

27. Mitsukami, Y.; Donovan, M.S.; Lowe, A.B.; McCormick, C.L. Water-soluble polymers. 81. Direct synthesis of hydrophilic styrenic-based homopolymers and block copolymers in aqueous solution via raft. *Macromolecules* **2001**, *34*, 2248–2256. [CrossRef]

28. Oae, S.; Yagihara, T.; Okabe, T. Reduction of semipolar sulphur linkages with carbodithioic acids and addition of carbodithioic acids to olefins. *Tetrahedron* **1972**, *28*, 3203–3216. [CrossRef]

29. Berne, B.; Pecora, R. *Dynamic Light Scattering*; John Wiley: New York, NY, USA, 1976; p. 174. ISBN 0486411559.

30. Zimm, B.H. Apparatus and methods for measurement and interpretation of the angular variation of light scattering; Preliminary results on polystyrene solutions. *J. Chem. Phys.* **1948**, *16*, 1099–1116. [CrossRef]

31. Huglin, M.B. *Light Scattering from Polymer Solutions*; Academic Press: New York, NY, USA, 1972.

32. Inayama, R.; Nakamura, S.; Tatsumi, N.; Otsuka Electronics Co., Osaka, Japan. Private communication, 2013.

33. Jiang, W.; Fischer, G.; Girmay, Y.; Irgum, K. Zwitterionic stationary phase with covalently bonded phosphorylcholine type polymer grafts and its applicability to separation of peptides in the hydrophilic interaction liquid chromatography mode. *J. Chromatogr. A* **2006**, *1127*, 82–91. [CrossRef] [PubMed]

34. Katagiri, K.; Hashizume, M.; Kikuchi, J.; Taketani, Y.; Murakami, M. Creation of asymmetric bilayer membrane on monodispersed colloidal silica particles. *Colloids Surf. B* **2004**, *38*, 149–153. [CrossRef] [PubMed]

35. Parker, A.P.; Reynolds, P.A.; Lewis, A.L.; Kirkwood, L.; Hughes, L.G. Investigation into potential mechanisms promoting biocompatibility of polymeric biomaterials containing the phosphorylcholine moiety: A physicochemical and biological study. *Colloids Surf. B* **2005**, *46*, 204–217. [CrossRef] [PubMed]

36. Shrivastava, S.; Matsuoka, H. Photo-responsive block copolymer: Synthesis, characterization and surface activity control. *Langmuir* **2014**, *30*, 3957–3966. [CrossRef] [PubMed]

37. Corrin, M.L.; Harkins, W.D. The effect of salts on the critical concentration for the formation of micelles in colloidal electrolytes. *J. Am. Chem. Soc.* **1947**, *69*, 683–688. [CrossRef] [PubMed]

38. Khougaz, K.; Astafieva, I.; Eisenberg, A. Micellization in block polyelectrolyte solutions. 3. Static light scattering characterization. *Macromolecules* **1995**, *28*, 7135–7147. [CrossRef]

39. Israelachvili, J.N. *Intermolecular and Surface Forces*, 2nd ed.; Academic Press: London, UK; New York, NY, USA, 1991.

40. Nagarajan, R. Molecular packing parameter and surfactant self-assembly: The neglected role of the surfactant tail. *Langmuir* **2002**, *18*, 31–38. [CrossRef]

![polymers logo] **polymers**

Article

Acetal-Linked Paclitaxel Polymeric Prodrug Based on Functionalized mPEG-PCL Diblock Polymer for pH-Triggered Drug Delivery

Yinglei Zhai [1,2,†], Xing Zhou [3,†], Lina Jia [4], Chao Ma [5], Ronghua Song [1], Yanhao Deng [1], Xueyao Hu [1] and Wei Sun [1,*]

1   Department of Biomedical Engineering, School of Medical Devices, Shenyang Pharmaceutical University, Shenyang 110016, China; zyllll1001@126.com (Y.Z.); srh.123@163.com (R.S.); dengyanhao88@163.com (Y.D); huxueyaosyphu@163.com (X.H.)
2   State Key Laboratory for Marine Corrosion and Protection, Luoyang Ship Material Research Institute (LSMRI), Qingdao 266101, China
3   Hainan Institute of Materia Medica, Haikou 570311, China; beyondyme@163.com
4   Department of Pharmacology, School of Life Science and Biopharmaceutics, Shenyang Pharmaceutical University, Shenyang 110016, China; frankjln@126.com
5   College of Food & Pharmaceutical Engineering, Guizhou Institute of Technology, Guizhou 550003, China; 13765046535@163.com
*   Correspondence: sunwei@syphu.edu.cn; Tel.: +86-24-43520357
†   These two authors contributed equally to this work

Received: 14 October 2017; Accepted: 7 December 2017; Published: 11 December 2017

Abstract: The differences in micro-environment between cancer cells and the normal ones offer the possibility to develop stimuli-responsive drug-delivery systems for overcoming the drawbacks in the clinical use of anticancer drugs, such as paclitaxel, doxorubicin, and etc. Hence, we developed a novel endosomal pH-sensitive paclitaxel (PTX) prodrug micelles based on functionalized poly(ethylene glycol)-poly(ε-caprolactone) (mPEG-PCL) diblock polymer with an acid-cleavable acetal (Ace) linkage (mPEG-PCL-Ace-PTX). The mPEG-PCL-Ace-PTX$_5$ with a high drug content of 23.5 wt % was self-assembled in phosphate buffer (pH 7.4, 10 mM) into nanosized micelles with an average diameter of 68.5 nm. The in vitro release studies demonstrated that mPEG-PCL-Ace-PTX$_5$ micelles was highly pH-sensitive, in which 16.8%, 32.8%, and 48.2% of parent free PTX was released from mPEG-PCL-Ace-PTX$_5$ micelles in 48 h at pH 7.4, 6.0, and 5.0, respectively. Thiazolyl Blue Tetrazolium Bromide (MTT) assays suggested that the pH-sensitive PTX prodrug micelles displayed higher therapeutic efficacy against MCF-7 cells compared with free PTX. Therefore, the PTX prodrug micelles with acetal bond may offer a promising strategy for cancer therapy.

Keywords: prodrug; polymer micelles; pH-sensitive; acetal; paclitaxel

1. Introduction

Cytotoxic chemotherapeutics, which act by killing rapidly proliferating cancer cells, remains to be one of the most preferred approaches for treatment of various cancers. However, its application suffers some major limitations such as organ damage, infertility, immunosuppression, and nausea/vomiting which severely reduce patient quality of life [1,2]. Furthermore, many anticancer drugs have low aqueous solubility, which requires the use of organic solvents or surfactants. These drawbacks therefore urge scientists to develop improved chemotherapy formulations by which the use of solubilizer would no more be needed and the selectivity and efficacies toward cancer cells can be enhanced [3].

In the past decades, nanoparticles based on enhanced permeability and retention (EPR) effect such as micelles [4,5], liposomes [6,7], and dendrimers [8,9] have been intensively reported as drug delivery systems. Compared with free drugs, chemotherapeutics conjugated with nanoparticles have the advantages of passive selective accumulation at the cancer tissues and increase the solubility of drug molecules [10–12]. Among them, responsive polymeric nanoparticles have emerged as one of the most promising delivery systems for anticancer drugs [13–15]. Internal and external stimuli, such as temperature [16–19], pH [20–24], or enzyme [25–28] were applied to maximal drug release in target tissues. As reported, the extracellular of solid tumors (pH 6.0–7.0) is slightly acidic compared with normal tissues, and pH inside cancer cells is even lower, especially inside endosome (pH 5.0–6.0) and lysosome (pH 4.0–5.0) [29,30]. Based on this unusual tumor tissue environment, a great variety of pH responsive polymer prodrugs were designed with acid-labile chemical bonds such as acetal [31], amine [32], hydrozone [33], and so forth.

Acetals are promising candidates for the development of acid-sensitive connections with hydroxyl groups because the rate of hydrolysis is highly relative to the concentration of hydronium ion [34]. To the best of my knowledge, several groups developed pH sensitive polymeric prodrugs via acetal bond. However, most of them show the disadvantages such as non-biodegradable or low drug content [31,35,36].

Poly(caprolactone) (PCL) is one of the extensively employed aliphatic polyestesr due to its excellent biocompatibility and biodegradability. Owing to the appropriate hydrophilic and hydrophobic balance, amphiphilic di-block copolymers based on poly(ethylene glycol)-block-poly(caprolactone) (PEG-b-PCL) show great potential to delivering water-insoluble hydrophobic drugs due to their large solubilizing power and loading capacity [37–39]. However, these copolymers are not suitable for the preparation of pH sensitive anti-cancer prodrug because of the absence of functional groups. Surnar and Jayakannan reported a newly developed pH-responsive PEG-b-PCL with carboxylic groups, which makes the synthesis of an acid-labile anti-cancer prodrug possible [40].

In this work, we report acetal-linked PTX prodrugs with different PTX content. The prodrug micelles offer two advantages: (i) the pH sensitive acetal bond offers prodrug micelles possessing specific tumor targetability. Moreover, the firmer conjugation between PEG-b-PCL and PTX may reduce the undesired drug release during circulation in the bloodstream; and (ii) the superior drug content can be easily achieved by varying PTX-to-polymer molar ratio. The structure and physiochemical characteristics of PTX prodrug and its micelles were determined using ^{1}H NMR, gel permeation chromatography (GPC), dynamic light scattering (DLS) and transmission electron microscopy (TEM). We also investigated pH-dependent drug release from PTX prodrug micelles and in vitro cytotoxicity of PTX-conjugated micelles to MCF-7 cell.

2. Experimental Section

2.1. Materials

1,4-Cyclohexanediol, potassium *t*-butoxide, *t*-butyl acrylate, pyridinium chlorochromate (PCC), metachloroperbenzoic acid (MCPBA), tin (II) 2-ethylhexanoate (Sn(Oct)$_2$), trifluoroacetic acid (TFA), Dicyclohexylcarbodiimide (DCC), and 4-dimethylaminopyridine (DMAP) were purchased from Aladdin Industrial Corporation (Shanghai, China). Paclitaxel (PTX) and Molecular sieves (4 Å) were obtained from Meilun Biotechnology Co., Ltd. (Dalian, China). Polyethylene glycol monomethyl ether (MW = 5000, here after referred as mPEG, Sigma-Aldrich, Shanghai, China), 2-(ethenyloxy) ethanol (Nanjing Chemlin, Nanjing, China) and *p*-toluenesulfonic acid monohydrate (*p*-TSA, Acros, Beijing, China) were used directly without any purification. *N*,*N*-dimethylformamide (DMF) was firstly dried by MgSO$_4$ and consequently distilled under reduced pressure, and 1,4-dioxane underwent the process of refluxing with sodium wire and distilling under nitrogen atmosphere prior to use. All other reagents of analytical grade were used without further purification. Human breast adenocarcinoma cell line (MCF-7 cell) was from American Type Culture Collection (ATCC) (Beijing, China). RPMI Medium 1640, Dulbecco's Modified Eagle Medium (DMEM) and 0.25% Trypsin-EDTA were purchased from Thermo

Fisher Scientific (Shanghai, China). Fetal Bovine Serum was obtained from Biological Industries (Shanghai, China).

2.2. Synthesis of Carboxylic Functionalized mPEG-PCL (mPEG-CPCL)

All steps for the synthesis of the carboxylic functionalized mPEG-PCL (mPEG-CPCL) were carried out as reported previously [40] but modifying some certain reactions' conditions, for example, the addition amount of some reagents, the lasting time of some reactions, et al.

2.2.1. Synthesis of 3-(4-Hydroxy-cyclohexyloxy)-propionic Acid *t*-Butyl Ester (1)

To the stirred solution of 1,4-cyclohexanediol (10.0 g, 86.0 mmol) in THF (150 mL), potassium *t*-butoxide (10 g, 89.1 mmol) was added portion wise, and the mixture was stirred for 30 min under nitrogen protection. Subsequently, *t*-butyl acrylate (5.5 g, 43.0 mmol) in THF (25 mL) was added slowly by using a dropping funnel, and then the reaction mixture was refluxed under nitrogen protection for another 48 h. The solvent was evaporated under reduced pressure, and the content was neutralized with 1 N HCl (10 mL). It was extracted with ethyl acetate, and the organic layer was dried over anhydrous MgSO$_4$. The solvent was concentrated and the residue was further purified by silica gel column chromatography using ethyl acetate and hexane (1:5 v/v) as eluent. Yield: 5.4 g (51.4%). ^{1}H NMR (600MHz, CDCl$_3$) δ 3.64 (m, 3H, O–CH$_2$– and CHO), 3.27 (m, 1H, CH–OH), 2.44 (t, 2H, –CH$_2$CO–), 1.96 (m, 2H, –CH(CH$_2$)$_2$), 1.79 (m, 2H, –CH(CH$_2$)$_2$), 1.64–1.32 (m, 4H, –CH(CH$_2$)$_2$), 1.45 (s, 9H, –C(CH$_3$)$_3$) (Supplementary Materials, Figure S1).

2.2.2. Synthesis of 3-(4-Oxo-cyclohexyloxy)-propionic Acid *t*-Butyl Ester (2)

PCC (18.8 g, 87.4 mmol) was added to the solution of compound **1** (10.7 g, 43.7 mmol) in DCM (200 mL) under nitrogen protection and then the reaction mixture was stirred at room temperature overnight. The reaction mixture was filtered through a Buchner funnel with the celite to remove PCC salts. The filtrate was concentrated, and the residue was purified by silica gel column chromatography using ethyl acetate and hexane (1:4 v/v) as eluent. Yield: 10.2 g (96%). ^{1}H NMR (600 MHz, CDCl$_3$) δ 3.70 (m, 3H, O–CH$_2$ and O–CH), 2.55 (m, 2H, CO–CH$_2$), 2.49 (t, 2H, CO–CH$_2$), 2.22 (m, 2H, CO–CH$_2$), 2.08 (m, 2H, CH(CH$_2$)), 1.87 (m, 2H, CH(CH$_2$)), 1.45 (s, 9H, C(CH$_3$)$_3$) (Supplementary Materials, Figure S2).

2.2.3. Synthesis of 3-(7-Oxo-oxepan-4-yloxy)-propionic Acid *t*-Butyl Ester (3)

To a stirred solution of compound **2** (10.5 g, 43.3 mmol) in DCM (100 mL), 75% metachloroperbenzoic acid (12.0 g, 52.2 mmol) was added slowly at 0 °C. Subsequently, anhydrous NaHCO$_3$ (10.9 g, 130.0 mmol) was added to the above reaction mixture, and the reaction mixture was allowed to warm to room temperature and stirred for another 12 h. The reaction was diluted with DCM (50 mL) and quenched by the addition of solid sodium thiosulphate (9.3 g, 58.8 mmol). The mixture was stirred for 1 h, filtered and the solid residues were washed with DCM (50 mL). The filtrate was evaporated under reduced pressure and the crude product was purified by silica gel column chromatography using ethyl acetate and hexane (1:5 v/v) as eluent. Yield: 10.1 g (90%). ^{1}H NMR (600 MHz, CDCl$_3$) δ 4.45 (m, 1H, COOCH$_2$), 4.02 (m, 1H, COOCH$_2$), 3.63 (m, 3H, OCH$_2$ and OCH), 2.94 (t, 1H, CH$_2$CO), 2.46 (t, 2H, COCH$_2$), 2.38 (m, 1H, CH$_2$CO), 1.99 (m, 2H, OCH–(CH$_2$)$_2$), 1.90 (t, 1H, OCH–(CH$_2$)$_2$), 1.78 (t, 1H, OCH–(CH$_2$)$_2$),1.45 (s, 9H, C(CH$_3$)$_3$) (Supplementary Materials, Figure S3).

2.2.4. Synthesis of mPEG-Bupcl Diblock Polymer

mPEG-BuPCL was synthesized by ring-opening polymerization (ROP). Initiator mPEG5000 (400 mg, 0.08 mmol), substituted caprolactone monomer compound **3** (1032 mg, 4 mmol) and Sn(Oct)$_2$ (16.2 mg, 0.04 mmol) were taken in a dry Schlenk flask. After vacuumed and charged in nitrogen, the reaction mixture was immersed in preheated oil bath at 130 °C, and the polymerization was continued for 48 h with constant stirring. After cooling to room temperature, DCM (20 mL) was added to the sticky polymer solution. The polymer was purified by pouring into icediethyl ether and then filtered to gain the off-white viscous solid of mPEG-BuPCL 930 mg (conversion rate: 80%). ^{1}H NMR (600 MHz, CDCl$_3$) δ 4.11 (m, OCH$_2$), 3.66–3.63 (m, PEG and OCH$_2$), 3.44 (m, OCH), 2.43 (t, COCH$_2$), 2.36 (m, COCH$_2$), 1.90–1.73 (m, CH(CH$_2$)$_2$), 1.45 (s, C(CH$_3$)$_3$) (Supplementary Materials, Figure S4) .

2.2.5. Synthesis of mPEG-CPCL

Trifluoroacetic acid (5.2 mL) was added dropwise into the solution of mPEG-BuPCL (2.7 g) in DCM at 0 °C. After finishing the addition, the reaction mixture was allowed to warm to room temperature and stirred for 8 h. The solvents were evaporated, and the polymer was redissolved in DCM (5 mL) and precipitated in freezing-cold ether. The purification was repeated at least twice to get pure mPEG-CPCL 2.2 g (yield: 96%). ^{1}H NMR (600 MHz, CDCl$_3$) δ 4.15 (s, OCH$_2$), 3.71–3.65 (d, PEG and OCH$_2$), 3.50 (s, OCH), 2.58 (s, COCH$_2$), 2.39 (s, COCH$_2$), 1.88–1.79 (d, CH(CH$_2$)$_2$).

2.3. Synthesis of PTX Prodrug (mPEG-PCL-Ace-PTX)

2.3.1. Synthesis of Vinyl Ether-Functionalized mPEG-PCL (mPEG-VPCL)

To the solution of mPEG-CPCL (242 mg, equiv. 0.74 mmol COOH) in anhydrous 1,4-dioxane (10 mL), DMAP (90 mg, 0.74 mmol) and DCC (458 mg, 2.22 mmol) were added at 0 °C, respectively. Subsequently, the flask was vacuumed and charged in nitrogen gas. The reaction mixture was allowed to warm to room temperature and proceeded under stirring for 24 h at 25 °C. Thereafter, 2-(ethenyloxy) ethanol (664 μL, 7.4 mmol) was added and the mixture continued stirring for another 72 h at 25 °C in the darkness. The cloudy solution was filtered, and the filtrate was purified by extensive dialysis against 1000 mL DI water with 10 changes of water for 48 h and collected by freeze-drying. Yield: (148 mg, 58%). ^{1}H NMR (600 MHz, CDCl$_3$) δ 6.45 (q, CH=CH$_2$), 4.32 (t, OCH$_2$), 4.20–4.18 (dd, CH=CH$_2$), 4.12 (m, OCH$_2$), 4.03 (dd, CH=CH$_2$), 3.88 (t, OCH$_2$), 3.71–3.63 (m, PEG and OCH$_2$), 3.45 (s, OCH), 2.56 (t, COCH$_2$), 2.34 (t, COCH$_2$), 1.86–1.76 (m, CH(CH$_2$)$_2$).

2.3.2. Synthesis of mPEG-PCL-Ace-PTX

mPEG-PCL-Ace-PTX was synthesized by using the procedure described below. To a 25 mL Schlenk flask, mPEG-VPCL (150 mg, equiv. 119.0 μmol –CH=CH$_2$), PTX (51 mg, 59.5 μmol), *p*-TSA (1.5 mg, catalytic amount) and 4 Å molecular sieve (1 g) were added into anhydrous DMF (5 mL) under a nitrogen flow. The flask was vacuumed and sealed with parafilm. Then the reaction was carried out under stirring at 25 °C. After 5 days, the mixture passed through a Buchner funnel and the filtrate was collected and dialyzed against 500 mL DMF (MWCO = 3500 Da) for 24 h with 4 times change of DMF to remove the unconjugated PTX, and then the resulting mixture was further dialyzed against 1000 mL DI water with 10 times change of water for another 48 h, and then the desired product was collected by freeze-drying. Yield: (147 mg, 75%). ^{1}H NMR (600 MHz, CDCl$_3$) PTX (δ 8.12, 7.74, 7.60, 7.52–7.47, 7.42–7.38, 7.33, 7.07, 6.26, 6.21, 5.77, 5.66, 4.93, 4.79, 3.79, 2.58–2.53, 2.38–2.23, 1.90–1.76, 1.68, 1.24, 1.14), acetal bond (–OCH(CH$_3$)O–: δ 4.38; –OCH(CH$_3$)O–: δ 1.24), 2-(ethenyloxy) ethyl (CH=CH$_2$: δ 6.45, 4.18, 4.03; –OCH$_2$CH$_2$O–: δ 4.32, 3.88) and PEG-CPCL (δ 4.12, 3.45, 3.70–3.64, 2.56, 2.35 and 1.90–1.76).

2.4. Structural Characterization

All compounds and polymers obtained at each step were characterized by ^{1}H NMR. A Bruker AVANCE III HD instrument equipped with a 600 MHz magnetic field was employed to record ^{1}H NMR spectra. The proton signals of all the products were measured in CDCl$_3$ solution. The $\overline{M}_n$ of conjugates and PTX loading contents in the prodrug were determined by the characteristic peaks of PEG, PCL and PTX.

The number-averaged molecular weight ($\overline{M}_n$) and molecular weight distribution (PDI) was measured by a gel permeation chromatography (Waters GPC system, Shanghai, China) equipped with 1515 isocratic HPLC pump (Waters Corp, Shanghai, China), 2414 Refractometer Detector and a GPC column (Waters Styragel HT3 THF, Shanghai, China).The polymers were dissolved in THF at a concentration of 5 mg/mL. All tests were carried out at ambient temperature using THF as eluent at a flow rate of 1 mL/min and the polystyrene as standards for calibration.

2.5. Preparation of mPEG-PCL-Ace-PTX Micelles

The polymer prodrug was dissolved in THF, and then the resultant solution was dialyzed against phosphate buffer (pH 7.4, 10 mM) (MWCO = 3500 Da). After 24 h, the water phase in dialysis bag was centrifuged and further purified by passing through 0.22 μm-microfiltration membrane. Subsequently, it was lyophilized and stored in a dry environment.

2.6. Particle Size and Surface Morphology Assessment

Dynamic light scattering was employed to determine the hydrodynamic diameters of mPEG-PCL-Ace-PTX micelles. In brief, 10 mg lyophilized mPEG-PCL-Ace-PTX micellar sample were dissolved in 10 mL Millipore water followed by ultrasonic treatment to obtain the uniform micelles dispersion. Both the hydrodynamic size and zeta potential of micelles were evaluated by ZetasizerNano ZS (MalvernInstruments) at 25 °C ($n = 3$).

Likewise, aqueous micellar solution (1 mg/mL) was put dropwise onto the clean copper grids and dried in the air; the size and morphology of micelles was determined by transmission electron microscopy (TEM) (JEOL JEM-100CX II, Tokyo, Japan). The image was calculated with reference to the ruler for getting micelle core size and size distribution.

2.7. In Vitro Drug Release

The micelle solution (1 mL) in a dialysis bag (MWCO = 3500 Da) was suspended in three different media, i.e., pH 5.0 (acetate buffer), pH 6.0 (acetate buffer) and pH 7.4 (phosphate buffer). The in vitro release was carried out under the sink condition. Temperature was held constant at 37 °C throughout the experiments. 3 mL of the release medium was withdrawn and replenished with 3 mL of fresh media at designated intervals. The amount of PTX released was analyzed by Shimadzu UV-2100 spectrophotometer (λ_{max} = 228 nm).

2.8. In Vitro Cellular Uptake

The polymer prodrug and coumarin-6 were co-dissolved in THF, and then the formed solution was transferred into a dialysis bag (MWCO = 3500 Da) against phosphate buffer (pH 7.4, 10 mM). After 24 h, the water solution in dialysis bag was centrifuged to obtain a supernatant which was further purified by passing through 0.22 μm filter. Then it was lyophilized and stored in a dry environment.

MCF-7 cells were seeded in 24-well plates at a density of 5×10^4 cells per well in DMEM medium and incubated at 37 °C under a 5% CO_2 atmosphere for 24 h. Then the cells were incubated with micelles encapsulating coumarin-6 for 1, 2 and 4 h. The MCF-7 cells were washed three times with PBS and fixed with 100% ethanol for 10 min at room temperature. The cell nuclei were stained with Hoechst 33342. Finally, the cells were monitored by Nikon C3 confocal laser scanning microscope.

2.9. In Vitro Cellular Viability

The cytotoxicity of PTX prodrug nanoparticles against human breast cancer cells (MCF-7) was studied using the MTT assay. MCF-7 cells were seeded in a 96-well plate (5×10^3) and incubated for 24 h. The media was aspirated and charged with fresh media containing various concentrations of PTX prodrug micelles. The MCF-7 cells were cultured in an atmosphere containing 5% CO_2 at 37 °C for 48 h. Then, 10 μL of MTT solution (5 mg/mL) was added. The MCF-7 cells were incubated for another 4 h. After discarding the culture medium, the MTT-formazan generated by live cells was dissolved in

100 µL of DMSO, and the absorbance at a wavelength of 492 nm of each well was recorded using a microplate reader.

3. Results and Discussion

3.1. Synthesis and Characterization of mPEG-PCL-Ace-PTX

In order to acquire a functionalized mPEG-PCL diblock polymer with pH-sensitive acetal groups, we employed a new functional caprolactone monomer with a protected carboxyl group as reported [40]. According to the synthetic route shown in Scheme 1, substituted caprolactone monomer compound **3** was synthesized in modest yield by using commercially available 1,4-cyclohexanediol as the starting material via Michael addition, PCC oxidation, and Baeyer-Villiger oxidation, respectively. All of the above intermediates were characterized by ^{1}H NMR. Subsequently, compound **3** was subjected to the ROP employing mPEG 5000 as macroinitiator. From ^{1}H NMR spectra of mPEG-BuPCL (Supplementary Materials, Figure S4), it showed clearly characteristic peaks of PEG (δ 3.63) and BuPCL (δ 1.45, 1.7–1.9, 2.3–2.5, 3.44, 3.6–3.7, and 4.11).

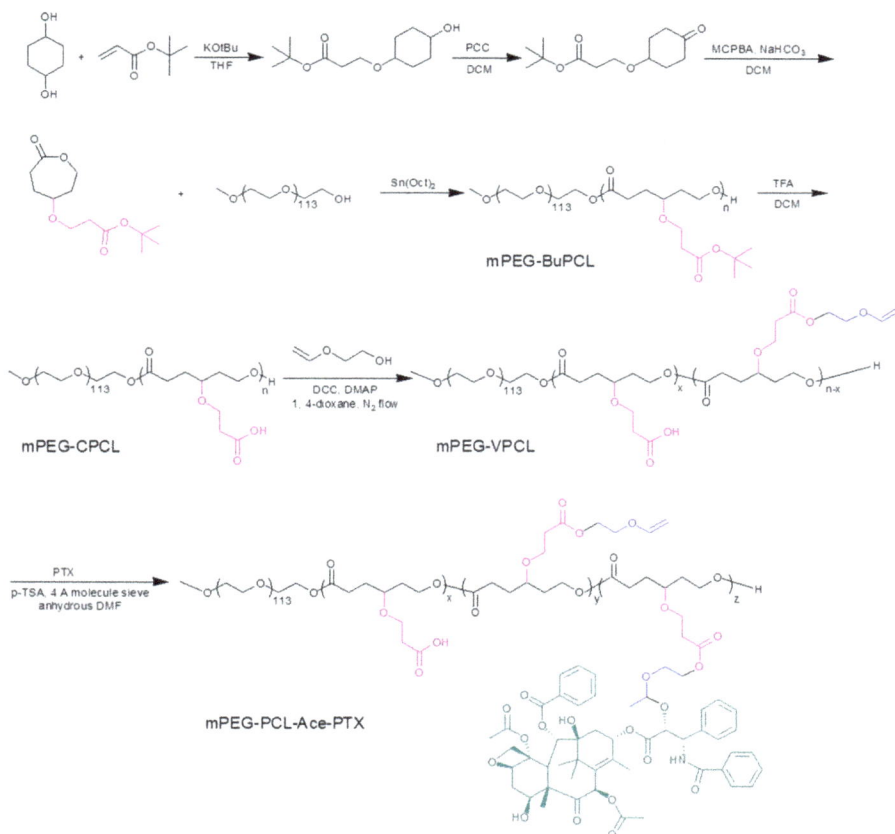

Scheme 1. Synthesis of mPEG-CPCL and mPEG-PCL-Ace-PTX.

The degree of polymerization of BuPCL block was estimated by comparing the intensities of signals at δ 4.11 and 3.63 to be 40. The *t*-butyl ester groups in the copolymer of mPEG-BuPCL were hydrolyzed by excessive trifluoroacetic acid to produce the corresponding carboxylic acid derivatives mPEG-CPCL. The peak assigned to be the methyl protons contributing to the *tert*-butyl groups at δ 1.45 ppm disappeared, which demonstrates the full hydrolysis of all the *tert*-butyl groups (Figure 1a). 2-(Ethenyloxy) ethanol was coupled with mPEG-CPCL employing DCC and DMAP in 1,4-dioxane for 72 h to acquire mPEG-VPCL. ^{1}H NMR spectra showed new peaks at δ 6.45/4.18/4.03 and δ 4.32/3.88 rather than peaks of mPEG-CPCL (Figure 1b). The number of vinyl ether coupled with per polymer chain was calculated to be 11 by comparing integrals of peaks at δ 6.45 with δ 3.63.

Finally, the acid-labile polymeric prodrug was obtained through a "click"-type conjugate reaction between 2'-hydroxyl group of PTX and the pendent vinyl ethers of mPEG-VPCL as reported before [31]. According to ^{1}H NMR (Figure 1c), it clearly showed the peaks attributed to PTX at δ 1.1–1.3, 1.6–1.9, 2.2–2.6, 4.7–5.0, 5.6–5.8, 6.2–6.3, and 7.0–8.2 besides the signals of mPEG-VPCL. Meanwhile, a peak assigned to acetal methine proton appeared at δ 4.38, which strongly supported the conjugation between PTX and mPEG-PCL via an acetal bond. The PTX content of prodrugs was determined according to the ^{1}H NMR spectra by comparing integrals of peaks at δ 7.0–8.2 with δ 3.64, which indicated that PTX prodrugs with 15.6 and 23.5 wt % PTX obtained by varying PTX-to-polymer molar ratios of 2/1 and 3/1. Notably, these PTX prodrugs have significantly higher drug contents compared with PTX-acetal-pDMA [35] and PTX-acetal-PEG without free PTX encapsulated [36] (only one molecular of PTX was chemically conjugated to the end of the polymers). GPC measurement showed a monomodal distribution of PTX prodrugs with a low polydispersity index (PDI) of 1.16 and 1.10 (Table 1, Supplementary Materials, Figure S5).

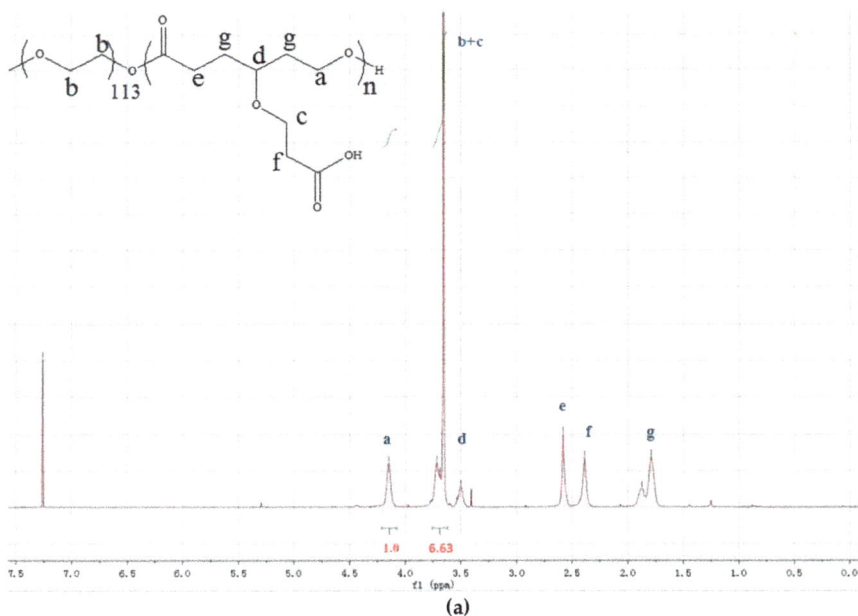

(a)

Figure 1. *Cont.*

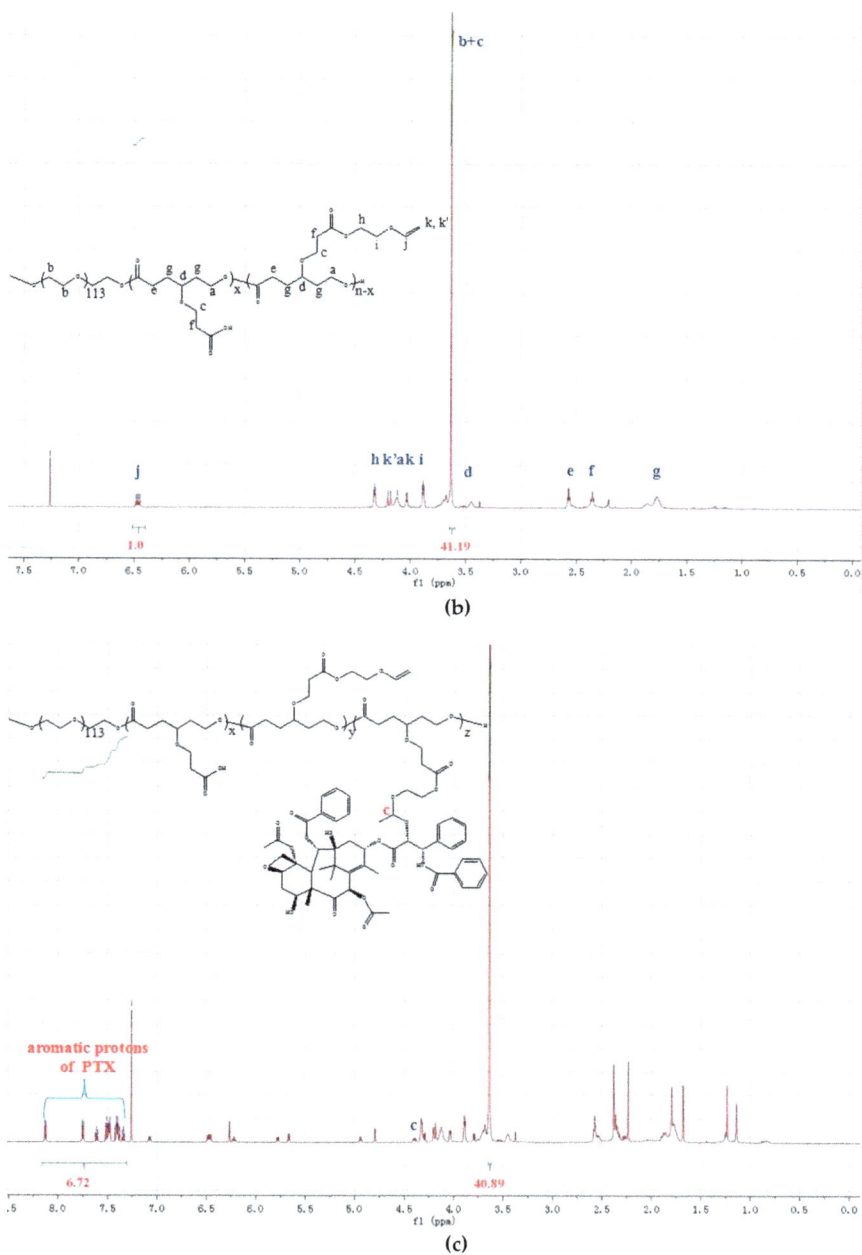

Figure 1. ^{1}H NMR spectra of mPEG-CPCL (**a**), mPEG-VPCL (**b**) and mPEG-PCL-Ace-PTX (**c**) in CDCl$_3$.

Table 1. Synthesis of mPEG-PCL-Ace-PTX.

PTX Prodrug	M_n [a] (kg/mol)	M_w/M_n [b] (PDI)	PTX [c] (wt %)
mPEG-PCL-Ace-PTX$_3$	16.4	1.16	15.6
mPEG-PCL-Ace-PTX$_5$	18.1	1.10	23.5

[a] Determined from ^{1}H NMR spectrum. [b] Determined from GPC measurements. [c] Calculated by ^{1}H NMR via comparing integrals of signals at δ 7.0–8.2 with δ 3.64.

3.2. Pharmaceutical Evaluation of mPEG-PCL-Ace-PTX Micelles

Due to the amphiphilic property, mPEG-PCL-Ace-PTX could self-assemble into nanosized micelles by dialyzing against phosphate buffer (pH 7.4, 10 mM), in which hydrophobic polyester and hydrophilic PEG form the core and the shell, respectively. Based on the higher PTX content, the size distribution and morphology of the mPEG-PCL-Ace-PTX$_5$ micelles was determined by DLS and TEM. The DLS measurement result showed the average particle size of mPEG-PCL-Ace-PTX$_5$ to be 68.5 nm (Figure 2a). TEM image indicated that the mPEG-PCL-Ace-PTX$_5$ micelles were well-dispersed in aqueous solution in the form of spherical morphology with an average diameter of about 50 nm (Figure 2b), which is somewhat smaller than that determined by DLS, likely due to shrinkage of the micelles upon drying.

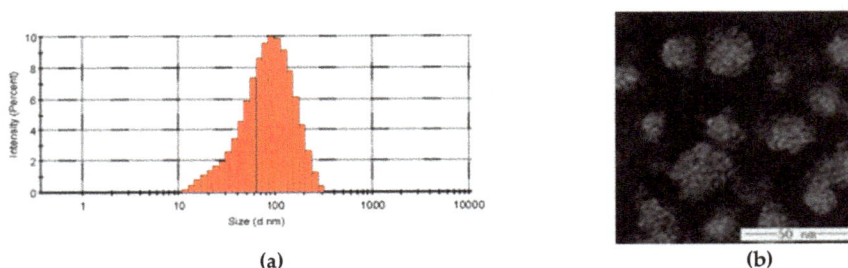

(a) (b)

Figure 2. Size distribution profiles for mPEG-PCL-Ace-PTX$_5$ micelles determined by dynamic light scattering (DLS) (**a**) and transmission electron microscopy (TEM) (**b**).

Because PTX was covalently conjugated to the mPEG-VPCL di-block copolymers through a pH-sensitive acetal linkage, the mPEG-PCL-Ace-PTX$_5$ micelles are expected to be pH-responsive. The in vitro drug release from the mPEG-PCL-Ace-PTX$_5$ micelles was investigated at 37 °C under three different pH levels, i.e., pH 5.0, 6.0, and 7.4. As shown in Figure 3, the cumulative release of PTX was 16.8%, 32.8%, and 48.2% in 48 h at pH 7.4, 6.0, and 5.0, respectively, which demonstrated that PTX release was remarkably accelerated under moderate acid environments. It should be noted that the drug release ratio of mPEG-PCL-Ace-PTX5 micelles is relatively lower than that of PEG-PAA-acetal-PTX [31] and PEG-acetal-PTX with free PTX encapsulated in 48 h at pH 7.4 [36]. However, a comparatively higher PTX release was observed for mPEG-PCL-Ace-PTX5 micelles compared with PTX-acetal-pDMA in 48 h under a mildly acidic environment. Moreover, as expected, no burst release was observed under both normal physiological and acidic conditions due to the covalent linkage between PTX and mPEG-VPCL. These results demonstrate that mPEG-PCL-Ace-PTX5 micelles possess the characteristics of better stability and lower undesired drug leakage in process of drug delievery.

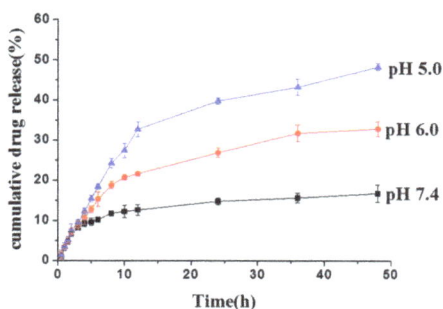

Figure 3. pH-dependent PTX release from mPEG-PCL-Ace-PTX$_5$ micelles. The data were presented as mean $\pm$ standard deviation (n = 3).

In order to investigate the cellular uptake behaviors by MCF-7 cells, confocal laser scanning microscopy was employed. Micelles encapsulating coumarin-6 were incubated for 1, 2, and 4 h, respectively. As shown in Figure 4, after 1 h incubation at 37 °C, the fluorescence intensity in the cytoplasm of MCF-7 cells was relatively very weak. When the incubation period elongated to 2 and 4 h, the quantity of green fluorescence increased dramatically, demonstrating efficient internalization of micelles.

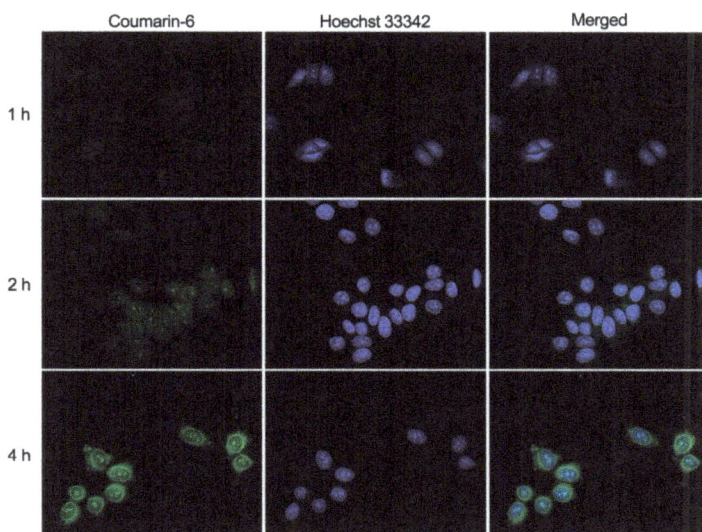

Figure 4. Confocal laser scanning microscopy images of MCF-7 cells incubated with micelles encapsulating coumarin-6 for 1, 2, and 4 h.

The in vitro anti-tumor activity of PTX prodrug was evaluated employing MTT assay. As shown in Figure 5a, the result indicated that PTX prodrug micelles displayed significant activity of inhibiting proliferation against MCF-7 cells. At drug concentrations lower than 1.0 μg·mL^{-1}, drug efficacy of PTX prodrug micelles was similar with free PTX or little higher against MCF-7 cells. In contrast, when drug concentrations was higher than 1.0 μg·mL^{-1}, antitumor activity of PTX prodrug micelles became slightly lower compared with free PTX. The half-maximal inhibitory concentrations (IC$_{50}$) were determined to be 0.49 and 0.65 μg·mL^{-1} for PTX prodrug micelles and free PTX, respectively.

It is clear that the PTX prodrug micelles shows an improvement of the cytotoxicity compared to the parent free PTX, which contributes to rapid cleavage of acetal linkage in response to the intracellular pH of MCF-7 cells [41–43].

Biocompatibility is one of the most important factors to be considered in the application of drug delivery system. Hence it is necessary to investigate the safety of the precursor of PTX prodrug. As shown in Figure 5b, vinyl ether-functionalized mPEG-PCL was actually nontoxic to MCF-7 cells, which tested concentrations varied from 0.05 to 1.0 mg/mL. The data of cell viability indicates that vinyl ether-functionalized mPEG-PCL possesses good biocompatibility and is safe as drug nanocarrier.

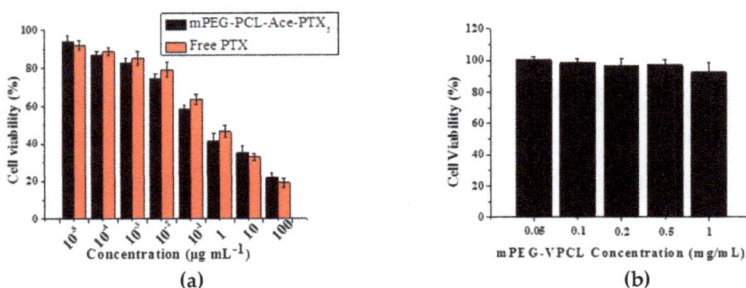

Figure 5. MTT assays of PTX mPEG-PCL-Ace-PTX$_5$ micelles (**a**) and Vinyl Ether-Functionalized mPEG-PCL (**b**) against MCF-7 cells. Data are presented as the average ± standard deviation ($n = 3$).

4. Conclusions

In conclusion, we have successfully developed a pH responsive polymer-drug conjugate for PTX delivery based on PEG-PCL, which exhibited excellent in vitro antitumor activity to MCF-7 cells. These smart PTX prodrug nanoparticles possess several unique features: (i) the mPEG-PCL-Ace-PTX micelles have a narrow size distribution and remarkable drug contents (23.5 wt % PTX); (ii) Owing to the acetal linkage between PTX and mPEG-CPCL, the PTX prodrug shows accelerated drug release in tumor-relevant acid micro-environment; (iii) They exhibit improved antitumor activity and good biocompatibility; (iv) They can readily be synthesized with controlled structures and molecular weight from PEG-PCL diblock polymer. All these results demonstrate that these acetal-linked PTX prodrug micelles have appeared as a highly promising alternative for cancer chemotherapy.

Supplementary Materials: The following are available online at www.mdpi.com/2073-4360/9/12/698/s1.

Acknowledgments: This work is supported by the National Natural Science Foundation of China (81603053 and 81502927), Hainan S&T Project (KYYS-2015-38), and the Scientific Research General Project of Liaoning Provincial Department of Education (201610163L29).

Author Contributions: Yinglei Zhai and Xing Zhou contributed equally to the synthesis of PTX prodrug; Lina Jia, Chao Ma, Ronghua Song, Yanhao Deng, and Xueyao Hu performed the evaluation; Wei Sun coordinated the study, analyzed the results, and wrote the paper.

Conflicts of Interest: The authors declare no conflict of interest.

References

1. Meirow, D.; Nugent, D. The effects of radiotherapy and chemotherapy on female reproduction. *Hum. Reprod. Update* **2001**, *7*, 535–543. [CrossRef] [PubMed]
2. Singal, P.K.; Iliskovic, N. Doxorubicin-induced cardiomyopathy. *N. Eng. J. Med.* **1998**, *339*, 900–905 [CrossRef] [PubMed]
3. Chao, T.-C.; Chu, Z.; Tseng, L.-M.; Chiou, T.-J.; Hsieh, R.-K.; Wang, W.-S.; Yen, C.-C.; Yang, M.-H.; Hsiao, L.-T.; Liu, J.-H.; et al. Paclitaxel in a novel formulation containing less Cremophor EL as first-line therapy for advanced breast cancer: A phase II trial. *Investig. New Drugs* **2005**, *23*, 171–177.

4. Tabatabaei Rezaei, S.J.; Sarbaz, L.; Niknejad, H. Folate-decorated redox/pH dual-responsive degradable prodrug micelles for tumor triggered targeted drug delivery. *RSC Adv.* **2016**, *6*, 62630–62639. [CrossRef]
5. Liu, C.; Guan, Y.; Su, Y.; Zhao, L.; Meng, F.; Yao, Y.; Luo, J. Surface charge switchable and core cross-linked polyurethane micelles as a reduction-triggered drug delivery system for cancer therapy. *RSC Adv.* **2017**, *7*, 11021–11029. [CrossRef]
6. Deng, Z.; Xiao, Y.; Pan, M.; Li, F.; Duan, W.; Meng, L.; Liu, X.; Yan, F.; Zheng, H. Hyperthermia-triggered drug delivery from iRGD-modified temperature-sensitive liposomes enhances the anti-tumor efficacy using high intensity focused ultrasound. *J. Control. Release* **2016**, *243*, 333–341. [CrossRef] [PubMed]
7. Meng, J.; Guo, F.; Xu, H.; Liang, W.; Wang, C.; Yang, X.-D. Combination Therapy using Co-encapsulated Resveratrol and Paclitaxel in Liposomes for Drug Resistance Reversal in Breast Cancer Cells in vivo. *Sci. Rep.* **2016**, *6*, 22390. [CrossRef] [PubMed]
8. Lv, T.; Yu, T.; Fang, Y.; Zhang, S.; Jiang, M.; Zhang, H.; Zhang, Y.; Li, Z.; Chen, H.; Gao, Y. Role of generation on folic acid-modified poly(amidoamine) dendrimers for targeted delivery of baicalin to cancer cells. *Mater. Sci. Eng. C* **2017**, *75*, 182–190. [CrossRef] [PubMed]
9. Öztürk, K.; Esendağlı, G.; Gürbüz, M.U.; Tülü, M.; Çalış, S. Effective targeting of gemcitabine to pancreatic cancer through PEG-cored Flt-1 antibody-conjugated dendrimers. *Int. J. Pharm.* **2017**, *517*, 157–167. [CrossRef] [PubMed]
10. Maeda, H.; Nakamura, H.; Fang, J. The EPR effect for macromolecular drug delivery to solid tumors: Improvement of tumor uptake, lowering of systemic toxicity, and distinct tumor imaging in vivo. *Adv. Drug Deliv. Rev.* **2013**, *65*, 71–79. [CrossRef] [PubMed]
11. Fang, J.; Nakamura, H.; Maeda, H. The EPR effect: Unique features of tumor blood vessels for drug delivery, factors involved, and limitations and augmentation of the effect. *Adv. Drug Deliv. Rev.* **2011**, *63*, 136–151. [CrossRef] [PubMed]
12. Maeda, H.; Wu, J.; Sawa, T.; Matsumura, Y.; Hori, K. Tumor vascular permeability and the EPR effect in macromolecular therapeutics: A review. *J. Control. Release* **2000**, *65*, 271–284. [CrossRef]
13. Mura, S.; Nicolas, J.; Couvreur, P. Stimuli-responsive nanocarriers for drug delivery. *Nat. Mater.* **2013**, *12*, 991–1003. [CrossRef] [PubMed]
14. Shim, M.S.; Kwon, Y.J. Stimuli-responsive polymers and nanomaterials for gene delivery and imaging applications. *Adv. Drug Deliv. Rev.* **2012**, *64*, 1046–1058. [CrossRef] [PubMed]
15. Zhuang, J.; Gordon, M.R.; Ventura, J.; Li, L.; Thayumanavan, S. Multi-stimuli responsive macromolecules and their assemblies. *Chem. Soc. Rev.* **2013**, *42*, 7421–7435. [CrossRef] [PubMed]
16. Ding, J.; Zhao, L.; Li, D.; Xiao, C.; Zhuang, X.; Chen, X. Thermo-responsive "hairy-rod" polypeptides for smart antitumor drug delivery. *Polym. Chem.* **2013**, *4*, 3345–3356. [CrossRef]
17. Abulateefeh, S.R.; Spain, S.G.; Thurecht, K.J.; Aylott, J.W.; Chan, W.C.; Garnett, M.C.; Alexander, C. Enhanced uptake of nanoparticle drug carriers via a thermoresponsive shell enhances cytotoxicity in a cancer cell line. *Biomater. Sci.* **2013**, *1*, 434–442. [CrossRef]
18. Li, Y.; Li, J.; Chen, B.; Chen, Q.; Zhang, G.; Liu, S.; Ge, Z. Polyplex Micelles with Thermoresponsive Heterogeneous Coronas for Prolonged Blood Retention and Promoted Gene Transfection. *Biomacromolecules* **2014**, *15*, 2914–2923. [CrossRef] [PubMed]
19. Roy, D.; Brooks, W.L.A.; Sumerlin, B.S. New directions in thermoresponsive polymers. *Chem. Soc. Rev.* **2013**, *42*, 7214–7243. [CrossRef] [PubMed]
20. Chen, C.-Y.; Kim, T.H.; Wu, W.-C.; Huang, C.-M.; Wei, H.; Mount, C.W.; Tian, Y.; Jang, S.-H.; Pun, A.K.Y.; Jen, S.H. pH-dependent, thermosensitive polymeric nanocarriers for drug delivery to solid tumors. *Biomaterials* **2013**, *34*, 4501–4509. [CrossRef] [PubMed]
21. Du, J.-Z.; Du, X.-J.; Mao, C.-Q.; Wang, J. Tailor-Made Dual pH-Sensitive Polymer-Doxorubicin Nanoparticles for Efficient Anticancer Drug Delivery. *J. Am. Chem. Soc.* **2011**, *133*, 17560–17563. [CrossRef] [PubMed]
22. Quan, C.; Chen, J.; Wang, H.; Li, C.; Chang, C.; Zhang, X.; Zhuo, R. Core-Shell Nanosized Assemblies Mediated by the alpha-beta Cyclodextrin Dimer with a Tumor-Triggered Targeting Property. *ACS Nano* **2010**, *4*, 4211–4219. [CrossRef] [PubMed]
23. Hocine, S.; Li, M. Thermoresponsive self-assembled polymer colloids in water. *Soft Matter* **2013**, *9*, 5839–5861. [CrossRef]
24. Issels, R.D. Hyperthermia adds to chemotherapy. *Eur. J. Cancer* **2008**, *44*, 2546–2554. [CrossRef] [PubMed]

25. Molla, M.R.; Rangadurai, P.; Pavan, G.M.; Thayumanavan, S. Experimental and theoretical investigations in stimuli responsive dendrimer-based assemblies. *Nanoscale* **2015**, *7*, 3817–3837. [CrossRef] [PubMed]

26. Harnoy, A.J.; Rosenbaum, I.; Tirosh, E.; Ebenstein, Y.; Shaharabani, R.; Beck, R.; Amir, R.J. Enzyme-Responsive Amphiphilic PEG-Dendron Hybrids and Their Assembly into Smart Micellar Nanocarriers. *J. Am. Chem. Soc.* **2014**, *136*, 7531–7534. [CrossRef] [PubMed]

27. Hu, Q.; Katti, P.S.; Gu, Z. Enzyme-responsive nanomaterials for controlled drug delivery. *Nanoscale* **2014**, *6*, 12273–12286. [CrossRef] [PubMed]

28. Ding, Y.; Kang, Y.; Zhang, X. Enzyme-responsive polymer assemblies constructed through covalent synthesis and supramolecular strategy. *Chem. Commun.* **2015**, *51*, 996–1003. [CrossRef] [PubMed]

29. Stubbs, M.; McSheehy, P.M.; Griffiths, J.R.; Bashford, C.L. Causes and consequences of tumour acidity and implications for treatment. *Mol. Med. Today* **2000**, *6*, 15–19. [CrossRef]

30. Ding, J.; Shi, F.; Xiao, C.; Zhuang, X.; He, C.; Chen, X. Facile preparation of pH and reduction responsive PEGylated polypeptide nanogel for efficient doxorubicin loading and intracellular delivery. *J. Control. Release* **2013**, *172*, E40–E41. [CrossRef]

31. Gu, Y.; Zhong, Y.; Meng, F.; Cheng, R.; Deng, C.; Zhong, Z. Acetal-Linked Paclitaxel Prodrug Micellar Nanoparticles as a Versatile and Potent Platform for Cancer Therapy. *Biomacromolecules* **2013**, *14*, 2772–2780. [CrossRef] [PubMed]

32. Yu, Y.; Chen, C.-K.; Law, W.-C.; Sun, H.; Prasad, P.N.; Cheng, C. A degradable brush polymer-drug conjugate for pH-responsive release of doxorubicin. *Polym. Chem.* **2015**, *6*, 953–961. [CrossRef]

33. Wang, C.E.; Wei, H.; Tan, N.; Boydston, A.J.; Pun, S.H. Sunflower Polymers for Folate-Mediated Drug Delivery. *Biomacromolecules* **2016**, *17*, 69–75. [CrossRef] [PubMed]

34. Fife, T.H.; Jao, L.K. Substituent Effects in Acetal Hydrolysis. *J. Org. Chem.* **1965**, *30*, 1492–1495. [CrossRef]

35. Louage, B.; Van Steenbergen, M.J.; Nuhn, L.; Risseeuw, M.D.P.; Karalic, I.; Winne, J.; Van Calenbergh, S.; Hennink, W.E.; De Geest, B.G. Micellar Paclitaxel-Initiated RAFT Polymer Conjugates with Acid-Sensitive Behavior. *ACS Macro Lett.* **2017**, *6*, 272–276. [CrossRef]

36. Huang, D.; Zhuang, Y.; Shen, H.; Yang, F.; Wang, X.; Wu, D. Acetal-linked PEGylated paclitaxel prodrugs forming free-paclitaxel-loaded pH-responsive micelles with high drug loading capacity and improved drug delivery. *Mater. Sci. Eng. C* **2018**, *82*, 60–68. [CrossRef] [PubMed]

37. Zupancich, J.A.; Batesr, F.S.; Hillmyer, M.A. Aqueous Dispersions of Poly(ethylene oxide)-*b*-poly(γ-methyl-ε-caprolactone) Block Copolymers. *Macromolecules* **2006**, *39*, 4286–4288. [CrossRef]

38. Geng, Y.; Discher, D.E. Hydrolytic Degradation of Poly(ethylene oxide)-block-Polycaprolactone Worm Micelles. *J. Am. Chem. Soc.* **2005**, *127*, 12780–12781. [CrossRef] [PubMed]

39. Gou, M.; Men, K.; Shi, H.; Xiang, M.; Zhang, J.; Song, J.; Long, J.; Wan, Y.; Luo, F.; Zhao, X.; Qian, Z. Curcumin-loaded biodegradable polymeric micelles for colon cancer therapy in vitro and in vivo. *Nanoscale* **2011**, *3*, 1558–1567. [CrossRef] [PubMed]

40. Surnar, B.; Jayakannan, M. Stimuli-Responsive Poly(caprolactone) Vesicles for Dual Drug Delivery under the Gastrointestinal Tract. *Biomacromolecules* **2013**, *14*, 4377–4387. [CrossRef] [PubMed]

41. Webb, B.A.; Chimenti, M.; Jacobson, M.P.; Barber, D.L. Dysregulated pH: A perfect storm for cancer progression. *Nat. Rev. Cancer* **2011**, *11*, 671–677. [CrossRef] [PubMed]

42. Zhong, Y.N.; Katharina, G.; Cheng, L.; Xie, F.; Meng, F.; Chao, D.; Zhong, Z.; Rainer, H. Hyaluronic acid-shelled acid-activatable paclitaxel prodrug micelles effectively target and treat CD44-overexpressing human breast tumor xenografts in vivo. *Biomaterials* **2016**, *84*, 250–261. [CrossRef] [PubMed]

43. Gao, C.; Tang, F.; Gong, G.; Zhang, J.; Hoi, M.P.M.; Lee, S.M.Y.; Wang, R. pH-Responsive prodrug nanoparticles based on a sodium alginate derivative for selective co-release of doxorubicin and curcumin into tumor cells. *Nanoscale* **2017**, *9*, 12533–12542. [CrossRef] [PubMed]

polymers

MDPI

Article

Photo Irradiation-Induced Core Crosslinked Poly(ethylene glycol)-*block*-poly(aspartic acid) Micelles: Optimization of Block Copolymer Synthesis and Characterization of Core Crosslinked Micelles

Kouichi Shiraishi [1], Shin-ichi Yusa [2], Masanori Ito [2], Keita Nakai [2] and Masayuki Yokoyama [1,*]

[1] Medical Engineering Laboratory, Research Center for Medical Sciences, The Jikei University School of Medicine, 163-1, Kashiwashita, Kashiwa, Chiba 277-0004, Japan; kshiraishi@jikei.ac.jp

[2] Department of Applied Chemistry, Graduate School of Engineering, University of Hyogo, 2167 Shosha, Himeji, Hyogo 671-2280, Japan; yusa@eng.u-hyogo.ac.jp (S.-i.Y.); Hatoyaito1@gmail.com (M.I.); good.west.56@hotmail.co.jp (K.N.)

* Correspondence: masajun2093ryo@jikei.ac.jp; Tel: +81-4-7164-1111 (ext. 6710)

Received: 27 November 2017; Accepted: 11 December 2017; Published: 14 December 2017

Abstract: We used photo irradiation to design core crosslinked polymeric micelles whose only significant physico-chemical change was in their physico-chemical stability, which helps elucidate poly(ethylene glycol) (PEG)-related immunogenicity. Synthetic routes and compositions of PEG-*b*-poly(aspartic acid) block copolymers were optimized with the control of *n*-alkyl chain length and photo-sensitive chalcone moieties. The conjugation ratio between *n*-alkyl chain and the chalcone moieties was controlled, and upon the mild photo irradiation of polymeric micelles, permanent crosslink proceeded in the micelle cores. In the optimized condition, the core crosslinked (CCL) micelles exhibited no dissociation while the non-CCL micelles exhibited dissociation. These results indicate that the photo-crosslinking reactions in the inner core were successful. A gel-permeation chromatography (GPC) measurement revealed a difference between the micellar-formation stability of CCL micelles and that of the non-CCL micelles. GPC experiments revealed that the CCL micelles were more stable than the non-CCL micelles. Our research also revealed that photo-crosslinking reactions did not change the core property for drug encapsulation. In conclusion, the prepared CCL micelles exhibited the same diameter, the same formula, and the same inner-core properties for drug encapsulation as did the non-CCL micelles. Moreover, the CCL micelles exhibited non-dissociable micelle formation, while the non-CCL micelles exhibited dissociation into single block copolymers.

Keywords: polymeric micelle; photo irradiation; core crosslinking; immunogenicity of PEG

---

1. Introduction

Nanoparticles, such as liposomes and polymeric micelles, have been explored as drug carriers for systemic chemotherapy in recent decades [1–3]. In drug targeting, nanoparticles can accumulate in solid tumor tissues through the enhanced permeability and retention (EPR) effect [4,5]. A long blood circulation time-period of the nanoparticles is a prerequisite for EPR-effect-based drug targeting. To achieve the long blood circulation, nanoparticles must evade interactions with plasma proteins and monocytes (such as macrophages and liver Kupffer cells) in blood. This evasion is obtained through surface coating with a hydrophilic layer, commonly poly(ethylene glycol) (PEG) [6,7], which is a water-soluble, non-toxic polymer of very low immunogenicity and which has many uses in pharmaceutics, cosmetics, and food. This PEG surface coating is generally called PEGylation. Most PEGylated nanocarriers for therapeutic or diagnostic purposes exhibit long-circulation characteristics in blood [8–15]. Among those nanoparticles, amphiphilic block copolymer micelles (polymeric micelles) have attracted substantial attention owing to their high potential for anticancer drug targeting [1,3]. Polymeric micelles tend to have relatively small hydrodynamic diameters (10–100 nm) and to exhibit stable micelle formation above the critical micelle concentration (CMC). We examined core-shell structures of polymeric micelles by means of small angle X-ray scattering (SAXS), and SAXS measurements revealed the precise hydrophobic core size of polymeric micelles [16,17]. Polymeric micelles' very small hydrophobic cores can encapsulate poorly water-soluble and water-insoluble drugs. Owing to the drug-encapsulation ability of polymeric micelles, clinical trials are now examining polymeric micelles loaded with anticancer drugs [18–20].

Polymeric micelles have shown great potential for drug targeting in terms of tumor accumulation. On the other hand, most injected polymeric micelles are excreted from blood mainly through renal filtration and spleen/liver capturing after systemic injection. A recent topic of PEGylated nanoparticles for drug targeting is the antibody responses of those PEGylated nanoparticles. Systemic injection of PEGylated nanoparticles induced the secretion of anti-PEG IgM antibodies (anti-PEG IgM), in turn changing the long blood-circulation characteristic of PEGylated nanoparticles. This phenomenon is called the accelerated blood clearance (ABC) phenomenon, which researchers have observed in mice injected with PEG-liposomes and polymeric micelles [21–25].

We have found that poly(ethylene glycol) chains do not strongly induce antibody responses, whereas poly(ethylene glycol) chains conjugating hydrophobic blocks become immunogenic for antibody responses [26]. Furthermore, poly(ethylene glycol) chains do not strongly bind to anti-PEG IgM, whereas poly(ethylene glycol) chains conjugating hydrophobic blocks can strongly bind to anti-PEG IgM. By contrast, PEGylation onto hydrophobic cores suppresses the binding of anti-PEG IgM to PEGylated nanoparticles. These results indicate the importance of core-forming hydrophobic blocks of hydrophilic (PEG)-hydrophobic block copolymers for anti-PEG IgM responses. In fact, even when the same PEG chain length (molecular weight = 5200) is used for polymeric-micelle formation, polymeric micelles with a 76-nm diameter induce the ABC phenomenon more effectively than polymeric micelles with a 34-nm diameter. This result implies that the size of polymeric micelles is an important factor for the ABC phenomenon.

We have uncovered evidence that PEG-block copolymers possessing hydrophobic blocks induce anti-PEG IgM responses, and that the hydrophobic block significantly enhances the immunogenicity of PEG chains. However, not only the size but also the state of block copolymers may affect the anti-PEG IgM responses related to hydrophobic blocks. Specifically, the ABC phenomenon may hinge on whether block copolymers have either a micellar form or a dissociated single polymer chain. As stated above, we have found that the larger a polymeric micelle is, the stronger the ABC phenomenon induction is. However, strong induction may result from the relative stability of large micelles. Therefore, polymeric micelles possessing discrete states are favored for elucidation of the ABC phenomenon.

One possible way to accurately investigate the anti-PEG IgM response is to compare non-dissociable polymeric micelles with dissociable polymeric micelles. The non-dissociable polymeric micelles are, as the name implies, free from dissociation; therefore, non-dissociable polymeric micelles may, in the micelle state, possess an induction mechanism of anti-PEG IgM responses. For the purposes of stable micellar carrier systems in drug delivery, inner core [27–36] and shell [37,38] crosslinking for non-dissociable micellar systems have been particularly attractive in recent years owing to the structural integrity of permanent covalent chemical bonding. In the current study, we focus on the core crosslinking of polymeric micelles, and the purpose of the study is to prepare photo-crosslinking polymeric micelles for evaluation of PEG-related immunogenicity. This approach can help reveal the in vivo behaviors, as well as the drug-release behaviors, of polymeric micelles in drug-delivery contexts.

Regarding efforts to achieve core crosslinking, there are synthetic limitations in the use of crosslinking agents. For example, a water-soluble crosslinking agent may have difficulty infiltrating a hydrophobic inner core to react with substrates. A water-insoluble agent might not be usable in aqueous media where micelles are formed. More important, crosslinking agents will change the chemical formula of block polymers, and this change might significantly affect PEG-related immune responses in vivo. We have chosen a photoreaction for the crosslinking of the polymeric-micelle cores. In the past decade, many research groups have developed various core crosslinked (CCL) approaches to enhance the stability of polymeric micelles for drug delivery. Studies have reported how they prepared photo-crosslinked micelles by using the photo-irradiated radical polymerization of poly(acrylate)s [27–30], the dimerization of cinnamoyl groups [31–33], and the dimerization of coumarin groups [35,36].

Herein, we report a successful development of CCL micelles by means of photo-irradiation. We used hydrocarbon chains for polymeric-micelle formation owing to their hydrophobic interactions and chalcone moieties for a photo-reactive species [39,40]. This study addresses the optimization of photo-crosslinkable block copolymer synthesis, photo-crosslinking of polymeric micelles in mild conditions, and the characterization of prepared CCL micelles. We succeeded in photo-crosslinking the inner core of polymeric micelles, and the obtained polymeric micelles exhibited high stability owing to their structural integrity.

2. Materials and Methods

2.1. Materials

α-Methoxy-ω-aminopropyl-poly(ethylene glycol) (PEG-NH$_2$, M_w = 5200) was purchased from NOF Corporation (Tokyo, Japan). PEG-NH$_2$ (M_w = 5200) was lyophilized from benzene before use. Sodium hydroxide, 6 N hydrochloric acid, trifluoroacetic acid (TFA), and 1,8-diazabicyclo[5.4.0]undec-7-ene (DBU) were purchased from Wako Pure Chemical Industries (Osaka, Japan). 3-Bromopropionyl chloride, 6-bromohexanoyl chloride, 9-bromononanoic acid, 1-iodopentane, and 1-iodononane, were purchased from Tokyo Kasei (Tokyo, Japan). Deuterium solvents were purchased from Sigma-Aldrich (Tokyo, Japan). Dehydrated *N*,*N*-dimethylformamide (DMF), dehydrated dimethylsulfoxide (DMSO), dehydrated dichloromethane, and thionyl chloride were purchased from Kanto Chemicals Co. (Tokyo, Japan). Dulbecco's PBS (D-PBS(-)) was purchased from Nakarai Tesque (Kyoto, Japan). We used all these commercial reagents as purchased, unless indicated otherwise. A dialysis membrane (Spectra/Por 6, molecular weight cut off (MWCO) = 1000) was purchased from Spectrum Laboratories (Tokyo, Japan).

2.2. Measurements

^{1}H NMR spectra were recorded on an Agilent Technologies, UNITY INOVA 400 MHz NMR spectrometer (Palo Alto, CA, USA). For ^{1}H NMR measurements, block copolymers were dissolved at 10.0 mg/mL in DMSO-d_6 containing 3 *v*/*v* % TFA. GPC measurements of the polymeric micelles (1.0 mg/mL in H$_2$O) were carried out with an HPLC system (LC 2000 series, Jasco, Tokyo, Japan) equipped with a TSK-gel G4000-PW$_{XL}$ column and a TSK-gel guard column PW$_{XL}$ (eluent = H$_2$O, flow rate = 1.0 mg/mL, detector = refractive index (RI)) at 40 °C. GPC measurements before and after the photo-irradiation of block copolymers were carried out by the use of a Shodex RI SE-61 refractive index detector (Showa Denko K.K, Tokyo, Japan) equipped with a Shodex Asahipak GF-7M HQ column and a GF-1G 7B guard column. Methanol containing 0.1 M LiClO$_4$ was used as eluent at 40 °C. The number-average molecular weight (M_n) and the weight-average molecular weight (M_w) were calibrated with standard PEG samples. UV-VIS spectra were recorded on either a JASCO V-550 or a JASCO V-530 spectrophotometer (JASCO, Tokyo, Japan) with a 1.0 cm path length cell. Static light scattering (SLS) and dynamic light scattering (DLS) measurements were carried out at 25 °C with a DLS-7000 instrument (Otsuka Electronics Co., Ltd., Tokyo, Japan) equipped with an ALV5000/EPP multi-t digital time correlator.

2.3. Synthesis of Bromo Alkylated Chalcone Amide (1a–c)

Synthesis of 1-(4-aminophenyl)-3-(2-naphthalenyl)-2-propen-1-one (4-amino-chalcone) is described elsewhere [39]. 4-amino-chalcone was stirred with an excess of 3-bromopropiponyl chloride (2–5 equivalent (eq)) in dehydrated THF at 0 °C, and the reaction mixture was gradually warmed up to r.t. After being stirred for 2 h at room temperature (r.t.), THF was removed under reduced pressure, and a yellow precipitate was obtained. The crude product was washed with diethyl ether several times. After filtration and drying under vacuum, a light-yellow powder was obtained. **1a** Yield = 77–79%. ^{1}H NMR (400 MHz, CDCl$_3$): δ 8.67 (d, *J* = 15.6 Hz, 2H), 8.26 (d, *J* = 8.4 Hz, 2H), 8.10 (d, *J* = 8.6 Hz, 2H), 7.92 (m, 3H), 7.71 (d, *J* = 8.6 Hz, 2H), 7.62 (d, *J* = 15.6 Hz, 2H), 7.56 (m, 4H), 7.44 (br, NH), 3.73 (t, *J* = 6.6 Hz, 2H), 3.01 (t, *J* = 6.6 Hz, 2H). In the same manner, two additional chalcone derivatives (chalcone-C$_x$-Br) were synthesized, as shown in Scheme 1. **1b** Yield = 77–78%. ^{1}H NMR (400 MHz, CDCl$_3$): δ 8.67 (d, *J* = 15.6 Hz, 2H), 8.26 (d, *J* = 8.4 Hz, 2H), 8.09 (d, *J* = 8.8 Hz, 2H), 7.92 (m, 3H), 7.70 (d, *J* = 8.4 Hz, 2H), 7.63 (d, *J* = 15.2 Hz, 2H), 7.57 (m, 4H), 7.32 (br, NH), 3.44 (t, *J* = 6.8 Hz, 2H), 2.44 (t, *J* = 7.2 Hz, 2H), 1.93 (m, 2H), 1.80 (m, 2H), 0.88 (t, *J* = 6.4 Hz, 3H). **1c** Yield = 81–85%. ^{1}H NMR (400 MHz, CDCl$_3$): δ 8.67 (d, *J* = 15.6 Hz, 2H), 8.26 (d, *J* = 8.4 Hz, 2H), 8.09 (d, *J* = 8.8 Hz, 2H), 7.92 (m, 3H), 7.70 (d, *J* = 8.8 Hz, 2H), 7.63 (d, *J* = 15.6 Hz, 2H), 7.56 (m, 4H), 7.34 (br, NH), 3.41 (t, *J* = 6.8 Hz, 2H), 2.41 (t, *J* = 7.2 Hz, 2H), 1.85 (m, 2H), 1.76 (m, 2H), 1.48–1.34 (m, 8H) (see Figure S1a–c of the Supplementary Materials).

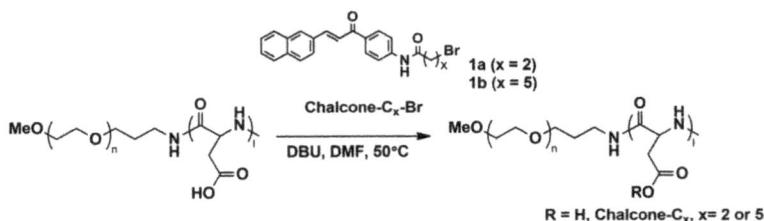

Scheme 1. The reaction of PEG-P(Asp) with chalcone-C$_x$-Br (Method A).

2.4. Esterification of PEG-b-P(Asp)

Synthesis of poly(ethylene glycol)-*b*-poly(aspartic acid), PEG-*b*-P(Asp) was reported in our previous report [17]. We synthesized poly(ethylene glycol)-*b*-poly(β-benzyl L-aspartate) (PEG-PBLA) by means of the ring-opening polymerization of β-benzyl L-aspartate N-carboxy anhydride (BLA-NCA) from α-methyl-ω-aminopropoxy poly(ethylene glycol). PEG-*b*-P(Asp) was obtained by alkaline (2.0 eq) hydrolysis of PEG-PBLA (see Figure S1d of the Supplementary Materials), followed by a hydrochloric acid treatment (2.5 eq) resulting in a free carboxylic acid form. For the esterification of PEG-*b*-P(Asp), PEG-*b*-P(Asp) was dissolved in dehydrated DMF, and a halogenated compound was added to the solution. DBU (1.05–1.20 eq/mol vs. mol of Asp) was added to the mixture, and the mixture was stirred overnight at 50 °C (16–20 h). The reaction mixture was dropped into a 10-fold volume of diethyl ether at 0 °C. The obtained precipitate was dissolved in DMSO, and hydrochloric acid (6N, 2.0 eq/mol vs. mol of DBU) was added to the solution. The resulting solution was dialyzed against H$_2$O (MWCO = 1000), and lyophilized. A ^{1}H NMR measurement (in DMSO-d_6 containing 3 *v/v* % TFA) was performed for characterization of the obtained block copolymers.

2.5. Micelle Preparation

The obtained block copolymer was dissolved in THF, which was evaporated at 40 °C under a dry N$_2$ flow. The obtained polymer film was further dried under reduced pressure for 1 h, and the film was sonicated with a VCX-750 sonicator (Sonic & Materials, Newtown, CT, USA) equipped with a 5-mm-diameter microtip in D-PBS(-). The obtained polymeric-micelle solutions were centrifuged

(3900 rpm, 10 min) for removal of possible insoluble precipitates. We used a filter unit equipped with an Amicon Ultra-15 (M_{WCO} = 100 k) to concentrate the polymeric-micelle solutions, which were passed through a 0.22 μm poly(vinylidene difluoride) (PVDF) filter unit (Nihon Millipore, K.K., Tokyo, Japan).

2.6. Photo-Crosslinking of Polymeric Micelles

Photo irradiation was applied to polymeric micelles at 5.0 mg/mL in a normal saline (N.S.) solution using an Asahi Spectra MAX-300 (Asahi Spectra Co., Ltd., Tokyo, Japan) equipped with a 300-W Xe-lamp and a 275-nm cutoff filter at 25 °C. The intensity was set to 8.0 mW/cm^2 at 350 nm for 3 h. For confirmation of the reaction's progress, a UV-VIS spectrum of the polymeric-micelle solution was monitored until an absorption peak at 350 nm disappeared.

2.7. GPC Elution Measurements of Polymeric Micelles

At a concentration range between 5.0 and 1.0 mg/mL, the CCL micelles and the non-CCL micelles (100 μL) were injected into the GPC, and the obtained peak area was normalized (μVsec/injected weight of polymeric micelles) in relation to either the CCL or the non-CCL micelles. We measured whole peak areas of both the CCL micelles and the non-CCL micelles by comparing the peak areas of micelles with one another upon injection of the micelles into the HPLC without the column (Figure S2 in the Supplementary Materials).

2.8. Preparation of Adriamycin-Encapsulated Polymeric Micelles

Block copolymers were dissolved in DMF, and Adriamycin (ADR) was dissolved in DMF containing trimethylamine (1.2 molar equivalent). The two solutions were mixed and dialyzed against H$_2$O by the use of a dialysis membrane with MWCO = 1000 (Spectra/Por 6). The obtained polymeric micelle solutions were centrifuged (3900 rpm, 10 min) for removal of possible aggregates. We used a filter unit equipped with an Amicon Ultra-15 (MWCO = 100,000) to concentrate the polymeric-micelle solutions. The obtained polymeric-micelle solutions were passed through a 0.22-μm PVDF filter unit.

2.9. Drug-Release Experiment

ADR-encapsulated CCL micelles and ADR-encapsulated non-CCL micelles were filled in a dialysis unit, with each micelle solution (100 μL) containing 13 μg ADR (Slide-A-Lyzer MINI Dialysis Unit (MWCO = 3500), Thermo Scientific, Japan), and were dialyzed against a 40-fold volume of either a 0.1-M phosphate buffer (pH = 7.4) or a 0.1-M acetate buffer (pH = 5.0) with stirring at r.t., respectively. An aliquot of the dialysate was freeze-dried, and its ADR content was determined by means of reverse-phased HPLC equipped with a μ-Bondasphere column (Waters, C4 bonded phase endcapped with 5-μm silica, pore size = 100 Å, eluent = acetonitrile containing 1% acetic acid, flow rate = 1.0 mg/mL, detector = UV at 485 nm) at 40 °C. This drug-release measurement was performed in triplicate.

3. Results and Discussion

3.1. Optimization of Photoreactive Chalcone-Conjugated Block Copolymers

Introduction of hydrophobic moieties into PEG-*b*-P(Asp) leads to the formation of polymeric micelles in aqueous media [16,17]. We selected *n*-alkyl chains, such as *n*-pentyl or *n*-nonyl, as hydrophobic moieties [24,25]. Together with these alkyl chains, we used alkylated photo-crosslinkable chalcone derivatives as a photo sensitive compound. We examined the following three methods for optimization of the conjugation reaction.

3.1.1. Method A

In our first examination, we performed esterification of PEG-*b*-P(Asp) with a chalcone derivative having an ethylene spacer, chal-C$_2$-Br (**1a**), at 50 °C in dehydrated DMF (see Scheme 1, method

A in Chart 1). However, we found that a very small number of chal-C_2 were conjugated to PEG-*b*-P(Asp) chains, and absorption at 350 nm, which corresponds to the π–π^* transition state of chalcone moieties [39], was decreased after the reaction (see Figure S3 of the Supplementary Materials). This decrease indicates that a side reaction, which changed an absorption spectrum of chal-C_2, occurred during the reaction. We confirmed that mixtures of **1a** with PEG-*b*-P(Asp) (without DBU) or **1a** with DBU (without PEG-*b*-P(Asp)) in the same reaction condition did not show a change in the π–π^* band (data not shown). The conjugation reaction occurred between carboxylate anion and alkyl halides in polar DMF. In this condition, the reactivity of alkyl halides in **1a** (in this case Br–CH_2CH_2–C(O)) was low owing to an electron-withdrawing carbonyl group attached to the ethylene group. However, reactive carboxylate anions remained in the mixture, and might have reacted with chalcone moieties instead of alkyl halides. We examined a reaction involving PEG-*b*-P(Asp) and 3-bromo-N-butyl-propanamide (BrCH_2CH_2C(O)NH*n*C_4H_9) (3-BNBPA), which is similar in chemical structure to **1a**, in the reaction condition identical to those of chal-C_2-Br and PEG-*b*-P(Asp) (2.0 eq/mol 3-BNBPA and 1.2 eq/mol DBU were applied; see Table S1 of the Supplementary Materials). This reaction yielded only 24% esterification for PEG-*b*-P(Asp) (see Figure S1e of the Supplementary Materials). From this result, we concluded that less reactive alkyl halides led to a smaller incidence of 3-BNBPA conjugation to PEG-*b*-P(Asp), and that the remaining carboxylate anions attacked chalcone π-conjugated moieties. Owing to this unfavorable change in the chalcone π-conjugated moieties in which a photo crosslinked reaction had to occur, we used long alkylated chalcone derivatives, such as pentyl spacers (chal-C_5-Br, **1b**) and octyl spacers (chal-C_8-Br, **1c**), instead of **1a**. These long alkylated chalcone derivatives were revealed to possess sufficient reactivity with carboxylate anions. Although the number of introduced chal-C_5 derivatives was not large enough (the number of introduced chalcone derivatives = 1.7, yield from feed mol was 22%; see Table S2 of the Supplementary Materials), we found that chal-C_5 was successfully conjugated into PEG-*b*-P(Asp) without a significant change in chalcone derivative's absorption band. The fact that methylene chains are, in general, longer than pentyl chains ($n > C_5$) in chalcone derivatives significantly reduced the change in chalcone derivatives' absorption peak in UV-VIS spectroscopy.

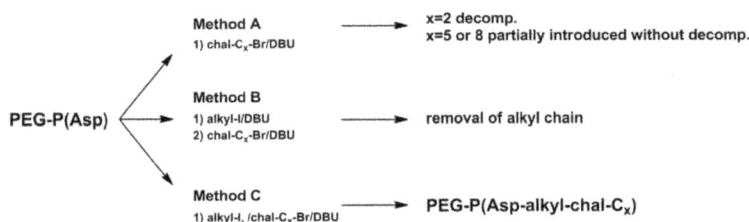

Chart 1. Summary of PEG-P(Asp-alkyl-chal-C_x) synthesis.

3.1.2. Method B

We performed two-step reactions as shown in part B of Chart 1. We conjugated pentyl groups to PEG-*b*-P(Asp) in the first esterification step (see Figure S1f of the Supplementary Materials), followed by conjugation of **1b** in the second step (see Scheme 2). The reaction of **1b** proceeded with PEG-P(Asp(26)-pentyl(76)), whose aspartic acid residues had been conjugated with pentyl groups (76% of the pentyl group in 26 aspartic-acid residues), under the same reaction condition as the first esterification step (see Scheme 2). The obtained block copolymer did not exhibit the change in chalcone derivatives' absorption peak, and we controlled the number of conjugated chal-C_5 by controlling the **1b** feed amount, as shown in Table 1. However, we found a decreasing number of pentyl groups with an increasing number of conjugated chal-C_5 in the PEG-P(Asp-pentyl). This result indicates that chal-C_5 conjugation took place with removal of the pentyl group.

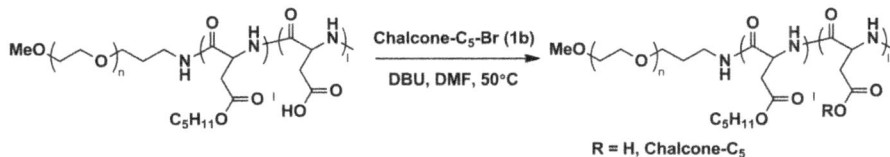

Scheme 2. The reaction of PEG-P(Asp-pentyl) with chalcone-C$_x$-Br (Method B).

Table 1. Chal-C$_5$ conjugation to PEG-P(Asp(26)-pentyl(76)).

Run	Chal-C$_5$-Br Feed		DBU Feed	Chal-C$_5$ Found	C$_5$H$_{11}$ Found
	Number	Ratio	/eq *	Number	Number
1	31.2	1.2	1.2	2.3	2.4
2	26.0	1.0	1.0	2.1	2.5
3	20.8	0.8	0.75	1.4	3.4
4	15.6	0.6	0.5	0.6	7.7

* eq indicates feed equivalent mol of 1,8-diazabicyclo[5.4.0]undec-7-ene (DBU) vs. 26 Asp residue in PEG-P(Asp(26)).

3.1.3. Method C

As the two-step reactions did not successfully provide a favorable composition, we performed an experiment involving a reaction of PEG-*b*-P(Asp) with alkyl halides and chal-C$_x$-Br (x = 5 or 8) simultaneously (see Scheme 3, method C in Chart 1). ^{1}H NMR was used for characterizations of PEG-P(Asp-alkyl-chal-C$_x$), as shown in Figure 1. We calculated the number of conjugated alkyl chains and chal-C$_x$ derivatives by comparing the integration among the PEG's methylene protons (degree of polymerization = 119), the terminal methyl protons of the conjugated alkyl chains at 0.77 ppm, and 3H aromatic protons of chal-C$_x$ derivatives at 7.58 ppm.

The reaction proceeded well without a change in absorption peak at 350 nm. As shown in Table 2 (Run 1–3), the obtained number of pentyl groups and chal-C$_5$ derivatives was well controlled by the feed ratio between 1-iodopentane and **1b**. In contrast, the chalcone having a C$_5$ alkyl chain (**1b**) conjugation reaction did not proceed in the mixture with *n*-C$_9$H$_{19}$-I (nonyl iodide), while conjugation of nonyl iodide took place for the block copolymer (see Run 4 in Table 2). The results indicate that the reaction of nonyl iodides to PEG-*b*-P(Asp) was faster than the reaction of **1b** to PEG-*b*-P(Asp). Once the nonyl chain was conjugated into PEG-*b*-P(Asp), a short pentyl chain having a bulky chalcone moiety possibly could not reach the aspartic acid residues at the PEG-*b*-P(Asp) backbone. From the above-mentioned results, we found that at least an equivalent number ($x = y$ − 1) of the methylene groups was necessary for the conjugation of chalcone derivatives to obtain PEG-P(Asp-C$_y$H$_{2y+1}$-chal-C$_x$) and long alkyl chains, such as a nonyl chain, exhibit better reactivity. As we expected, the reaction between PEG-P(Asp), *n*-C$_9$H$_{19}$-I, and a C$_8$ alkyl chain (**1c**) yielded good results (Table 2, Run 5–8). Results indicated that reaction efficiencies (% N(found)/N(feed)) of each substrate was 33–42% for **1c** and 28–47% for *n*-C$_9$H$_{19}$-I, respectively, therefore, we succeeded in preparing PEG-P(Asp-C$_y$H$_{2y+1}$-chal-C$_x$) in a well-controlled manner.

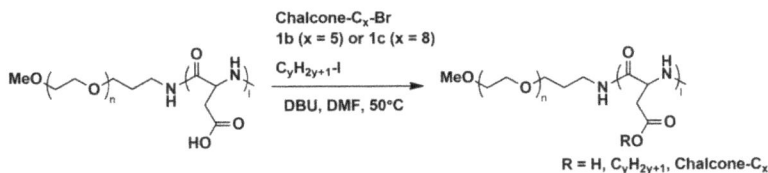

Scheme 3. The reaction of PEG-P(Asp) with chalcone-C$_x$-Br (Method C).

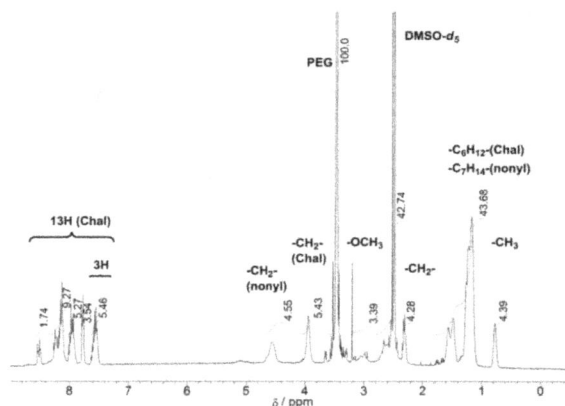

Figure 1. ^{1}H NMR spectrum of the PEG-P(Asp-nonyl-chal-C$_8$) block copolymer in DMSO-d_6 containing 3 v/v % TFA.

Table 2. Results of the conjugation of chal-C$_x$ and C$_y$H$_{2y+1}$ to PEG-P(Asp(26)).

Run	Chal-C$_x$-Br Feed			C$_y$H$_{2y+1}$-I Feed			Chal-C$_x$ Found	C$_y$H$_{2y+1}$ Found
	x	eq *	Number	y	eq *	Number	Number	Number
1	5	0.3	7.8	5	1.1	28.6	2.6	8.8
2	5	0.3	7.8	5	1.5	39.0	3.5	12.9
3	5	0.3	7.8	5	2.0	52.0	2.7	14.2
4	5	0.3	7.8	9	1.5	39.0	0.0	13.2
5	8	0.3	7.8	9	1.5	39.0	2.8	18.2
6	8	0.6	15.6	9	1.2	31.2	5.6	9.3
7	8	0.8	20.8	9	1.0	26.0	6.9	7.0
8	8	0.8	20.8	9	0.8	20.8	8.6	6.0
9	8	0.8	20.8	9	0.6	15.6	8.8	4.4

* equivalent vs. 26 Asp residues. DBU equivalent = 1.1 vs. Asp.

3.2. Photo-Crosslinking of Polymeric Micelles

We used a solvent evaporation–sonication method to prepare PEG-P(Asp-nonyl-chal-C$_8$) micelles [16,17]. In addition, we performed DLS and UV-VIS measurements to confirm the obtained PEG-P(Asp-nonyl-chal-C$_8$) micelles' characteristics. As shown in Table 3, diameters of the micelles were in a range of 20–40 nm, except in Run 1. The block copolymers having 18.2 nonyl chains and 2.8 Chal-C$_8$ groups exhibited a mixture of micelle and vesicle formation, which we determined by means of TEM (see Figure S4 of the Supplementary Materials). This helped clarify why we observed this exceptional diameter in Run 1 of Table 3.

Table 3. The number of introduced *n*-nonyl and chal-C$_8$ moieties in block copolymers and the size of the prepared polymeric micelles.

Run	C$_9$H$_{19}$/N	Chal-C$_8$/N	Diameter/nm
1	18.2	2.8	129 *1
2	9.3	5.6	37
3	7.0	6.9	17
4	7.0	8.6	29

*1 The diameter in run 1 was calculated by means of a cumulative method.

The obtained micelles exhibited a maximum π–π* absorption peak at 350 nm. We performed photo-irradiation on the micelle concentration at 5.0 mg/mL with an intensity of 8.0 mW/cm^2 at 350 nm of UV-VIS light for 3 h. In this mild light-intensity condition, no absorption peak around 325–360 nm was found after 3 h, and a new absorption peak at 300 nm appeared upon photo-irradiation, as shown in Figure 2. This spectrum change was due to single bond formation via [2 + 2] addition of the chalcone moieties (see Scheme S1 of the Supplementary Materials). The disappearance of the peak in the range extending from about 325 to 360 nm indicates that the photo reaction proceeded in polymeric micelles' inner cores. In other photo-crosslinking systems, many studies have applied high light intensities at shorter wavelengths to polymeric micelles [31–36]; however, in the current study, we have successfully performed a crosslinking reaction in polymeric micelles' core with photo irradiation at mild light intensity.

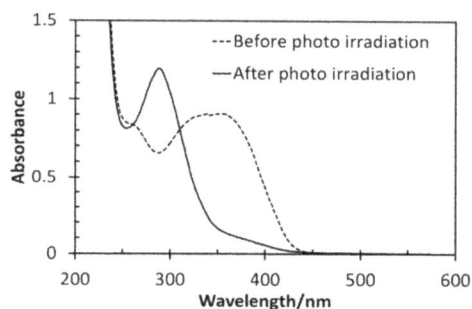

Figure 2. UV-VIS absorption spectra of PEG-P(Asp-nonyl-chal-C$_8$) (nonyl = 7.0, chal-C$_8$ = 8.6) micelles before photo irradiation (solid line) and after photo irradiation (dashed line) in a 0.1 mg/mL solution.

3.3. Measurement of Size and Molecular Weight upon Photo-Irradiation

We used GPC measurements in an organic solvent to compare the molecular weight of PEG-P(Asp-nonyl-Chal-C$_8$) block copolymers before and after photo irradiation. The CCL micelles exhibited an M_n of 1.6×10^5 with a M_w/M_n of 1.63, while the non-crosslinked block copolymer exhibited an M_n of 5.1×10^3 with a M_w/M_n of 1.29 (Run 4 of Table 3) in GPC (eluent = methanol containing 0.1 M LiClO$_4$) (see Figure S5 of the Supplementary Materials). The obtained M_n of non-crosslinked block copolymers was in good agreement with the ^{1}H NMR values, indicating that the non-crosslinked block copolymers did not form aggregates in methanol.

It should be noted that the M_n values of PEG-P(Asp-nonyl-Chal-C$_8$) block copolymers (nonyl = 7.0, chal-C$_8$ = 8.6) and the CCL micelles in DMF were 5.6×10^4 and 6.5×10^5 with M_w/M_n = 1.14 and 1.17, respectively (data not shown). The molecular weight of the PEG-P(Asp-nonyl-Chal-C$_8$) block copolymers, when measured in DMF, was greater than the corresponding molecular weight obtained in the context of ^{1}H NMR. This fact indicates that PEG-P(Asp-nonyl-Chal-C$_8$) block copolymers underwent aggregation before photo-crosslinking in DMF. Therefore, an appropriate choice of solvents is required for determining the actual molecular weight of PEG-P(Asp-nonyl-Chal-C$_8$) block copolymers in GPC. We have succeeded in preparing CCL micelles from PEG-P(Asp-nonyl-chal-C$_8$) block copolymers (nonyl = 7.0, chal-C$_8$ = 6.9; nonyl = 7.0, chal-C$_8$ = 8.6; Runs 3 and 4 of Table 3). By contrast, when in methanol, PEG-P(Asp-nonyl-chal-C$_8$) block copolymers (nonyl = 9.3, chal-C$_8$ = 5.6; Runs 2 of Table 3) exhibited several peaks on the GPC chart because the number of chalcone derivatives that had undergone incomplete photo-crosslinking was smaller than the number of chalcone derivatives that had undergone complete photo-crosslinking. The number of chalcone derivatives in the polymeric micelles' core was revealed to be an important factor in our effort to complete the photo-crosslinking. To confirm the formation of CCL micelles (nonyl = 7.0, chal-C$_8$ = 8.6), we characterized the obtained polymeric micelles by means of DLS and SLS.

Our DLS measurements yielded polymeric micelles whose diameters after photo-irradiation were nearly the same as the diameters before photo-irradiation in aqueous media (Table 4). This result indicates that photo-irradiation did not affect the radius of polymeric micelles in aqueous media. Figure 3 presents the Zimm plots for the CCL-micelles obtained by means of SLS measurements either in normal saline before the photo-crosslinking or in methanol containing 0.1 M LiCO$_4$ after the photo-crosslinking. Table 4 shows the SLS data and hydrodynamic radius (R_h) for the polymeric micelles. We could not obtain SLS data regarding PEG-P(Asp-nonyl-chal-C$_8$) block copolymers in methanol owing to the very weak scattering intensities of the unimer form of the block copolymers. To estimate the values of apparent M_w for polymeric micelles before and after photo-crosslinking, we extrapolated the polymer concentration (C_p) and scattering angle (θ) to zero, and to estimate the values of the radius of gyration (R_g) and the second virial coefficient (A_2), we referred respectively to the slope of the angular dependence and the slope of the concentration dependence in the Zimm plots. Polymeric micelles' M_w before photo-crosslinking was 3.93×10^6, and the aggregation number (N_{agg}) was 310, as estimated by the molecular weight of the block copolymer. After photo-crosslinking, we dialyzed the CCL-micelle solution against methanol to exchange the solvent and remove possible non-crosslinked block copolymers. The M_w of CCL micelles was 1.14×10^6, and the N_{agg} was 90. The R_g/R_h values of the polymeric micelles before and after photo-crosslinking were 0.837 and 0.982, respectively. The R_g/R_h value of the CCL micelles indicates that the structure of their inner core was less dense than the inner core of the non-CCL micelles in methanol.

Table 4. Diameter, weight-average molecular weight (M_w), second virial coefficient (A_2), radius of gyration (R_g), and aggregation number (N_{agg}) of non-core crosslinked (CCL) and CCL micelles.

Run	DLS Diameter/nm	SLS $M_w \times 10^{-6}$	A_2 (cm$^3 \cdot$mol/g^2)$^b \times 10^5$	R_g/R_h	N_{agg}	Media
Non-CCL micelle	29	3.93	1.55	0.837	310	in NS [1]
CCL micelle	30	1.14	1.29	0.982	90	Methanol [2]

[1] normal saline, [2] methanol containing 0.1 M LiClO$_4$. The composition of the PEG-P(Asp-nonyl-chal-C$_8$) used in this study was nonyl = 7.0, chal-C$_8$ = 8.6.

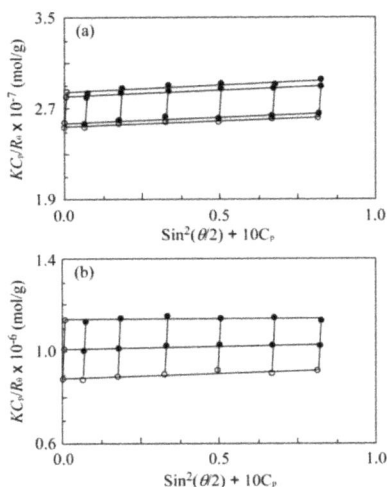

Figure 3. Zimm plots of PEG-P(Asp-nonyl-chal-C$_8$) (nonyl = 7.0, chal-C$_8$ = 8.6) micelles (**a**) before and (**b**) after photo-crosslinking; and (**a**) in normal saline and (**b**) in methanol containing 0.1 M LiClO$_4$.

3.4. DLS Measurements of CCL and Non-CCL Micelles

Owing to the aggregation of PEG-P(Asp-nonyl-chal-C_8) block copolymers as confirmed by GPC measurements in DMF, we dissolved the CCL micelles and block copolymers in methanol containing 0.1 M $LiClO_4$. Figure 4 exhibits the radius of CCL and non-CCL micelles before and after photo irradiation. We have observed that the CCL and the non-CCL micelles exhibited the same diameters in normal saline (Figure 4a). In methanol containing 0.1 M $LiClO_4$, the radius of the CCL micelles indicated the presence of an aggregate form, while the radius of the non-photo-irradiated block copolymers indicates solely the presence of a unimer form without the aggregation, as shown in Figure 4b. These findings suggest that photo-crosslinking of the cores proceeded successfully and that, consequently, the polymeric micelles did not dissociate into single block copolymers in methanol containing 0.1 M $LiClO_4$.

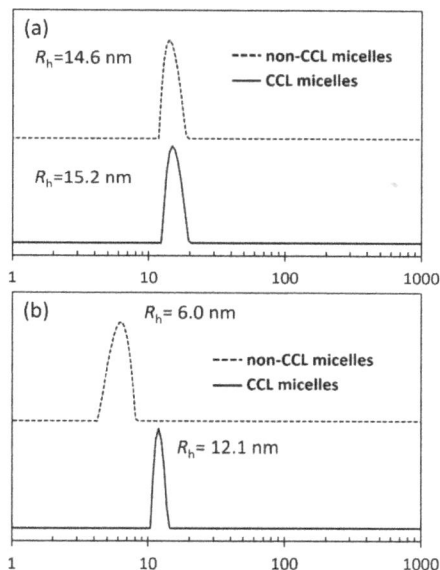

Figure 4. DLS charts of PEG-P(Asp-nonyl-chal-C_8) micelles (nonyl = 7.0, chal-C_8 = 8.6) before and after photo irradiation (**a**) in normal saline and (**b**) in methanol containing 0.1 M $LiClO_4$.

3.5. GPC Measurements of CCL and Non-CCL Micelles for Calculation of CMC

Measurements of the emission and excitation spectra of pyrene encapsulated in polymeric micelles provided critical micelle concentration (CMC) values of polymeric micelles [41,42]. We followed a pyrene-encapsulation method to determine the CMC values of the CCL and the non-CCL micelles (nonyl = 7.0, chal-C_8 = 8.6) [42]; however, emission spectra of the pyrene encapsulated polymeric micelles exhibited a new broad absorption peak at 480 nm in a concentration range extending from 0.1 to 2.0 mg/mL. This peak resulted from exciplex formation between pyrene and chalcone moieties in the inner core, since PEG-P(Asp-nonyl) micelles (without chalcone moieties) did not exhibit such a new peak. Owing to the exciplex formation between pyrene and chalcone moieties, we could not estimate the CMC values of the two types of polymeric micelles (see Figure S6 of the Supplementary Materials).

Rather than measure the emission and excitation spectra of pyrene, we measured concentration-dependent GPC elution peaks in order to observe the shear stress-mediated dissociation and adsorption behaviors of the CCL and the non-CCL micelles (nonyl = 7.0, chal-C_8 = 8.6). We have proved that this concentration-dependent GPC measurement deeply correlated with the stability of micelles in vivo [17]. Figure 5 exhibited plots of the normalized elution-peak areas (µVsec/injected weight of polymeric micelles) of the CCL and the non-CCL micelles at each concentration. The plots of non-CCL micelles dropped at a higher concentration than did the plots of CCL micelles. Under this GPC condition, the non-CCL micelles exhibited almost no elution peak below 0.01 mg/mL. These two kinds of micelles were identical to one another regarding diameter and chemical formula, but exhibited different plot profiles. The difference in GPC measurements was due mainly to the adsorption—stemming from hydrophobic interactions—of either the unimer form or the polymeric micelles onto the GPC column. If the polymeric micelles exhibited no adsorption on the column, the plots of the normalized peak areas should have been flat. Owing to the hydrophobic interaction between the column and unimer forms of the block copolymers, the normalized peak area of non-CCL micelles drastically decreased in the diluted conditions. However, as shown in Figure 5, the plots of the CCL micelles gradually dropped even after photo-crosslinking. It should be noted that adsorption behaviors of those micelles depend on the number of adsorption sites in a dynamic environment (flow rate and flow pressure). Therefore, a significant amount of injected micelles adsorbed onto the GPC column, and a low concentration of micelles exhibited significant decreases in elution-peak areas, whereas a high concentration of micelles exhibited negligible decreases in elution-peak areas. In fact, 88% of the non-CCL micelles were eluted at 1.0 mg/mL, whereas 98% of the CCL micelles were eluted at the same injection concentration. Therefore, this pattern was due not to dissociation, but perhaps to the hydrophobic interactions between the CCL micelle's inner core and the column's surface. These data show that there was a stability difference between the CCL micelles and the non-CCL micelles in the GPC measurements.

The detection limit of the non-CCL micelles' concentration was 0.01 mg/mL. Moreover, as shown in Figure 5, an obtained CMC value from a crossing point between the two approximation straight lines was 20 µg/mL. This CMC value was in a good agreement with the CMC of benzylated PEG-*b*-P(Asp) block copolymer micelles obtained in pyrene excitation and emission measurements (5–40 µg/mL) [16]. By contrast, the CCL micelles in the GPC measurements did not exhibit a drastic decrease in the elution-peak area. We observed the CCL micelles exhibiting high stability even in dilute conditions in the GPC experiment, as shown in Figure 5.

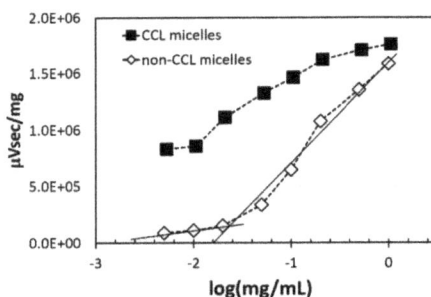

Figure 5. GPC measurements conducted before (diamond) and after (square) the photo-crosslinking of polymeric micelles. Concentrations of the polymers were in a range extending from 0.5 µg/mL to 1.0 mg/mL. Y axis indicates normalized elution-peak areas per injected weight gram of polymeric micelles in GPC measurements. The obtained normalized elution-peak area was plotted.

3.6. Comparisons of Two Micelles' Adriamycin Encapsulation and Release Behaviors

In this experiment, we performed an in vitro drug-release experiment to clarify the effects of the inner core's hydrophobic properties on the release of Adriamycin (ADR). We used two block copolymers for ADR encapsulation (in Table 2, see Run 6's block copolymer (nonyl = 9.3 and chal-C_8 = 5.6) and Run 7's block copolymer (nonyl = 7.0 and chal-C_8 = 6.9)). ADR was encapsulated in the two types of block copolymer micelles in good yields (75–80% in 2.0 wt % ADR feed to a block copolymer), and the obtained ADR-encapsulated micelles respectively had a 19-nm diameter (nonyl = 7.0 and chal-C_8 = 6.9) and a 33-nm diameter (nonyl = 9.3 and chal-C_8 = 5.6). We examined photo irradiation to these ADR-encapsulated polymeric micelles. It should be noted that ADR is a photo-sensitive compound, so we took additional precautions during the photo irradiation [43,44]. Upon photo irradiation, a 50% decrease in ADR absorption at 488 nm was observed in the UV-vis spectrum of the 19-nm-diameter polymeric micelles, while 72% of the ADR absorption remained for the 33-nm-diameter polymeric micelles. The smaller-diameter (19 nm) ADR-encapsulated polymeric micelles exhibited much faster ADR decomposition upon photo-irradiation than did the larger-diameter (33 nm) ADR-encapsulated polymeric micelles. Owing to the decomposition of ADR upon photo irradiation, only the 33-nm-diameter CCL or non-CCL micelles were employed for further examination in this study.

We examined the ADR-release behaviors of both the CCL and the non-CCL micelles in conditions either at pH = 7.4 or at pH = 5.0 by means of a dialysis method. ADR-encapsulated micelles were filled in a dialysis bag, and we monitored released-ADR in the outer medium until day 8. Results are shown in Figure 6. Both the CCL and the non-CCL micelles exhibited similar ADR-release behaviors in the two conditions. After 4 days, both the CCL micelles and the non-CCL micelles released about 20% of their ADR at pH = 7.4 and about 50% at pH 5.0. Owing to the higher solubility of ADR in the acidic medium, the two types of micelles exhibited faster ADR release at pH = 5.0 than at pH = 7.4. However, ADR was stably encapsulated in the two types of micelles, and photo irradiation did not change ADR-releasing behaviors. This was unexpected for us. We had expected that two micelles exhibit different ADR-release behaviors. In fact, polymeric micelles have emerged as a novel anticancer-drug carrier in drug-delivery systems in recent decades, research has not yet completely elucidated the drug-release mechanism involving polymeric micelles, particularly regarding whether the drug release is dominated by a mechanism of diffusion in the inner core, by a mechanism of the dissociation of polymeric micelles, or by both of these mechanisms. Results of ADR-release experiments show that a diffusion mechanism contributed to ADR release from polymeric micelles' inner core in vitro and that a mechanism of the dissociation of polymeric micelles did not. This result was probably due to two different hydrophobic characteristics of bulky aromatic rings of ADR and long alkyl chains. Previous study shown that chemically conjugated ADR to P(Asp) helps to maintain physically entrapped ADR in micelles' core [44]. Therefore, hydrophobic interactions between aromatic rings, as well as π–π interactions improve stability of ADR encapsulation, and these interactions may be more stable for crosslinked micelles, which neighboring hydrophobic blocks can interact easily, than the dissociable micelles. At least, the obtained results proved that photo-irradiation did not change the ADR-releasing behaviors.

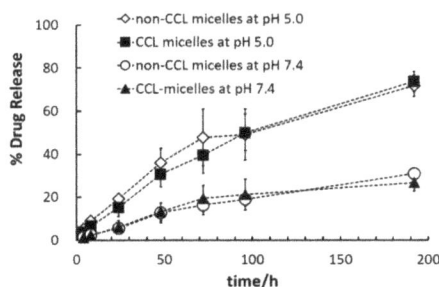

Figure 6. Release of Adriamycin from polymeric micelles in different media where either pH = 7.4 (D-PBS) or pH = 5.0 (acetate buffer). Diamonds and squares indicate the after and before photo-irradiation contexts, respectively. The release assay was performed in triplicate.

4. Conclusions

In the current study, we succeeded in preparing two distinct types of polymeric micelles: those with photo-crosslinking and those without photo-crosslinking. We focused on photo-crosslinking the polymeric micelles' cores so that the two types of polymeric micelles would possess the same chemical formula. We succeeded in syntheses of photo-crosslinkable block copolymers—PEG-P(Asp-alkyl-chal-C_x). We found that reactions among photo-crosslinkable chalcone moieties, alkyl chains, and PEG-*b*-P(Asp) successfully proceeded when we mixed all the substrates at once. The obtained block copolymers formed polymeric micelles with diameters ranging from 20 to 40 nm in aqueous media. We optimized the composition of PEG-P(Asp-nonyl-chal-C_8) micelles for successful photo-crosslinking in the cores. Mild photo-irradiation led to the crosslinking of chalcone moieties in the polymeric micelles' inner cores. The core crosslinked micelles having a diameter of between 25 and 30 nm were present in both aqueous media and methanol, while the non-CCL micelles exhibited dissociation in methanol. We examined GPC experiments to evaluate dissociation and adsorption behaviors of the two types of micelles. The CCL micelles exhibited greater elution-peak areas than the non-CCL micelles owing to either the complete absence of—or the very small presence of—dissociation-related adsorption onto the GPC column. No release-profile difference emerged between the CCL and the non-CCL micelles when the encapsulated Adriamycin was released from the micelles' inner cores. These results indicate that we obtained our desired property associated with the polymeric micelles: namely, the CCL micelles and the non-CCL micelles shared identical diameters, chemical formulas, and hydrophobic properties, even though the CCL micelles exhibited no dissociation into a unimer of block copolymers. We observed cmc of the non-CCL micelles and no cmc-related dissociation behavior of the CCL micelle. By using these two micelles, we can evaluate PEG-related immunogenicity in both forms: a micelle form and a dissociated polymer form. This is a big challenge to elucidate mechanisms of polymeric micelles' immunogenicity in vivo.

Supplementary Materials: The following are available online at www.mdpi.com/2073-4360/9/12/710/s1, Additional data: Figure S1: ^{1}H NMR of (a–c) chalcone derivatives, (d) ^{1}H NMR of PEG-P(Asp) block copolymer, (e–f) esterified PEG-P(Asp) block copolymers. Figure S2: GPC experiments for estimating the non-adsorption behaviors of CCL micelles and non-CCL micelles. Figure S3: UV-VIS spectrum of PEG-P(Asp-chal-C_2). Figure S4: TEM images of the prepared PEG-P(Asp-nonyl-chal-C_8) micelles. Figure S5: GPC traces before and after photo-irradiation in MeOH containing 0.1 M LiClO$_4$. Figure S6: emission spectra of pyrene-encapsulated PEG-P(Asp-nonyl-chal-C_8) micelles before and after photo-irradiation.

Acknowledgments: This research was financially supported by a Grant-in-Aid for Scientific Research (B) from the Ministry of Education, Culture, Sports, Science and Technology (MEXT), Japan.

Author Contributions: Kouichi Shiraishi and Masayuki Yokoyama conceived and designed the experiments. Kouichi Shiraishi performed experiments (synthesis and evaluation) and wrote the paper. Masanori Ito, Keita Nakai, and Shin-ichi Yusa performed experiments (photo-irradiation and SLS). All the authors helped with the data analysis and revision of the paper.

Conflicts of Interest: The authors declare no conflict of interest.

References

1. Aliabadi, H.M.; Lavasanifar, A. Polymeric micelles for drug delivery. *Expert Opin. Drug Deliv.* **2006**, *3*, 139–162. [CrossRef] [PubMed]

2. Torchilin, V.P. Recent advances with liposomes as pharmaceutical carriers. *Nat. Rev. Drug Discov.* **2005**, *4*, 145–160. [CrossRef] [PubMed]

3. Yokoyama, M. Polymeric micelles as a new drug carrier system and their required considerations for clinical trials. *Expert Opin. Drug Deliv.* **2010**, *7*, 145–158. [CrossRef] [PubMed]

4. Matsumura, Y.; Maeda, H. A new concept for macromolecular therapeutics in cancer chemotherapy: Mechanism of tumoritropic accumulation of proteins and the antitumor agent SMANCS. *Cancer Res.* **1986**, *46*, 6387–6392. [PubMed]

5. Torchilin, V. Tumor delivery of macromolecular drugs based on the EPR effect. *Adv. Drug Deliv. Rev.* **2011**, *63*, 131–135. [CrossRef] [PubMed]

6. Harris, J.M.; Chess, R.B. Effect of pegylation on pharmaceuticals. *Nat. Rev. Drug Discov.* **2003**, *2*, 214–221. [CrossRef] [PubMed]

7. Veronese, F.M.; Pasut, G. PEGylation, successful approach to drug delivery. *Drug Discov. Today* **2005**, *10*, 1451–1458. [CrossRef]

8. Klibanov, A.L.; Maruyama, K.; Torchilin, V.P.; Huang, L. Amphipathic polyethyleneglycols effectively prolong the circulation time of liposomes. *FEBS Lett.* **1990**, *268*, 235–237. [CrossRef]

9. Yokoyama, M. Polymeric micelles as drug carriers: Their lights and shadows. *J. Drug Target.* **2014**, *22*, 576–583. [CrossRef] [PubMed]

10. Shi, J.; Kantoff, P.W.; Wooster, R.; Farokhzad, O.C. Cancer nanomedicine: Progress, challenges and opportunities. *Nat. Rev. Cancer* **2017**, *17*, 20–37. [CrossRef] [PubMed]

11. Nishiyama, N.; Matsumura, Y.; Kataoka, K. Development of polymeric micelles for targeting intractable cancers. *Cancer Sci.* **2016**, *107*, 867–874. [CrossRef] [PubMed]

12. Torchilin, V.P.; Omelyanenko, V.G.; Papisov, M.I.; Bogdanov, A.A., Jr.; Trubetskoy, V.S.; Herron, J.N.; Gentry, C.A. Poly(ethylene glycol) on the liposome surface: On the mechanism of polymer-coated liposome longevity. *Biochim. Biophys. Acta* **1994**, *1195*, 11–20. [CrossRef]

13. Yokoyama, M.; Okano, T.; Sakurai, Y.; Ekimoto, H.; Shibazaki, C.; Kataoka, K. Toxicity and antitumor activity against solid tumors of micelle-forming polymeric anticancer drug and its extremely long circulation in blood. *Cancer Res.* **1991**, *51*, 3229–3236. [PubMed]

14. Yokoyama, M.; Okano, T.; Sakurai, Y.; Fukushima, S.; Okamoto, K.; Kataoka, K. Selective delivery of Adriamycin to a solid tumor using a polymeric micelle carrier system. *J. Drug Target.* **1999**, *7*, 171–186. [CrossRef] [PubMed]

15. Shiraishi, K.; Wang, Z.; Kokuryo, D.; Aoki, I.; Yokoyama, M. A polymeric micelle magnetic resonance imaging (MRI) contrast agent reveals blood–brain barrier (BBB) permeability for macromolecules in cerebral ischemia-reperfusion injury. *J. Control. Release* **2017**, *253*, 165–171. [CrossRef] [PubMed]

16. Sanada, Y.; Akiba, I.; Hashida, S.; Sakurai, K.; Shiraishi, K.; Yokoyama, M.; Yagi, N.; Shinohara, Y.; Amemiya, Y. Composition dependence of the micellar architecture made from poly(ethylene glycol)-block-poly(partially benzyl-esterified aspartic acid). *J. Phys. Chem. B* **2012**, *116*, 8241–8250. [CrossRef] [PubMed]

17. Shiraishi, K.; Sanada, Y.; Mochizuki, S.; Kawano, K.; Maitani, Y.; Sakurai, K.; Yokoyama, M. Determination of polymeric micelles' structural characteristics, and effect of the characteristics on pharmacokinetic behaviors. *J. Control. Release* **2015**, *203*, 77–84. [CrossRef] [PubMed]

18. Matsumura, Y. Poly (amino acid) micelle nanocarriers in preclinical and clinical studies. *Adv. Drug Deliv. Rev.* **2008**, *60*, 899–914. [CrossRef] [PubMed]

19. Hamaguchi, T.; Matsumura, Y.; Suzuki, M.; Shimizu, K.; Goda, R.; Nakamura, I.; Nakatomi, I.; Yokoyama, M.; Kataoka, K.; Kakizoe, T. NK105, a paclitaxel-incorporating micellar nanoparticle formulation, can extend in vivo antitumour activity and reduce the neurotoxicity of paclitaxel. *Br. J. Cancer* **2005**, *92*, 1240–1246. [CrossRef] [PubMed]

20. Koizumi, F.; Kitagawa, M.; Negishi, T.; Onda, T.; Matsumoto, S.; Hamaguchi, T.; Matsumura, Y. Novel SN-38–incorporating polymeric micelles, NK012, eradicate vascular endothelial growth factor–secreting bulky tumors. *Cancer Res.* **2006**, *66*, 10048–10056. [CrossRef] [PubMed]

21. Dams, E.T.M.; Laverman, P.; Oyen, W.J.; Storm, G.; Scherphof, G.L.; van der Meer, J.W.; Corstens, F.H.; Boerman, O.C. Accelerated blood clearance and altered biodistribution of repeated injections of sterically stabilized liposomes. *J. Pharmacol. Exp. Ther.* **2000**, *292*, 1071–1079. [PubMed]

22. Laverman, P.; Carstens, M.G.; Boerman, O.C.; Dams, E.T.M.; Oyen, W.J.; van Rooijen, N.; Corstens, F.H.; Storm, G. Factors affecting the accelerated blood clearance of polyethylene glycol-liposomes upon repeated injection. *J. Pharmacol. Exp. Ther.* **2001**, *298*, 607–612. [PubMed]

23. Wang, X.Y.; Ishida, T.; Kiwada, H. Anti-PEG IgM elicited by injection of liposomes is involved in the enhanced blood clearance of a subsequent dose of PEGylated liposomes. *J. Control. Release* **2007**, *119*, 236–244. [CrossRef] [PubMed]

24. Koide, H.; Asai, T.; Hatanaka, K.; Urakami, T.; Ishii, T.; Kenjo, E.; Nishihara, M.; Yokoyama, M.; Ishida, T.; Kiwada, H.; Oku, N. Particle size-dependent triggering of accelerated blood clearance phenomenon. *Int. J. Pharm.* **2008**, *362*, 197–200. [CrossRef] [PubMed]

25. Koide, H.; Asai, T.; Kato, H.; Ando, H.; Shiraishi, K.; Yokoyama, M.; Oku, N. Size-dependent induction of accelerated blood clearance phenomenon by repeated injections of polymeric micelles. *Int. J. Pharm.* **2012**, *432*, 75–79. [CrossRef] [PubMed]

26. Shiraishi, K.; Kawano, K.; Maitani, Y.; Aoshi, T.; Ishii, K.J.; Sanada, Y.; Mochizuki, S.; Sakurai, K.; Yokoyama, M. Exploring the relationship between anti-PEG IgM behaviors and PEGylated nanoparticles and its significance for accelerated blood clearance. *J. Control. Release* **2016**, *234*, 59–67. [CrossRef] [PubMed]

27. Kadam, V.S.; Nicol, E.; Gaillard, C. Synthesis of flower-like poly(ethylene oxide) based macromolecular architectures by photo-cross-linking of block copolymers self-assemblies. *Macromolecules* **2012**, *45*, 410–419. [CrossRef]

28. Wu, Y.Y.; Chen, W.; Meng, F.; Wang, Z.; Cheng, R.; Deng, C.; Liu, H.; Zhong, Z. Core-crosslinked pH-sensitive degradable micelles: A promising approach to resolve the extracellular stability versus intracellular drug release dilemma. *J. Control. Release* **2012**, *164*, 338–345. [CrossRef] [PubMed]

29. Shunai, X.; Merdan, T.; Schaper, A.K.; Xi, F.; Kissel, T. Core-cross-linked polymeric micelles as paclitaxel carriers. *Bioconj. Chem.* **2004**, *15*, 441–448. [CrossRef] [PubMed]

30. Iijima, M.; Nagasaki, Y.; Okada, T.; Kato, M.; Kataoka, K. Core-polymerized reactive micelles from heterotelechelic amphiphilic block copolymers. *Macromolecules* **1999**, *32*, 1140–1146. [CrossRef]

31. Yusa, S.; Sugahara, M.; Endo, T.; Morishima, Y. Preparation and characterization of a pH-responsive nanoge based on a photo-cross-linked micelle formed from block copolymers with controlled structure. *Langmuir* **2009**, *25*, 5258–5265. [CrossRef] [PubMed]

32. Lin, W.; Kim, D. pH-Sensitive micelles with cross-linked cores formed from polyaspartamide derivatives for drug delivery. *Langmuir* **2011**, *27*, 12090–12097. [CrossRef] [PubMed]

33. Zhao, Y.; Bertrand, J.; Tong, X.; Zhao, Y. Photo-cross-linkable polymer micelles in hydrogen-bonding-built layer-by-layer films. *Langmuir* **2009**, *25*, 13151–13157. [CrossRef] [PubMed]

34. Jiang, X.; Luo, S.; Armes, S.P.; Shi, W.; Liu, S. UV irradiation-induced shell cross-linked micelles with pH-responsive cores using ABC triblock copolymers. *Macromolecules* **2006**, *39*, 5987–5994. [CrossRef]

35. Jiang, J.; Qi, B.; Lepage, M.; Zhao, Y. Polymer micelles stabilization on demand through reversible photo-cross-linking. *Macromolecules* **2007**, *40*, 790–792. [CrossRef]

36. He, J.; Tong, X.; Zhao, Y. Corona-cross-linked polymer vesicles displaying a large and reversible temperature-responsive volume transition. *Macromolecules* **2009**, *42*, 4845–4852. [CrossRef]

37. Sun, X.; Rossin, R.; Turner, J.L.; Becker, M.L.; Joralemon, M.J.; Welch, M.J.; Wooley, K.L. An assessment of the effects of shell cross-linked nanoparticle size, core composition, and surface PEGylation on in vivo biodistribution. *Biomacromolecules* **2005**, *6*, 2541–2554. [CrossRef] [PubMed]

38. Qi, K.; Ma, Q.; Remsen, E.E.; Clark, C.G.; Wooley, K.L. Determination of the bioavailability of biotin conjugated onto shell cross-linked (SCK) nanoparticles. *J. Am. Chem. Soc.* **2004**, *126*, 6599–6607. [CrossRef] [PubMed]

39. Selvam, P.; Nanjundan, S. Synthesis and characterization of new photoresponsive acrylamide polymers having pendant chalcone moieties. *React. Funct. Polym.* **2005**, *62*, 179–193. [CrossRef]

40. Allcock, H.R.; Cameron, C.G. Synthesis of photo-cross-linkable chalcone-bearing polyphosphazenes. *Macromolecules* **1994**, *27*, 3131–3135. [CrossRef]

41. Zhao, C.L.; Winnik, M.A. Fluorescence probe techniques used to study micelle formation in water-soluble block copolymers. *Langmuir* **1990**, *6*, 514–516. [CrossRef]

42. Kwon, G.; Naito, M.; Yokoyama, M.; Okano, T.; Sakurai, Y.; Kataoka, K. Micelles based on AB block copolymers of poly(ethylene oxide) and poly(β-benzyl L-aspartate). *Langmuir* **1993**, *9*, 945–949. [CrossRef]

43. Kataoka, K.; Matsumoto, T.; Yokoyama, M.; Okano, T.; Sakurai, Y.; Fukushima, S.; Okamoto, K.; Kwon, G.S. Doxorubicin-loaded poly(ethylene glycol)–poly(β-benzyl-L-aspartate) copolymer micelles: Their pharmaceutical characteristics and biological significance. *J. Control. Release* **2000**, *64*, 143–153. [CrossRef]

44. Yokoyama, M.; Fukushima, S.; Uehara, R.; Okamoto, K.; Kataoka, K.; Sakurai, Y.; Okano, T. Characterization of physical entrapment and chemical conjugation of adriamycin in polymeric micelles and their design for in vivo delivery to a solid tumor. *J. Control. Release* **1998**, *50*, 79–92. [CrossRef]

polymers

MDPI

Article

PLMA-b-POEGMA Amphiphilic Block Copolymers as Nanocarriers for the Encapsulation of Magnetic Nanoparticles and Indomethacin

Athanasios Skandalis [1], Andreas Sergides [2,3,4], Aristides Bakandritsos [5] and Stergios Pispas [1,*]

[1] Theoretical and Physical Chemistry Institute, National Hellenic Research Foundation, 48 Vassileos Constantinou Avenue, 11635 Athens, Greece; thanos.skan@gmail.com

[2] Department of Pharmacy, University of Patras, 26504 Rio Patras, Greece; andreas.sergides.15@ucl.ac.uk

[3] Department of Materials Science, University of Patras, 25604 Rio Patras, Greece

[4] UCL Healthcare Biomagnetic and Nanomaterials Laboratories, Department of Physics and Astronomy, The Royal Institution of Great Britain, 21 Albemarle street, London W1S 4BS, UK

[5] Regional Centre of Advanced Technologies and Materials, Department of Physical Chemistry, Faculty of Science, Palacky University, Šlechtitelů 27, 783 71 Olomouc, Czech Republic; aristeidis.bakandritsos@upol.cz

* Correspondence: pispas@eie.gr; Tel.: +30-210-727-3824

Received: 24 November 2017; Accepted: 20 December 2017; Published: 23 December 2017

Abstract: We report here on the utilization of poly(lauryl methacrylate)-b-poly(oligo ethylene glycol methacrylate) (PLMA-b-POEGMA) amphiphilic block copolymers, which form compound micelles in aqueous solutions, as nanocarriers for the encapsulation of either magnetic iron oxide nanoparticles or iron oxide nanoparticles, and the model hydrophobic drug indomethacin in the their hydrophobic core. The mixed nanostructures were characterized using dynamic light scattering (DLS) and transmission electron microscopy (TEM) in terms of their structure and solution properties. Magnetophoresis experiments showed that the mixed solutions maintain the magnetic properties of the initial iron oxide nanoparticles. Results indicate that the cumulative hydrophilic/hydrophobic balance of all components determines the colloidal stability of the nanosystems. The effect of salt and bovine serum albumin (BSA) protein concentration on the structure of the mixed nanostructures was also investigated. Disintegration of the mixed nanostructures was observed in both cases, showing the importance of these parameters in the structure formation and stability of such complex mixed nanosystems.

Keywords: amphiphilic block copolymers; micelles; self-assembly; encapsulation; magnetic nanoparticles; indomethacin

1. Introduction

Polymer-based nanomedicine has gained vast attention in the worldwide scientific community because of the promising potential it provides in the fields of bioimaging and the therapy of various diseases and disorders. Theranostics is a new scientific term describing nanosystems used for applications in diagnosis and therapy at the same time [1,2]. The polymeric systems used for theranostics usually consist of the polymeric material, the pharmaceutical substance that is used for therapy purposes and a chemical group/component that is used for diagnostic purposes [3–5]. Amphiphilic block copolymers consist of two or more covalently connected macromolecular chains with different hydrophobicity. Therefore, these polymers are able to self-assemble into various complex nanostructures, such as spherical or cylindrical micelles and vesicles, where the hydrophobic polymer chains form the core and the hydrophilic polymer chains form the corona when inserted in aqueous solutions [6,7]. The morphology of the self-assemblies depends on the hydrophobic/hydrophilic ratio.

This property renders them ideal materials for the encapsulation and delivery of hydrophobic drugs and imaging agents in the hydrophobic core [8–14]. The encapsulation of nanoparticles (NPs) in the self-assembled nanostructures of block copolymers can provide the resulting hybrid nanostructures with enhanced properties such as improved stability, easy multi-functionalization and reduced toxicity, combining the characteristics of both the nanoparticles and the polymers [15,16]. Nowadays, magnetic nanoparticles are key materials, as far as biomedical applications are concerned, in fields such as magnetic resonance imaging (MRI), magnetic hyperthermia and drug delivery systems for triggered release of the encapsulated drug [15,17–21]. Encapsulation of magnetic NPs in block copolymers micelles can lead to improved stability and biocombatibility [22].

The aim of this work was to utilize amphiphilic poly(lauryl methacrylate)-*b*-poly (oligo ethylene glycol methacrylate), (PLMA-*b*-POEGMA) micelles as nanocarriers for the encapsulation of iron oxide magnetic NPs (synthesized by thermal decomposition) and both NPs and indomethacin (IND, a model hydrophobic drug) simultaneously in their hydrophobic PLMA core. The structure and properties of the hybrid nanostructures are studied by a gamut of physicochemical techniques including dynamic light scattering (DLS) and transmission electron microscopy (TEM).

2. Materials and Methods

2.1. Materials

Tetrahydrofuran (THF, 99.9%, Sigma-Aldrich, Athens, Greece), *n*-Hexane (95%, Carlo-Erba, Athens, Greece), 1-octadecene (Alfa Aesar, Athens, Greece, 90%,), iron chloride hexahydrate (FeCl$_3$·6H$_2$O, Acros Organics, Athens, Greece, 99%,), sodium oleate (Sigma-Aldrich, Athens, Greece, 82%,), oleic acid (Alfa Aesar, Athens, Greece, 90%,), bovine serum albumin (BSA, 99%, Sigma-Alrdrich, Athens, Greece), sodium chloride (99.5%, Sigma-Arldrich, Athens, Greece), indomethacin (IND, 99%, Fluka, Athens, Greece).

2.2. PLMA-b-POEGMA Block Copolymers Synthesis

The PLMA-*b*-POEGMA block copolymers were synthesized in various compositions by a two-step Reversible Addition-Fragmentation Chain Transfer (RAFT) polymerization process using 2,2′-Azobis (Isobutyronitrile), (AIBN) as the radical initiator and 4-cyano-4-(phenyl-carbonothioylthio) pentanoic acid (CPAD) as the chain transfer agent. The synthetic procedure has been described in detail in our previous work [23].

2.3. Synthesis of Iron Oxide Nanoparticles

Iron oxide NPs were synthesized by thermal decomposition of iron-oleate complex (precursor) in the presence of oleic acid and 1-octadecene.

(a) Synthesis of the precursor: The synthesis of the precursor was performed according to a procedure reported in the literature [24]. In a typical synthesis, iron chloride hexahydrate (FeCl$_3$·6H$_2$O, 8.1 g, 30 mmol) and sodium oleate (27.4 g, 90 mmol) were dissolved in a mixture of 60 mL absolute ethanol and 45 mL distilled water, followed by the addition of 105 mL hexane. The mixture was then refluxed (70 °C) for 4 h. Thereafter, the upper organic layer containing the iron-oleate complex (red-brown) was washed three times with 50 mL distilled water in a separatory funnel. In order to remove any residual water, sodium sulphate anhydrous (0.5–1.0 g) was added to the solution of the precursor, followed by filtration. Hexane was evaporated off in a rotary evaporator and the resulting waxy-solid iron-oleate complex was further dried for 2 h.

(b) Synthesis of Fe$_2$O$_3$ NPs: (3.6 g, 4.0 mmol) iron-oleate complex and (3.4 g, 12 mmol) oleic acid were dissolved in 30 g, 38 mL 1-octadecene. The mixture was heated to 100 °C for 30 min and then to reflux (318 °C) for 1 h (heating rate: 7 °C/min). The resulting black dispersion containing the NPs was left to cool down to room temperature and washed four times with a mixture of chloroform and acetone (1:3–1:1). At every wash, the nanocrystals were first dispersed in chloroform (CHCl$_3$) and then precipitated by addition of acetone and centrifugation. Finally, the NPs were dispersed in $CHCl_3$. The procedure described above, led to the formation of monodisperse nanocrystals with an average diameter of 9.4 nm determined by TEM (JEOL, Akishima, Tokyo, Japan).

2.4. Preparation of PLMA-b-POEGMA Micelles Loaded with Magnetic Nanoparticles

The amphiphilic block copolymer PLMA-*b*-POEGMA was dissolved in THF at a concentration of 5×10^{-3} g/mL. A dispersion of iron oxide nanoparticles in $CHCl_3$, stabilized with oleic acid, was added to the solution. The concentration of magnetic nanoparticles in the mixture was set to 10 or 20 wt % relative to the PLMA block. Then the mixture was injected fast in 10 mL distilled water under vigorous stirring. Thereafter, the mixed solution was placed in a rotary evaporator in order to evaporate the organic solvents (THF and $CHCl_3$). The mixed micelles with encapsulated NPs were generated during the evaporation of the organic solvents. The loading percentage in all cases was more than 95%, since no precipitate was observed, at least for the stable solutions.

2.5. Preparation of PLMA-b-POEGMA Micelles Loaded with Magnetic Nanoparticles and Indomethacin

The amphiphilic block copolymer PLMA-*b*-POEGMA was dissolved in THF at a concentration of 5×10^{-3} (g/mL). The appropriate amount of indomethacin was dissolved in THF. A dispersion of iron oxide nanoparticles in $CHCl_3$, stabilized with oleic acid and indomethacin in THF, were added to the solution. The concentration of magnetic nanoparticles in the mixture was 10 wt % and the concentration of indomethacin was 20 wt % relative to the PLMA block. Then, the mixture was injected quickly in 10 mL distilled water under vigorous stirring. Thereafter, the mixed solution was placed in a rotary evaporator in order to evaporate the organic solvents. The mixed micelles with encapsulated NPs and IND were generated during the evaporation of the organic solvents. The loading percentage in all cases was more than 95%, as evidenced by the absence of precipitates and UV–Vis (Perkin Elmer, Waltham, MA, USA) characterization.

2.6. Methods

Dynamic light scattering measurements were conducted on an ALV/CGS-3 compact goniometer system (ALVGmbH, Hessen, Germany), equipped with an ALV 5000/EPP multi-τ digital correlator with 288 channels and an ALV/LSE-5003 light scattering electronics unit for stepper motor drive and limit switch control. A JDS Uniphase 22 mW He-Ne laser ($\lambda = 632.8$ nm) was used as the light source. Measurements were carried out five times for each concentration and angle and were averaged for each angle. The solutions were filtered through 0.45 μm hydrophilic PTFE filters (Millex-LCR from Millipore, Billerica, MA, USA) before measurements. The angular range for the measurements was 30–150°. Obtained correlation functions were analyzed by the cumulants method and CONTIN software (ALVGmbH, Hessen, Germany). The size data and figures shown below are from measurements at 90°.

The magnetophoretic experiments were performed using a Perkin Elmer (Lambda 19) UV–Vis–NIR spectrophotometer (Waltham, MA, USA) and by inserting next to the cuvette holder a cylindrical Nd-Fe-B magnet (dimensions: diameter = 20 mm, thickness = 10 mm, magnetization unit: N45, attraction/repulsion strength: max 16 kg). The wavelength of the measurements was set at 450 nm and the absorbance of the solution containing the hybrid nanostructures with encapsulated NPs was measured for 1 h under the influence of the magnet. It should be noted that absorption in this wavelength comes from the encapsulated NPs and not from the copolymers.

TEM micrographs were obtained by a JEOL, JEM-2100 instrument (Akishima, Tokyo, Japan) operating at 200 kV. Samples for TEM were prepared by casting a droplet of a dilute aqueous suspension (ca. 1×10^{-4} g·mL^{-1}) of the mixed micelles on copper grids coated by Formvar carbon film.

3. Results and Discussion

The self-assembly of PLMA-*b*-POEGMA amphiphilic block copolymers has been studied extensively in our previous work [23]. In aqueous solutions, the block copolymers have been found to form compound micelles where the PLMA block is forming the hydrophobic domains and POEGMA the hydrophilic corona.

In order to impart magnetic properties to the polymeric micelles, small size ($D = 9.4$ nm) iron oxide nanoparticles stabilized with oleic acid were encapsulated in the hydrophobic core of the micelles. It was expected that the oleic acid corona of NPs is compatible and miscible with the lauryl side chains of the PLMA block and, in this way, the encapsulation of the magnetic NPs would be greatly facilitated. Moreover, both indomethacin and NPs could be simultaneously encapsulated in the micellar core in order to create nanocarrier systems suitable for both imaging and therapy. Thus, PLMA is the interacting functional block that promotes the miscibility of the components. On the other hand, POEGMA blocks provide solubility/colloidal stability and stealth properties to the mixed nanostructures. Table 1 shows the molecular characteristics of the PLMA-*b*-POEGMA block copolymers, as well as the DLS results for all mixed nanostructures. In all cases, mixed structures within the nanoscale range, with a relatively narrow size distribution were obtained. Scheme 1 depicts the procedure followed for the preparation of the micelles loaded only with magnetic NPs.

Table 1. Molecular characteristics and dynamic light scattering (DLS) results for poly(lauryl methacrylate)-*b*-poly(oligo ethylene glycol methacrylate) (PLMA-*b*-POEGMA) block copolymers and solutions of their micelles loaded with nanoparticles (NPs) and both indomethacin (IND) and NPs.

	Block copolymers					NPs loaded		NPs and IND loaded	
Sample	M_w [a] ($\times 10^4$) (g/mol)	M_w/M_n [a]	wt % PLMA [b]	D_h [c] (nm)	PDI [c]	D_h [c] (nm)	PDI [c]	D_h [c] (nm)	PDI [c]
PLMA$_{22}$-*b*-POEGMA$_{58}$	3.3	1.30	16	162	0.28	-	-	166	0.33
PLMA$_{22}$-*b*-POEGMA$_{32}$	2.07	1.31	27	122	0.35	198	0.37	194	0.39
PLMA$_{22}$-*b*-POEGMA$_{13}$	1.17	1.18	52	140	0.31	200	0.35	-	-

[a] Determined by size exclusion chromatography (SEC), [b] Determined by nuclear magnetic resonance (^{1}H NMR), [c] Determined by dynamic light scattering (DLS) at 90°.

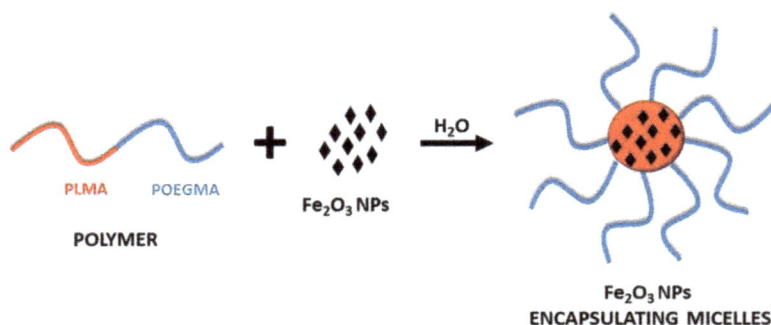

Scheme 1. Simplified schematic illustration depicting the preparation of the PLMA-*b*-POEGMA/NPs mixed nanostructures.

3.1. Encapsulation of Fe$_2$O$_3$ Nanoparticles in the PLMA-b-POEGMA Micelles

A major concern when it comes to the preparation of such mixed nanosystems is the colloidal stability of the mixed solutions. This was possible only for PLMA$_{22}$-*b*-POEGMA$_{32}$ and PLMA$_{22}$-*b*-POEGMA$_{13}$ containing 10 wt % NPs with regard to PLMA mass. In the case of PLMA$_{22}$-*b*-POEGMA$_{58}$, the magnetic NPs precipitated, probably due to the low PLMA hydrophobic content relative to POEGMA, which apparently is not enough to accommodate a significant amount of the NPs. Trials for the encapsulation of greater amounts of NPs were carried out, but the mixed solutions were not colloidally stable, with only a small fraction of the NPs being encapsulated in the form of an aqueous suspension, while most of the materials existed as precipitates at the bottom of the vials (Figure 1).

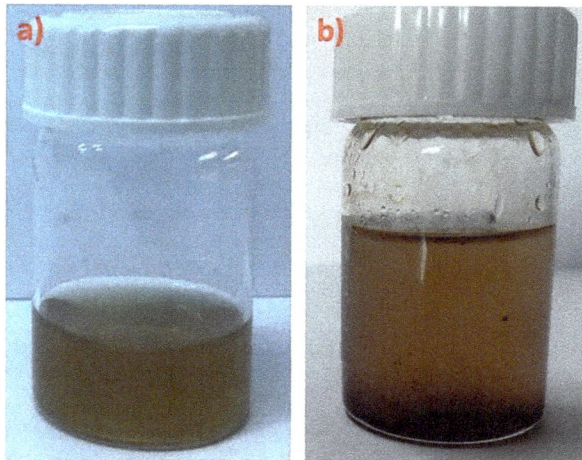

Figure 1. Colloidally stable (**a**) and unstable (**b**) mixed solutions of PLMA$_{22}$-*b*-POEGMA$_{32}$ micelles containing 10 and 20 wt % of magnetic nanoparticles (NPs) relative to PLMA, respectively.

In order to investigate the size of the NPs encapsulating micelles, dynamic light scattering was utilized. In Figure 2, a typical size distribution graph from CONTIN analysis before and after the encapsulation of the magnetic NPs is shown. The results indicate that the Fe$_2$O$_3$ NPs affect the size of the mixed nanostructures, and the hydrodynamic diameter (D_h) is significantly larger after the encapsulation of the magnetic NPs (Table 1). This is a result of the fact that a larger amount of hydrophobic components exists in the mixed nanostructures, and a larger amount of block copolymer participates in the organization of the materials in order to provide a colloidally stable nanosystem.

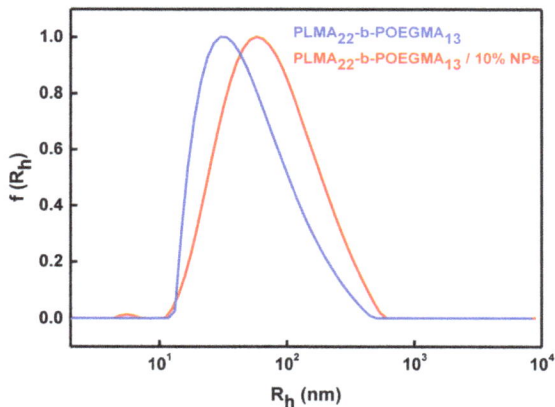

Figure 2. Comparative size distribution graphs from CONTIN analysis before and after the encapsulation of the magnetic NPs within the PLMA$_{22}$-*b*-POEGMA$_{13}$ micelles.

Magnetophoretic experiments were performed in order to investigate whether the iron oxide NPs containing mixed nanostructures maintain the magnetic properties of the inorganic part after the encapsulation in the polymeric micelles. As can be seen in Figure 3, the magnetic mixed colloidal nanostructures are gathered in the side of the measuring cell in the place where the magnetic field was applied.

In addition, in the magnetophoretic experiments performed with the aid of a UV–Vis spectrophotometer, there is a decrease in the absorbance at 450 nm because the magnetic material gathers in the side where the magnet is placed. The magnetophoresis graphs shown in Figure 4 prove that the NPs loaded micelles have a strong response to the application of the external magnetic field, since the absorbance of the solution decreases rapidly within the first ten minutes of the measurement and then shows a small and more gradual decrease up to approximately 20 min. No further decrease in the absorbance can be observed in the remaining time of the experiment. It seems that the phenomenon is faster for the PLMA$_{22}$-*b*-POEGMA$_{13}$/10% NPs mixed nanosystem, which obviously contains the higher amount of the magnetic material (as can be deduced by the stoichiometry of the solutions and from the initial value of the absorbance).

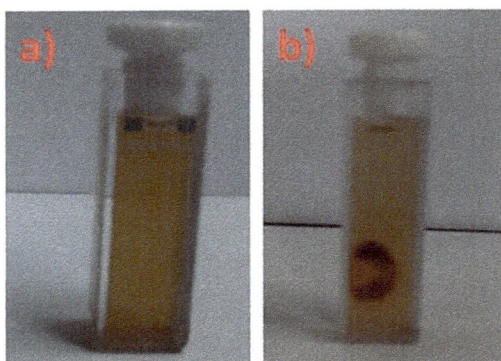

Figure 3. PLMA$_{22}$-*b*-POEGMA$_{13}$/10% NPs mixed solution before (**a**) and after (**b**) the application of magnetic field (using a magnet) for the magnetophoresis experiments.

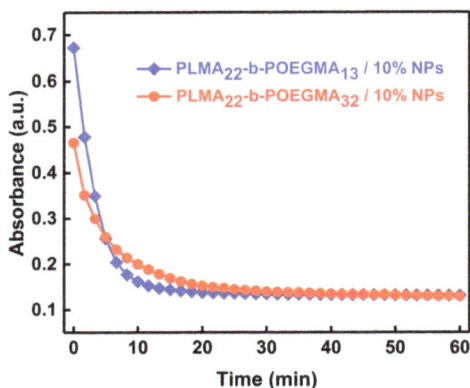

Figure 4. Magnetophoresis graphs measuring absorbance as a function of time (λ = 450 nm) for PLMA$_{22}$-*b*-POEGMA$_{32}$ and PLMA$_{22}$-*b*-POEGMA$_{58}$ micelles loaded with 10 wt % magnetic NPs in regard to PLMA mass in the mixed nanosystem.

In order to have a more complete picture of the morphology of the iron oxide NPs before and after their loading in the polymeric micelles, TEM was utilized as a nanoscale imaging technique.

At this point, it should be noted that the contrast in the mixed systems studied is mainly due to the inorganic NPs, and the polymeric material is barely visible in the micrographs. In Figure 5a, it can be observed that the iron oxide NPs in the absence of copolymer are in some way oriented to the carbon coated grid after the evaporation of the solvent, and this ordering is determined by the oleic acid ligands on the NPs surface. The mean distance between the particles is approximately 1.9 nm, which corresponds rather well with double the space of an oleic acid tail. In Figure 5b, where the block copolymer PLMA$_{22}$-b-POEGMA$_{32}$ is involved, it can be observed that the mean distance between the particles is significantly larger (ca. 4.6 nm); a fact that leads to the conclusion that the surface of the NPs is decorated with the block copolymer chains. This idea is enhanced from the observation of white shadows in the TEM images (noted with red arrows) which must be areas with a higher concentration of polymeric chains (or parts of polymeric spherical micelles). In the case of a PLMA$_{22}$-b-POEGMA$_{13}$-based mixed nanosystem, similar areas can be observed, but also areas with large aggregation of magnetic NPs (red circles); an observation that indicates both the encapsulation of the magnetic NPs within polymeric micelles and the decoration of NPs surface with block copolymer chains. Most probably, the length of the PLMA block and its content dictate the interactions between block copolymer chains and the NPs, as well as the formation of mixed nanostructures in aqueous media. On the other hand, a restructuring of the mixed nanostructures on the carbon surface of the TEM grid in the dry state cannot be ruled out due to the low T_g of the PLMA block. Therefore, there is a possibility that the structures observed by DLS and TEM are not exactly the same, due to the different state of the material during the measurements (suspension vs. dry state). In any case, TEM micrographs also show the presence of interactions between PLMA-b-POEGMA copolymers and magnetic NPs and their nanoscale organization. However, the general picture for the nanosystems obtained from DLS measurements is closer to the environments found in biomedical applications such as drug delivery, since they are mostly related to solutions or suspensions.

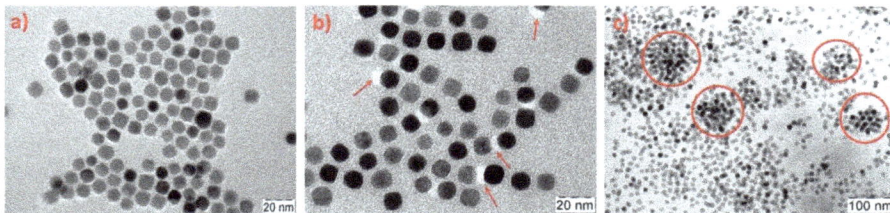

Figure 5. TEM images for (**a**) iron oxide NPs; (**b**) PLMA$_{22}$-b-POEGMA$_{32}$/10% NPs (red arrows indicate areas with a higher concentration of polymeric chains) and (**c**) PLMA$_{22}$-b-POEGMA$_{13}$/10% NPs mixed nanosystems after solvent evaporation (red circles indicate areas with large aggregation).

3.2. Encapsulation of Both Iron Oxide Nps and IND into PLMA-b-POEGMA Micelles

The next step was to investigate whether the polymeric micelles could encapsulate a hydrophobic drug together with the magnetic NPs in their PLMA core forming a three component colloidal nanosystem. The procedure followed for the preparation of PLMA-b-POEGMA micelles loaded with iron oxide NPs and the non-steroidal, anti-inflammatory drug indomethacin is depicted in Scheme 2.

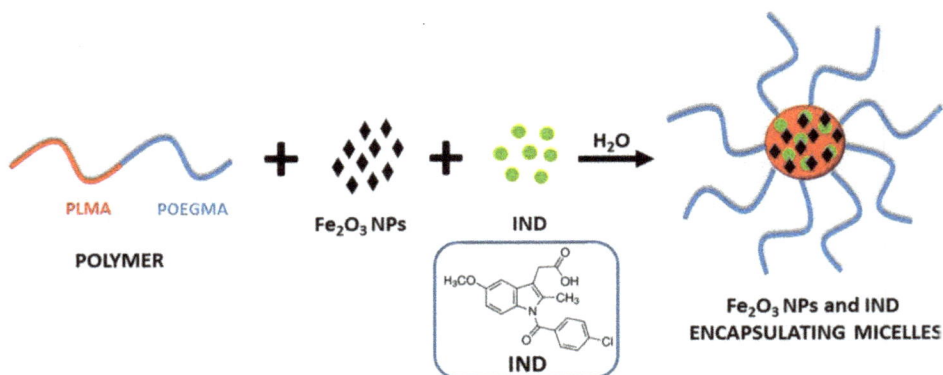

Scheme 2. Simplified schematic illustration for the preparation of the PLMA-*b*-POEGMA/NPs/IND mixed nanostructures.

This was achieved only for sample PLMA$_{22}$-*b*-POEGMA$_{58}$ in terms of colloidal stability. As stated above, this block copolymer was not able to form stable solutions with encapsulated iron oxide NPs alone. Probably, the polar groups contained in the hydrophobic IND molecule make it act as a low molecular surfactant, affecting the structural characteristics and colloidal stability of the mixed self-assembled nanostructures formed [23,25–27]. Regarding the rest of the mixed systems, PLMA$_{22}$-*b*-POEGMA$_{32}$/NPs/IND was stable for 72 h and PLMA$_{22}$-*b*-POEGMA$_{13}$/NPs/IND precipitated almost immediately after mixing the initial solutions. This fact indicates that the hydrophilic/hydrophobic balance, taking into account all components utilized in the preparation of the mixed nanostructures, is responsible for the stability of the final solution.

DLS was utilized for the determination of the size distribution of the mixed PLMA-*b*-POEGMA/NPs/IND nanostructures. For the PLMA$_{22}$-*b*-POEGMA$_{58}$ system, the size is significantly larger than before the encapsulation (Figure 6a), whereas for PLMA$_{22}$-*b*-POEGMA$_{32}$ it is almost the same with the system where only magnetic NPs were encapsulated in the copolymer. This may be another proof for the surfactant-like activity of indomethacin in the cases under investigation. TEM observations (Figure 6b) show the dominating contrast of magnetic NPs and no substantial changes in the morphology compared with the systems where IND is absent (Figure 5b,c).

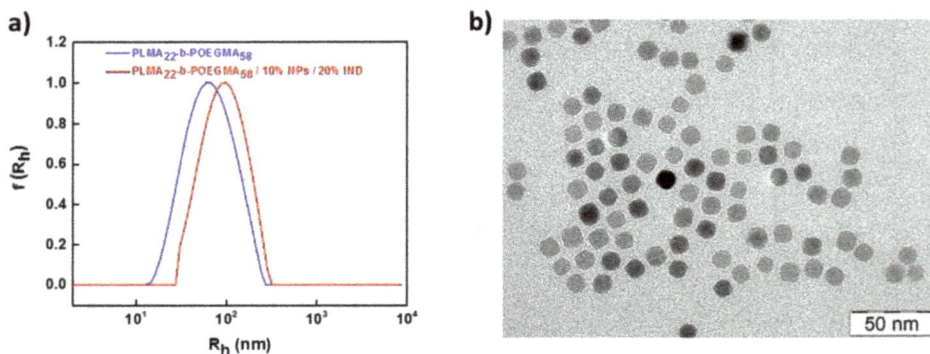

Figure 6. (a) Comparative size distribution graphs from CONTIN analysis before and after the encapsulation of magnetic NPs and IND in PLMA$_{22}$-*b*-POEGMA$_{58}$ micelles; (b) TEM micrograph for PLMA$_{22}$-*b*-POEGMA$_{58}$/10% NPs/20% IND.

Further proof for the incorporation of IND in the polymeric micelles is given by the UV–Vis measurements shown in Figure 7. The increase in absorbance at 450 nm is attributed to the existence of the magnetic NPs in the micellar core, and the one at 321 nm is attributed to the presence of encapsulated indomethacin[27].

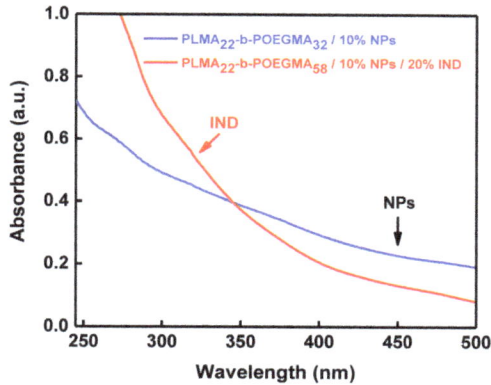

Figure 7. UV–Vis spectra of PLMA$_{22}$-POEGMA$_{32}$/10% NPs (blue line) and PLMA$_{22}$-*b*-POEGMA$_{58}$/10% NPs/20% IND (red line).

Magnetophoretic measurements were also performed in order to investigate the behavior of the three-component mixed nanostructures under the application of a magnetic field (Figure 8). The obtained results are similar to those for nanosystems that have only NPs encapsulated. This shows that the presence of IND does not influence the magnetic properties of the mixed nanostructures.

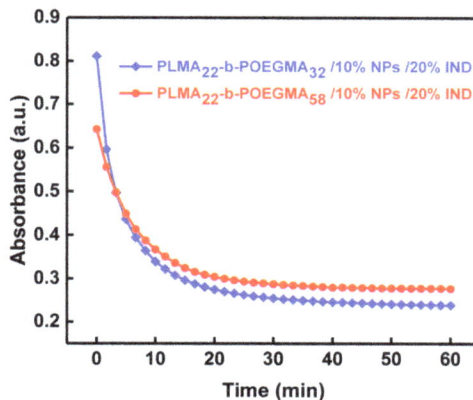

Figure 8. Magnetophoresis graphs showing absorbance as a function of time (λ = 450 nm) for PLMA$_{22}$-*b*-POEGMA$_{58}$ and PLMA$_{22}$-*b*-POEGMA$_{32}$ loaded with 10% magnetic NPs and 20% IND in respect to PLMA mass.

Solution ionic strength is another parameter that affects the properties of aqueous solutions of nanoparticles. The effect of ionic strength in the structure of the PLMA-*b*-POEGMA/NPs/IND mixed nanosystems in aqueous solutions was also investigated by gradually increasing the concentration of NaCl (by addition of a 1 M NaCl stock solution) and subsequent DLS measurements on the resulting solutions. As can be observed in Figure 9, scattering intensity initially remains constant for ionic strengths up to ca. 0.1 M NaCl and then has the tendency to decrease as the concentration of salt increases. This observation supports a decrease in the mass of the mixed nanostructures by increasing ionic strength above 0.1 M NaCl. However, the determined hydrodynamic radius remains practically the same in the whole range investigated, revealing an absence of aggregation phenomena and pointing towards disintegration of the nanostructures accompanied by swelling (increase in the volume of the nanostructures). This may be related to phenomena taking place at the surface of the magnetic NPs, where an increase of ionic strength may result in a decrease of the oleic acid/Fe_2O_3 interactions and a decrease of the bound amount of stabilizing ligands, which in turn affects copolymer/NPs interactions, since the gluing components/interactions are eventually diminished.

Having in mind the potential application of the mixed nanostructures in biological environments and most probably that of blood serum, the interaction of BSA with the mixed nanoassemblies was investigated by titration with aqueous BSA solution (1×10^{-3} g/mL stock solution concentration) and subsequent DLS measurements. The results are presented in Figure 10b. The scattered intensity decreases as the concentration of BSA increases, and there are no significant changes in the R_h values. The behavior depicted in Figure 10b indicates a decrease in the mass of the nanostructures accompanied by swelling of the structures, as was the case for the variation of the solution ionic strength.

In Figure 10a, the behavior of neat PLMA$_{22}$-*b*-POEGMA$_{58}$ micelles is depicted in the presence of BSA. It can be seen that both scattering intensity and size of the pure polymer micelles solutions increase rapidly after the first addition of BSA in the micellar solution. This means that BSA is adsorbed/incorporated in the polymeric micelles, resulting to the formation of higher aggregates. As the concentration of BSA increases, the values of R_h remain relatively constant but there is a decrease in the values of scattered intensity. This decrease in the intensity can be related to a decrease in the mass of the PLMA$_{22}$-*b*-POEGMA$_{58}$ micelles along with nanostructure swelling and that may explain the non-significant changes in the size after the initial addition of BSA. The phenomenon of the rapid increase in the size and intensity is not observed after the encapsulation of NPs and IND in the polymeric micelles, probably because, in the latter case, the structure of PLMA$_{22}$-*b*-POEGMA$_{58}$/10% NPs/20% IND assemblies is more compact and more stable, due to increased mainly hydrophobic interactions of the components, compared to that of neat PLMA$_{22}$-*b*-POEGMA$_{58}$ compound micelles. Apparently, BSA leads to partial disintegration of the mixed nanostructures, through incorporation of the protein within the newly formed PLMA$_{22}$-*b*-POEGMA$_{58}$/NPs/IND/BSA assemblies, due to the amphiphilic character of BSA, and its ability to interact with and to encapsulate hydrophobic compounds in the blood environment [28].

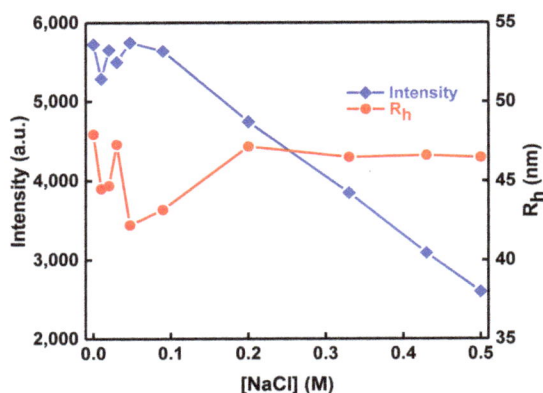

Figure 9. R_h and scattered intensity as a function of ionic strength ([NaCl]) for PLMA$_{22}$-*b*-POEGMA$_{58}$ /10% NPs/20% IND mixed aqueous solution.

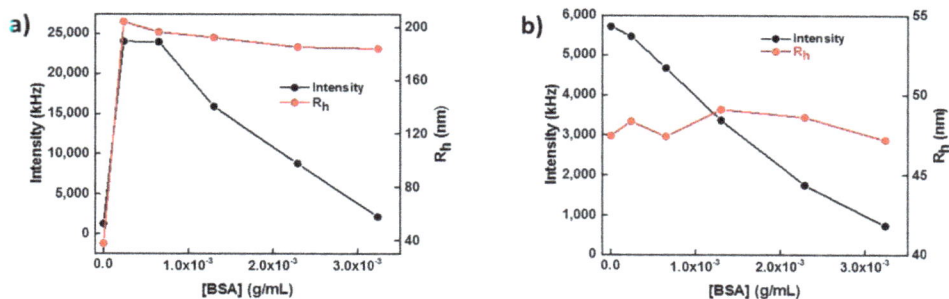

Figure 10. R_h and scattered intensity as a function of BSA concentration for PLMA$_{22}$-*b*-POEGMA$_{58}$ (**a**) and PLMA$_{22}$-*b*-POEGMA$_{58}$/10% NPs/20% IND (**b**) mixed nanostructures in water.

4. Conclusions

To summarize, the ability of poly(lauryl methacrylate)-*b*-poly(oligo ethylene glycol methacrylate) (PLMA-*b*-POEGMA) amphiphilic block copolymer self-assemblies in aqueous solutions to encapsulate magnetic iron oxide nanoparticles was investigated. These block copolymers self-assemble in compound micelles in aqueous solutions and can load up to 10 wt % iron oxide nanoparticles (NPs) in their PLMA cores forming mixed nanostructures. The size distribution of the mixed micelles is significantly broader after the loading of the magnetic NPs, as dynamic light scattering (DLS) results have shown. The NPs maintain their magnetic properties after the encapsulation in the micelles as it was proved by magnetophoretic measurements. The mixed solutions are colloidally stable for copolymers where the hydrophobic ratio is larger than 30 wt %.

In the case where both indomethacin (IND) and magnetic NPs were simultaneously loaded in the micellar core, only the lowest hydrophobic ratio copolymer was able to assure colloidal stability of the mixed three-component aggregates. The magnetic properties of the iron oxide NPs were also maintained after the encapsulation of the drug, resulting in magnetically active mixed nanostructures with encapsulated IND. The mixed aggregates seem to be affected by the presence of increased concentrations of salt and BSA but without loss of their nanosized dimensions and colloidal stability.

The colloidal stability, drug encapsulation ability and magnetic properties of the nanosystems based on biocompatible PLMA-*b*-POEGMA amphiphilic block copolymers prepared in this study hold potential for utilization of these hybrid nanostructures as drug delivery and triggered release systems, as well as for bioimaging applications.

Acknowledgments: The authors gratefully acknowledge support from the Ministry of Education, Youth and Sports of the Czech Republic (project LO1305 and CZ.1.05/2.1.00/19.0377) and Starska Jana for the transmission electron microscopy (TEM) facilities and TEM measurements respectively.

Author Contributions: Athanasios Skandalis and Stergios Pispas conceived and designed the experiments; Athanasios Skandalis and Andreas Sergides performed the experiments; Athanasios Skandalis, Andreas Sergides, Aristides Bakandritsos and Stergios Pispas analyzed the data; Stergios Pispas and Aristides Bakandritsos contributed reagents/materials/analysis tools; Athanasios Skandalis, Andreas Sergides, Aristides Bakandritsos and Stergios Pispas wrote the paper.

Conflicts of Interest: The authors declare no conflict of interest.

References

1. Lammers, T.; Kiessling, F.; Hennink, W.E.; Storm, G. Nanotheranostics and Image-Guided Drug Delivery: Current Concepts and Future Directions. *Mol. Pharm.* **2010**, *7*, 1899–1912. [CrossRef] [PubMed]
2. Kelkar, S.S.; Reineke, T.M. Theranostics: Combining Imaging and Therapy. *Bioconj. Chem.* **2011**, *22*, 1879–1903. [CrossRef] [PubMed]

3. Krasia-Christoforou, T.; Georgiou, T.K. Polymeric theranostics: Using polymer-based systems for simultaneous imaging and therapy. *J. Mater. Chem. B* **2013**, *1*, 3002–3025. [CrossRef]

4. Talelli, M.; Rijcken, C.J.F.; Lammers, T.; Seevinck, P.R.; Storm, G.; van Nostrum, C.F.; Hennink, W.E. Superparamagnetic Iron Oxide Nanoparticles Encapsulated in Biodegradable Thermosensitive Polymeric Micelles: Toward a Targeted Nanomedicine Suitable for Image-Guided Drug Delivery. *Langmuir* **2009**, *25*, 2060–2067. [CrossRef] [PubMed]

5. Zahraei, M.; Marciello, M.; Lazaro-Carrillo, A.; Villanueva, A.; Herranz, F.; Talelli, M.; Costo, R.; Monshi, A.; Shahbazi-Gahrouei, D.; Amirnasr, M.; et al. Versatile theranostics agents designed by coating ferrite nanoparticles with biocompatible polymers. *Nanotechnology* **2016**, *27*, 255702. [CrossRef] [PubMed]

6. Mai, Y.; Eisenberg, A. Self-assembly of block copolymers. *Chem. Soc. Rev.* **2012**, *41*, 5969–5985. [CrossRef] [PubMed]

7. Tritschler, U.; Pearce, S.; Gwyther, J.; Whittell, G.R.; Manners, I. 50th Anniversary Perspective: Functional Nanoparticles from the Solution Self-Assembly of Block Copolymers. *Macromolecules* **2017**, *50*, 3439–3463. [CrossRef]

8. Fairbanks, B.D.; Gunatillake, P.A.; Meagher, L. Biomedical applications of polymers derived by reversible addition—Fragmentation chain-transfer (RAFT). *Adv. Drug Deliv. Rev.* **2015**, *91* (Suppl. C), 141–152. [CrossRef] [PubMed]

9. Duncan, R. The dawning era of polymer therapeutics. *Nat. Rev. Drug Discov.* **2003**, *2*, 347–360. [CrossRef] [PubMed]

10. Aliabadi, H.M.; Lavasanifar, A. Polymeric micelles for drug delivery. *Expert Opin. Drug Deliv.* **2006**, *3*, 139–162. [CrossRef] [PubMed]

11. Gaucher, G.; Dufresne, M.-H.; Sant, V.P.; Kang, N.; Maysinger, D.; Leroux, J.-C. Block copolymer micelles: Preparation, characterization and application in drug delivery. *J. Control. Release* **2005**, *109*, 169–188. [CrossRef] [PubMed]

12. Miyata, K.; Christie, R.J.; Kataoka, K. Polymeric micelles for nano-scale drug delivery. *React. Funct. Polym.* **2011**, *71*, 227–234. [CrossRef]

13. Jones, M.-C.; Leroux, J.-C. Polymeric micelles—A new generation of colloidal drug carriers. *Eur. J. Pharm. Biopharm.* **1999**, *48*, 101–111. [CrossRef]

14. Pinto Reis, C.; Neufeld, R.J.; Ribeiro, J.A.; Veiga, F. Nanoencapsulation I. Methods for preparation of drug-loaded polymeric nanoparticles. *Nanomedicine NBM* **2006**, *2*, 8–21. [CrossRef] [PubMed]

15. Wang, J.; Li, W.; Zhu, J. Encapsulation of inorganic nanoparticles into block copolymer micellar aggregates: Strategies and precise localization of nanoparticles. *Polymer* **2014**, *55*, 1079–1096. [CrossRef]

16. Yuan, J.; Müller, A.H.E. One-dimensional organic–inorganic hybrid nanomaterials. *Polymer* **2010**, *51*, 4015–4036. [CrossRef]

17. Kim, J.; Lee, J.E.; Lee, S.H.; Yu, J.; Lee, J.H.; Park, T.G.; Hyeon, T. Designed Fabrication of a Multifunctional Polymer Nanomedical Platform for Simultaneous Cancer—Targeted Imaging and Magnetically Guided Drug Delivery. *Adv. Mater.* **2008**, *20*, 478–483. [CrossRef]

18. Zhang, L.; Gu, F.; Chan, J.; Wang, A.; Langer, R.S.; Farokhzad, O.C. Nanoparticles in Medicine: Therapeutic Applications and Developments. *Clin. Pharmacol. Ther.* **2008**, *83*, 761–769. [CrossRef] [PubMed]

19. Laurent, S.; Forge, D.; Port, M.; Roch, A.; Robic, C.; Vander Elst, L.; Muller, R.N. Magnetic Iron Oxide Nanoparticles: Synthesis, Stabilization, Vectorization, Physicochemical Characterizations, and Biological Applications. *Chem. Rev.* **2008**, *108*, 2064–2110. [CrossRef] [PubMed]

20. Kim, J.; Piao, Y.; Hyeon, T. Multifunctional nanostructured materials for multimodal imaging, and simultaneous imaging and therapy. *Chem. Soc. Rev.* **2009**, *38*, 372–390. [CrossRef] [PubMed]

21. Hua, X.; Yang, Q.; Dong, Z.; Zhang, J.; Zhang, W.; Wang, Q.; Tan, S.; Smyth, H.D.C. Magnetically triggered drug release from nanoparticles and its applications in anti-tumor treatment. *Drug Deliv.* **2017**, *24*, 511–518. [CrossRef] [PubMed]

22. Mahmoudi, M.; Hosseinkhani, H.; Hosseinkhani, M.; Boutry, S.; Simchi, A.; Journeay, W.S.; Subramani, K.; Laurent, S. Magnetic Resonance Imaging Tracking of Stem Cells in Vivo Using Iron Oxide Nanoparticles as a Tool for the Advancement of Clinical Regenerative Medicine. *Chem. Rev.* **2011**, *111*, 253–280. [CrossRef] [PubMed]

23. Skandalis, A.; Pispas, S. PLMA-b-POEGMA amphiphilic block copolymers: Synthesis and self-assembly in aqueous media. *J. Polym. Sci. A* **2017**, *55*, 155–163. [CrossRef]

24. Park, J.; An, K.; Hwang, Y.; Park, J.G.; Noh, H.J.; Kim, J.Y.; Park, J.H.; Hwang, N.M.; Hyeon, T. Ultra-large-scale syntheses of monodisperse nanocrystals. *Nat. Mater.* **2004**, *3*, 891–895. [CrossRef] [PubMed]

25. Pispas, S.; Hadjichristidis, N. Aggregation Behavior of Poly(butadiene-b-ethylene oxide) Block Copolymers in Dilute Aqueous Solutions: Effect of Concentration, Temperature, Ionic Strength, and Type of Surfactant. *Langmuir* **2003**, *19*, 48–54. [CrossRef]

26. Raffa, P.; Wever, D.A.Z.; Picchioni, F.; Broekhuis, A.A. Polymeric Surfactants: Synthesis, Properties, and Links to Applications. *Chem. Rev.* **2015**, *115*, 8504–8563. [CrossRef] [PubMed]

27. Nagy, M.; Szöllösi, L.; Kéki, S.; Faust, R.; Zsuga, M. Poly(vinyl alcohol)-based Amphiphilic Copolymer Aggregates as Drug Carrying Nanoparticles. *J. Macromol. Sci. A* **2009**, *46*, 331–338. [CrossRef]

28. Zhao, Y.; Marcel, Y.L. Serum Albumin Is a Significant Intermediate in Cholesterol Transfer between Cells and Lipoproteins. *Biochemistry* **1996**, *35*, 7174–7180. [CrossRef] [PubMed]

polymers

MDPI

Article

Micellization Thermodynamics of Pluronic P123 ($EO_{20}PO_{70}EO_{20}$) Amphiphilic Block Copolymer in Aqueous Ethylammonium Nitrate (EAN) Solutions

Zhiqi He and Paschalis Alexandridis *

Department of Chemical and Biological Engineering, University at Buffalo, The State University of New York (SUNY), Buffalo, NY 14260-4200, USA; zhiqihe@buffalo.edu
* Correspondence: palexand@buffalo.edu; Tel.: +1-716-645-1183

Received: 30 November 2017; Accepted: 25 December 2017; Published: 28 December 2017

Abstract: Poly(ethylene oxide)-poly(propylene oxide)-poly(ethylene oxide) (PEO-PPO-PEO) block copolymers (commercially available as Pluronics or Poloxamers) can self-assemble into various nanostructures in water and its mixtures with polar organic solvents. Ethylammonium nitrate (EAN) is a well-known protic ionic liquid that is expected to affect amphiphile self-assembly due to its ionic nature and hydrogen bonding ability. By proper design of isothermal titration calorimetry (ITC) experiments, we determined the enthalpy and other thermodynamic parameters of Pluronic P123 ($EO_{20}PO_{70}EO_{20}$) micellization in aqueous solution at varied EAN concentration. Addition of EAN promoted micellization in a manner similar to increasing temperature, e.g., the addition of 1.75 M EAN lowered the critical micelle concentration (CMC) to the same extent as a temperature increase from 20 to 24 °C. The presence of EAN disrupts the water solvation around the PEO-PPO-PEO molecules through electrostatic interactions and hydrogen bonding, which dehydrate PEO and promote micellization. At EAN concentrations lower than 1 M, the PEO-PPO-PEO micellization enthalpy and entropy increase with EAN concentration, while both decrease above 1 M EAN. Such a change can be attributed to the formation by EAN of semi-ordered nano-domains with water at higher EAN concentrations. Pyrene fluorescence suggests that the polarity of the mixed solvent decreased linearly with EAN addition, whereas the polarity of the micelle core remained unaltered. This work contributes to assessing intermolecular interactions in ionic liquid + polymer solutions, which are relevant to a number of applications, e.g., drug delivery, membrane separations, polymer electrolytes, biomass processing and nanomaterial synthesis.

Keywords: micelle; self-assembly; surfactant; block copolymer; Pluronic; Poloxamer; poly(ethylene oxide); ionic liquid; water; calorimetry

1. Introduction

Poly(ethylene oxide)-poly(propylene oxide)-poly(ethylene oxide) (PEO-PPO-PEO) block copolymers are commercially available amphiphiles (known as Pluronics or Poloxamers) that have been studied widely and applied in many fields [1–5]. Their ability to self-assemble in solvents [6,7] and interact with particles and surfaces [8] renders PEO-PPO-PEO block copolymers useful amphiphiles for applications ranging from paints and coatings to pharmaceutical formulations [9].

The micellization of amphiphilic block copolymers in water has been widely studied [2]. The spontaneous micellization process of Pluronics is entropy-driven and endothermic. Temperature can alter the thermodynamics and structure of Pluronic micellization in water [10]. Following the micellization, water molecules predominantly form hydrogen bonds with the PEO blocks, while the PPO micelle core is mostly dehydrated [11]. At higher temperatures, the hydration of the PEO blocks decreases [12], and the segregation between the PEO and PPO blocks increases [10].

The hydrophobic PPO block plays an important role in the assembly of the core-shell micelle structure [5,13]. The different composition and length of the PEO and PPO blocks of the PEO-PPO-PEO block copolymers lead to varied micelle corona and core sizes and aggregation numbers [14,15].

The modification of the solvent properties of water by the addition of electrolytes or polar organic solutes or solvents [16–19] affects the thermodynamics of PEO-PPO-PEO block copolymer micellization and the micelle structure [20]. Alkali-halide salts act as cosolutes and promote the PEO-PPO-PEO micellization by lowering the critical micelle temperature (CMT) [2]. Some polar organic solvents, such as ethanol and 1-propanol, act as cosolvents that are able to decrease the cohesive forces in the solvent mixture and increase the solubility of the PEO-PPO-PEO block copolymer molecules, resulting in increased critical micelle concentration (CMC). Other solvents, such as glycerol, favor the micellization and decrease CMC due to a contraction (dehydration) of PEO blocks [17], depending on the solvent quality in the mixed solvents [12,16,19,21,22].

Ionic liquids (ILs), low-melting salts with combinatorially high chemical diversity, are being considered for applications in various fields [23], such as biomass processing [24,25], pharmaceuticals [26,27], separations and extraction [28,29], catalysis and batteries [23,30] due to their unique properties and intermolecular interactions [23,31]. Some amphiphiles can self-assemble in neat ionic liquids and in ionic liquid aqueous solutions into ordered structures including micelles and lyotropic liquid crystals [31–34]. The same types of amphiphile self-assembled structures can be formed in select ionic liquids as in aqueous solutions; however, a higher portion of the solvophobic part is frequently required for an amphiphile to form similar structures in ionic liquids due to a greater solubility of hydrocarbons in many ionic liquids [32].

One of the better studied protic ionic liquids, which have available proton(s) on their cations as hydrogen bond donors [32], is ethylammonium nitrate (EAN), which forms an intermolecular hydrogen bonding network similar to water. This provides an opportunity for EAN to replace or be mixed with water in many uses [35]. Homopolymer PEO (molecular weight 38 kDa) can dissolve in EAN with a more contracted conformation (8.1-nm radius of gyration) compared to that in water (9.6-nm radius of gyration), which implies that EAN is a less good solvent for PEO than water [36]. The shell of Pluronic P123 ($EO_{20}PO_{70}EO_{20}$) spherical micelles shrank by 40% in pure EAN compared to that in water, while the micelle core radii remained similar (1 wt % Pluronic P123 at 25 °C) [37]. The ethyl moiety on the EAN cation leads to the formation of polar and nonpolar segregated domains, which increases the solubility of hydrocarbons and, as a result, can increase the CMC of PEO-PPO-PEO block copolymers [37]. Surface tension measurements [38] of the ionic low-molecular weight surfactant Aerosol-OT (AOT) in EAN indicated a spontaneous, endothermic and entropy-driven micellization process similar to that observed in water [39]. The surface activity of AOT in EAN is weaker than that in water due to the relatively higher dielectric constant and viscosity of EAN [38].

Ionic liquids play one or more roles, such as additives and/or solvents, during the self-assembly process in IL-containing water. The role of ionic liquids as additives in decreasing the critical micellization temperature (CMT) of Pluronic F108 ($EO_{128}PO_{54}EO_{128}$) aqueous solution [40] is mainly affected by the anion and the concentration of the IL [41]. EAN is totally miscible with water and forms hydrogen bonds in the mixture [42]. It was reported that the Pluronic P123 ($EO_{20}PO_{70}EO_{20}$) lyotropic liquid crystal (LLC) phases formed in EAN and in EAN aqueous solution are similar [43].

Whereas there are several studies on PEO-PPO-PEO block copolymer self-assembly in neat EAN [33,34,37,44,45], most have focused on the structure rather than the thermodynamics. Studies on PEO-PPO-PEO self-assembly in aqueous EAN solutions are quite limited and are concerned with concentrated systems that form lyotropic liquid crystals [43]. To our best knowledge, there is no study available on PEO-PPO-PEO block copolymer micellization in aqueous EAN solutions.

The thermodynamics and structure of PEO-PPO-PEO block copolymer micellization in water are well established [5,13]. The addition of classic salt cosolutes and/or organic cosolvents leads to various effects on Pluronic micellization, depending on the change of solvent quality. While the ionic liquid EAN exhibits interesting and useful properties when mixed with water [46], the lack of research on

PEO-PPO-PEO micellization in aqueous EAN solutions leaves a gap in our fundamental knowledge of this self-assembly process. Quantifying the effect of EAN on PEO-PPO-PEO micellization in aqueous solution in terms of thermodynamics and interactions merits further work.

Isothermal titration calorimetry (ITC) can directly measure enthalpy changes during (physical) mixing or binding or (chemical) reaction; this provides a way to quantify the thermodynamics of PEO-PPO-PEO micellization process in the presence of EAN. ITC has been previously used to study the physicochemical properties and interactions of Pluronic micellization [47–50]. With relatively large molecular weight and a gradual transition taking place at the CMC [5], the quantification of the thermodynamics during PEO-PPO-PEO block copolymer micellization can be difficult. Moreover, EAN absorbs a large amount of heat when mixing with water [51], which can obstruct the observation of the enthalpy changes caused by PEO-PPO-PEO micellization in such mixtures. To study the IL addition effect on PEO-PPO-PEO micellization in aqueous solution, we designed experiments to exclude the significant enthalpy change resulting from IL dilution by keeping the EAN concentration constant during titration. Moreover, in order to evaluate the interactions between IL, Pluronic and water, we designed a set of ITC titrations to separately determine the IL dilution enthalpy and the enthalpy of IL-Pluronic interactions. Upon subtracting the enthalpy of EAN dilution and of Pluronic interacting with water from the total observed enthalpy during micellization, the enthalpy associated with interactions between EAN and Pluronic has been assessed and is discussed here as a function of Pluronic concentration at varied EAN concentration. In this paper, the thermodynamics (including CMC, free energy, enthalpy and entropy) of Pluronic P123 micellization are determined in EAN aqueous solution with varied concentrations. The effect of EAN is compared to the temperature effect. The interactions within the aqueous EAN Pluronic solution are explored during micellization. This is the first report that quantifies the effect of EAN addition on the change of thermodynamic parameters of PEO-PPO-PEO micellization in aqueous solution.

2. Materials and Methods

2.1. Materials

Pluronic P123 poly(ethylene oxide)-poly(propylene oxide)-poly(ethylene oxide) (PEO-PPO-PEO) block copolymer can be represented by the formula $EO_{20}PO_{70}EO_{20}$ on the basis of its 30 wt % PEO and 5750 g/mol nominal molecular weight. Pluronic P123 was obtained as a gift from BASF Corp. (Vandalia, IL, USA) and was used as received. With a large portion of PPO, Pluronic P123 is relatively hydrophobic. The micellization of Pluronic P123 has a relatively low CMC in water (0.313 mM at 20 °C [13]) and is sensitive to the change of conditions such as temperature [13], which makes this a good system to observe the effects of ionic liquid addition.

Ethylammonium nitrate (EAN, $C_2H_5NH_3NO_3$, melting point = 12 °C) was obtained from Iolitech (Tuscaloosa, Alabama, AL, USA). It was stored in a desiccator and vacuum dried for at least 8 h before use. Milli-Q purified water was used to prepare aqueous solutions.

Pyrene purchased from Fluka (Buchs, Switzerland) was used to probe the micropolarity of the Pluronic block copolymer-ionic liquid-water systems. Two microliters of 1 mM pyrene in ethanol mixture were added to 3-g solution samples for fluorescence spectroscopy. The resulting pyrene and ethanol concentrations were about 0.7 μM and 6.7×10^{-4} vol %.

2.2. Sample Preparation

Ionic liquid aqueous solutions: A specified mass of EAN was mixed with a specified volume of water in a volumetric flask (EAN concentration in water ranging from 0 to 3 M), under rolling at room temperature for at least 8 h before use. Each EAN concentration was prepared separately.

Samples for ITC experiments: Pluronic P123 in ionic liquid aqueous solution was prepared by dissolving Pluronic P123 in the ionic liquid aqueous solution prepared as discussed above.

These samples were allowed to equilibrate at least 8 h under rolling at room temperature. The samples were measured within two days after preparation.

Samples for fluorescence experiments: 0.1 wt % Pluronic P123 solution was prepared by mixing neat Pluronic P123 with ionic liquid aqueous solution. This stock solution was allowed to equilibrate for around 24 h under rolling at room temperature. Further samples were prepared by diluting this stock solution to achieve Pluronic concentrations in the range 1×10^{-6} to 1×10^{-2} wt %. The samples were measured within two days following preparation.

2.3. Isothermal Titration Calorimetry

ITC experiments were performed using a Microcal ITC200 calorimeter (Malvern, Worcestershire, UK). (Figure 1) The sample cell (200 µL) was filled with water or water/EAN solution. The syringe (38.4 µL) was loaded with micellar Pluronic P123 in water or in water/EAN solution. Aliquots (1.5 µL per injection) of a micellar P123 solution were injected into the sample cell at 120-s intervals between successive injections. The syringe rotated at a constant speed of 750 rpm. The temperature inside the sample cell was kept constant during the experiment.

In order to assess the enthalpy changes during the Pluronic micellization process, we successively titrated micellar Pluronic solution into the solvent present in the sample cell. At first, all the Pluronic micelles dissociate into unimers (non-associated block copolymer molecules); but after the Pluronic concentration in the sample cell increased to around CMC, not all titrated micelles dissociate beyond this point. The appropriate experimental parameters (injection volume, interval and titrant concentration) were determined following pre-experiments. We selected the Pluronic P123 titrant concentration at 3.2 mM, so that the P123 concentration in the sample cell changes to span the micellization process as the titration proceeds. We also considered that the heat release/absorption of each titration recorded by ITC should not exceed the maximum measurement capability of the ITC instrument and is not too small compared to the background noise.

By integrating the enthalpy of each injection and then standardizing by injected amphiphile molar amount, the observed enthalpy change with amphiphile concentration can be obtained. For the first several titrations, the Pluronic P123 concentration in the sample cell is lower than the CMC, and the Pluronic P123 micelles titrated into the sample cell dissociate into unimers. As more injections occur, the Pluronic P123 concentration in the sample cell increases. When the Pluronic P123 concentration in the sample cell reaches the CMC, the magnitude of the observed enthalpy change quickly decreases, indicating that the micellized P123 gradually stops dissociating into unimers. We define the concentration range at which micelles gradually stop dissociating as a "transition region". For the last few titrations, the Pluronic P123 concentration in the sample cell is well above CMC, the micellar Pluronic P123 no longer dissociates into unimers, and only dilution of the micelles takes place. The micellization enthalpy was calculated as the difference between the enthalpy at the end of the micelle dissociation concentration region and the enthalpy at the start of the micelle dilution region. Extrapolation was conducted by a linear fit of the first/last several points on the enthalpogram. The position where the extrapolation line diverged from the enthalpy curve was designated as the end of the micelle dissociation region/start of the micelle dilution region and was used for the calculation of the micellization enthalpy (as indicated in Figure 1c) [52]. The inflection point of the enthalpogram is assigned as the CMC, which can be calculated by the maximum of the first derivative of the corresponding enthalpy change versus amphiphile concentration [53,54]. In the case of higher temperatures (>22 °C) or higher EAN concentrations (>1 M), the CMC is too small to be quantified by titrating 3.2 mM of Pluronic P123; thus, we used a lower Pluronic P123 concentration (around 1 mM, depending on the experiment) titrant in order to probe a more detailed change at the concentrations around CMC.

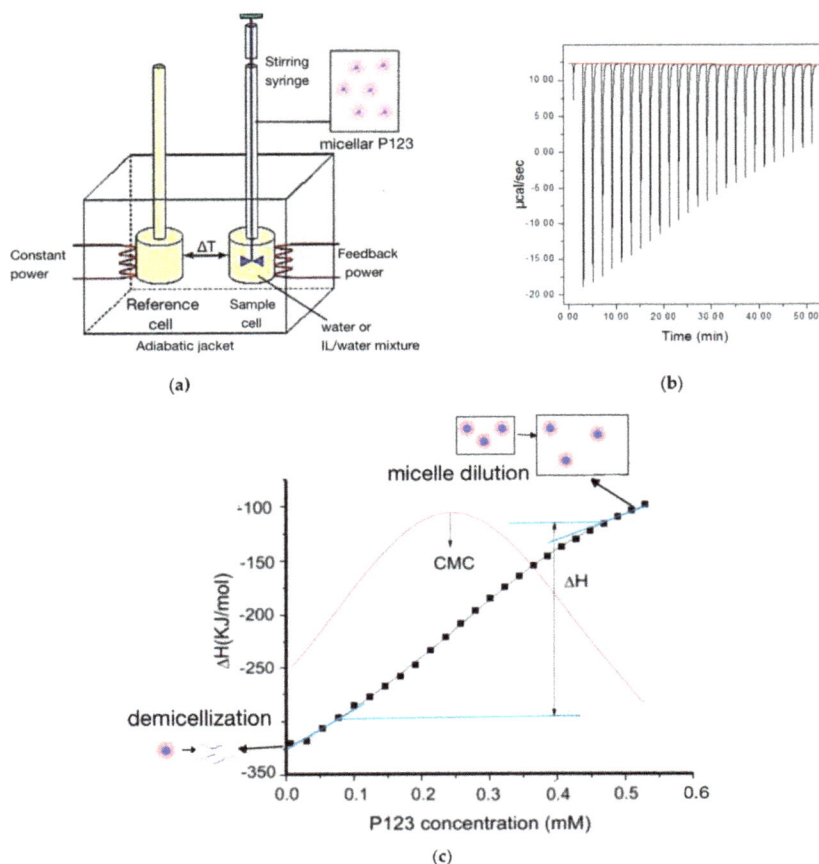

Figure 1. (**a**) Schematic of the titration process for micellization in an isothermal titration calorimeter. Micellar Plurnic P123 inside the syringe is titrated into the sample cell, which contains water or water + ethylammonium nitrate (EAN) solution. ITC records the heat difference between the sample cell and reference cell caused by the dilution and dissociation of injected micelles at constant temperature. (**b**) Raw ITC data of aqueous micellar Pluronic P123 solution titrated into water at 20 °C. (**c**) Enthalpy change during micellar Pluronic P123 solution titrated into water plotted versus Pluronic P123 concentration in the sample cell at 20 °C. The large heat release for the initial injections indicated that all the titrated micelles dissociate into unimers; the relative small heat release over the last injections indicated that the injected micelles only diluted without dissociation. The enthalpy of micellization is determined by the difference between the enthalpy value of extrapolation at the end of the first few injections and the start of the last few injections.

2.4. Pyrene Fluorescence

A Hitachi 2500 fluorescence spectrophotometer (Tokyo, Japan) was used to record fluorescence emission intensity in the 340 to 460 nm range from pyrene-containing Pluronic P123 solutions at 23 °C. The excitation wavelength was $\lambda = 335$ nm. The ratio of the first to the third vibronic peak (I_1/I_3) of the pyrene emission spectrum is a measure of the polarity of the pyrene microenvironment [12]. The micropolarity in EAN aqueous solutions was probed by the I_1/I_3 intensity ratio of the pyrene vibronic band of fluorescence emission spectra. The fluorescence emission of blank (without pyrene addition) EAN and water mixture was measured and used as a control to adjust the baseline. With the Pluronic P123 concentration increase, pyrene partitions into hydrophobic sites (micelle core), which results in a decrease of the I_1/I_3 value due to the lower polarity of a hydrophobic medium compared to that of an aqueous medium. At Pluronic P123 concentrations above the CMC, the value of the I_1/I_3 ratio no longer changes as most pyrene has accumulated in the hydrophobic moieties.

3. Results and Discussion

3.1. Thermodynamics of Pluronic P123 Micellization in Water

In order to quantify the Pluronic micellization in plain water that serves as a reference system for investigating EAN effects, and to establish the methodology for ITC analysis of Pluronic micellization, micellar Pluronic P123 aqueous solution was successively titrated into water.

Enthalpy changes recorded during titrations of micellar 3.2 mM Pluronic P123 aqueous solution into water are shown in Figure 2 at various temperatures. The enthalpy change during the whole titration process is negative (exothermic process) at all temperatures considered here. The possible contributions to such exothermic enthalpy profiles include the dilution of micelles and the dissociation of micelles (demicellization) [55].

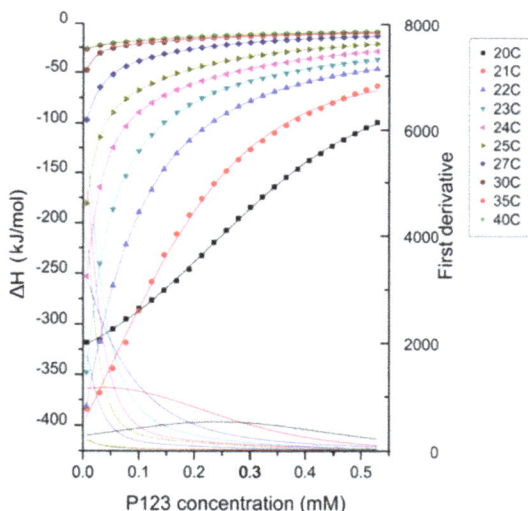

Figure 2. Enthalpy changes generated by titrating 3.2 mM Pluronic P123 in water at varied temperatures (shown in the legend) in the range 20 to 40 °C. The titrant concentration is well above the critical micelle concentration (CMC) (0.313 mM at 20 °C [13]). The enthalpy change for each injection was obtained by ITC and then was standardized by dividing by the amount of the injected Pluronic for each injection. The maximum of the first derivative of the enthalpogram (solid lines) indicates the CMC.

The enthalpogram has a sigmoidal shape at 20 °C, while it is non-sigmoidal at higher temperatures. This shape transition of the enthalpogram shows that at higher temperatures, the concentration region at which all micelles dissociate into unimers is not observed. As the temperature increased, the CMC and the transition region shifted to lower Pluronic P123 concentration, indicating that higher temperature promoted the micellization process.

The Gibbs free energy of micellization (ΔG_{mic}) and the entropy of micellization (ΔS_{mic}) can be calculated from ΔH_{mic} and CMC by the following equations [56]:

$$\Delta G_{mic} = RT\ln(X_{cmc})$$

$$\Delta G_{mic} = \Delta H_{mic} - T\Delta S_{mic}$$

where X_{cmc} is the amphiphile concentration (mole fraction) at the CMC.

The CMC, enthalpy, Gibbs free energy and entropy of micellization extracted and calculated from the enthalpograms in Figure 2 are plotted as a function of temperature in Figure 3. The more negative free energy of micellization values at higher temperatures reflect that Pluronic P123 molecules are thermodynamically easier to assemble into micelles at higher temperature, consistent with previous findings [13]. The positive value of micellization enthalpy ΔH_{mic} over the entire temperature range considered here indicates that Pluronic P123 micellization in water is endothermic. The positive ΔS_{mic} observed at all the temperatures considered here contributes to the negative free energy and drives the micellization. At lower temperatures (20 to 22 °C), the entropy increase plays an important role to drive Pluronic P123 micellization; while at higher temperatures (>22 °C), the decrease of ΔH_{mic} primarily contributes to the decrease of ΔG_{mic} and further lowers the CMC. It is noted that at temperatures above 35 °C, the enthalpy of micellization decreased to a very small value according to ITC measurements. These enthalpy values may be caused by the micellization region shifting to very low Pluronic concentrations at this temperature, and the extrapolation method (described in Section 2) is not precise due to the steep change of enthalpy during the initial few injections. The change of micellization enthalpy with temperature is in agreement with studies reporting an increase of micellization enthalpy at lower temperature and a decrease of micellization enthalpy at higher temperature [57]; however, the lowest micellization enthalpy reported in aqueous solution was around 60 kJ/mol at a CMT of 60 °C [58]. The decrease of micellization free energy with temperature increase can be attributed to the temperature-dependent difference in the solvation of PEO and PPO blocks in aqueous solutions. With increasing temperature, PPO blocks experience a large degree of dehydration [59]. The ITC technique provides quantitative information on the change of enthalpy and entropy as the temperature increases. The micellization enthalpy is often assumed to be independent of temperature [13]. However, this assumption is not proper for the temperature-sensitive PEO-PPO-PEO block copolymers.

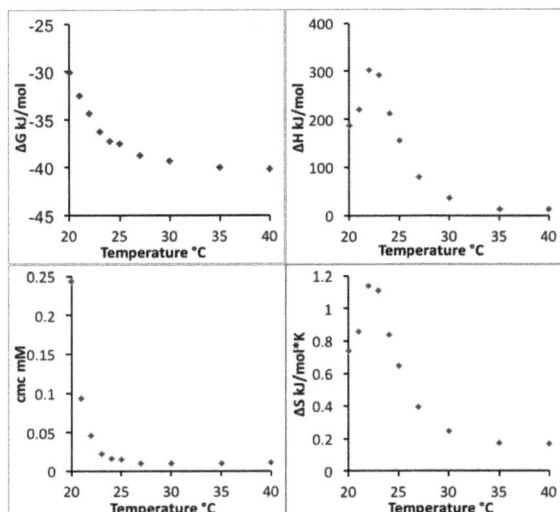

Figure 3. Micellization free energy, enthalpy, CMC and entropy of Pluronic P123 in water plotted as a function of temperature.

As the temperature increases, the micellization entropy and enthalpy change in the same trend, i.e., they both either increase or decrease together. A linear relationship between the micellization enthalpy and entropy at varied temperature (shown in Figure 4) indicates that changes in ΔH_{mic} are compensated by changes in ΔS_{mic} for the micellization process [52], which keeps the free energy negative and results in a spontaneous process. The slope of the linear regression of micellization enthalpy versus entropy provides the compensation temperature, which characterizes solvent-solute interactions [60]. The compensation temperature determined here is 298.4 K, which is close to the compensation temperature reported for other Pluronics block copolymers in aqueous solutions characterized by ITC and surface tension (298.1 K for Pluronic 127 ($EO_{100}PO_{65}EO_{100}$) [52], and 294.9 K for Pluronics F88 ($EO_{104}PO_{39}EO_{104}$) and F68 ($EO_{79}PO_{28}EO_{79}$) [57]).

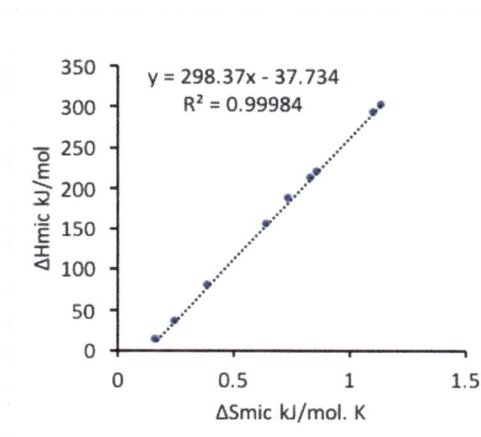

Figure 4. Enthalpy-entropy compensation of Pluronic P123 micellization process. The slope is defined as the compensation temperature (298.37 K) characterizing solvent-solute interaction.

3.2. EAN Effects on Pluronic P123 Micellization in Aqueous Solutions

Having characterized PEO-PPO-PEO micellization in aqueous solution with ITC, we assess here the effect on Pluronic micellization of EAN addition to water. The enthalpy changes of titrating micellar Pluronic P123 in (EAN + water) into (EAN + water) solution were measured by ITC. (Figure 5) Significant heat is absorbed upon mixing EAN and water [51], which indicates an energetically unfavorable (endothermic) mixing resulting from the reduction of strong electrostatic attractions and hydrogen bonds between the EAN ions in the presence of water. Because the large heat absorption during EAN dilution cannot be ignored, it would hinder the assessment of the "real" effect of the EAN presence on Pluronic P123 micellization in aqueous solution.

To study the EAN effect on Pluronic P123 micellization, 3.2 mM Pluronic P123 in EAN aqueous solution was successively titrated into EAN aqueous solution (having the same EAN content as in the titrant). By keeping the EAN concentration the same in the titrant as in the sample cell, the significant heat change caused by possible EAN dilution during the titration can be excluded, and the results thus obtained reflect the actual enthalpy change of Pluronic P123 micellization in EAN + water mixture. Various EAN concentrations were used in order to assess the effect of EAN concentration on PEO-PPO-PEO micellization.

The observed enthalpy change is negative during all the titration process with EAN present at varying concentrations, suggesting that the EAN presence does not alter the exothermic aspect of Pluronic P123 demicellization in aqueous solutions. With higher EAN concentration, the CMC significantly shifted to a lower P123 concentration. For example, upon 1 M EAN addition to water, the CMC is 0.06 mM, much lower than the 0.24 mM CMC in plain water. From the enthalpograms of Figure 5, the CMC and micellization enthalpy were extracted through the inflection point and the extrapolation respectively (discussed in Section 2). The Gibbs free energy and entropy of micellization were then calculated based on the value of CMC and micellization enthalpy. ΔH_{mic} of Pluronic P123 are measured over the EAN concentration range 0 to 3 M; however, other parameters related to CMC can only be assessed at EAN concentrations up to 2 M. At higher EAN concentrations, the CMC is too low to be measured with this experiment, as the micelles no longer completely dissociate into unimers even at the very first injection. The inability to obtain the enthalpy of micelle dissociation makes it difficult to accurately determine the micellization enthalpy at EAN >2 M. By extrapolating the enthalpy change of the first few injections to infinite dilution (zero Pluronic concentration), an estimation of the micelle dissociation enthalpy was obtained and then used to calculate the micellization enthalpy. With an EAN concentration higher than 3 M, the heat release from each injection is too small to be accurately determined, but the trend from the experiments reported here indicates that more concentrated EAN leads to lower micellization enthalpy.

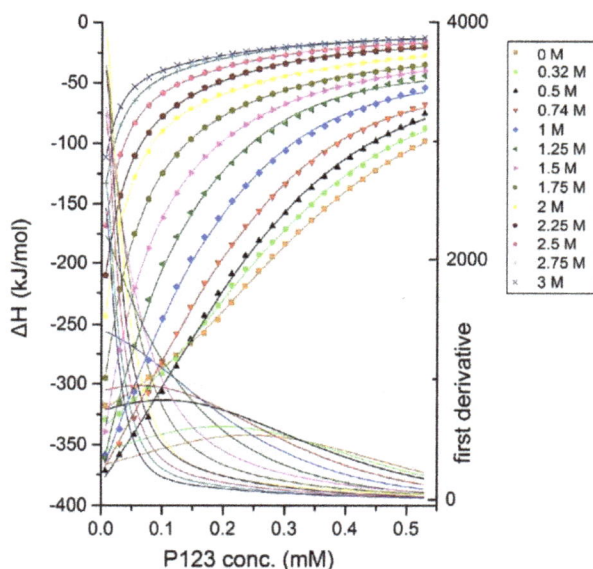

Figure 5. Enthalpy changes of titrating micellar 3.2 mM Pluronic P123 in EAN aqueous solution into aqueous solution with the same EAN content at 20 °C. The enthalpy change for each injection was obtained by ITC and then was standardized by dividing by the amount of the injected Pluronic for each injection. The maximum of the first derivative of the enthalpogram (solid lines) indicates the CMC. The EAN concentration at each titration is shown in the legend.

The CMC, Gibbs free energy, enthalpy and entropy of Pluronic P123 micellization in EAN aqueous solution are plotted as a function of EAN concentration in Figure 6. The Gibbs free energy of Pluronic P123 micellization decreased monotonically with increasing EAN concentration in water, suggesting that EAN promotes Pluronic P123 micellization in aqueous solution, which is similar to the effect of increasing temperature. For example, the addition to water of 1.75 M EAN lowers the CMC to the same extent as a temperature increase from 20 to 24 °C. The entropy of micellization increased with EAN addition. This can be rationalized in terms of the water molecules arranged in the proximity of the Pluronic hydrophobic chains being released into the bulk solution and becoming more disorganized during micellization. However, when the EAN concentration is higher than around 1 M, the micellization entropy no longer increases significantly with EAN concentration. The properties of the EAN and water mixture depend on the ratio of EAN and water. At dilute EAN aqueous solution (<0.02 mole fraction, 1 M, 10.6 wt % EAN), water molecules solvate around the ions of EAN, which behaves like a typical 1:1 electrolyte in water [61]. However, at higher EAN concentrations, distinct semi-ordered nano-domains of EAN and water have been reported by molecular dynamics simulation even though EAN and water are completely miscible in the macro-scale, which is different from traditional ion-water interactions [62]. Concentrated EAN aqueous solutions exhibit different properties, e.g., lower conductivity, from dilute EAN aqueous solutions [63]. As the EAN concentration in water increases, instead of EAN being released into the bulk during micellization, EAN forms ion clusters, which hinder the entropy increase [64]. With higher EAN concentration in water, the enthalpy of micellization decreases further to contribute to a lower free energy and facilitate micellization.

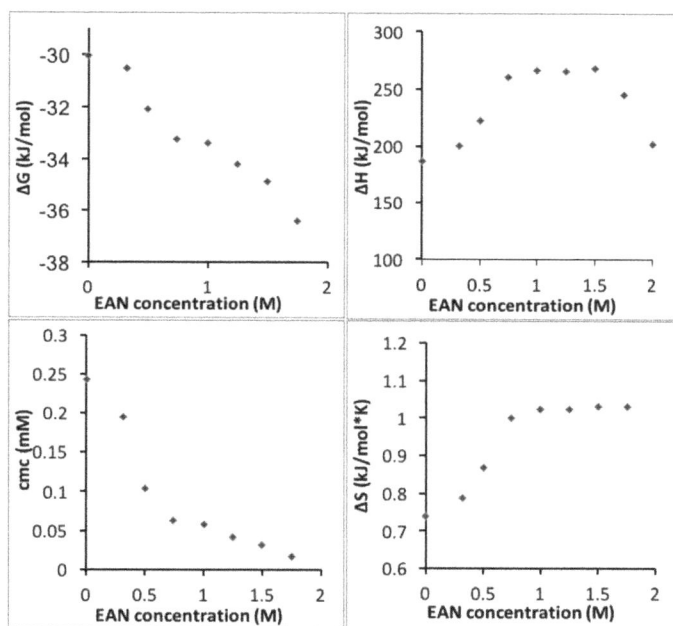

Figure 6. ΔG_{mic}, ΔH_{mic}, ΔS_{mic} and CMC of Pluronic P123 micellization in EAN aqueous solution plotted as a function of EAN concentration.

Both temperature increase and EAN addition promote Pluronic P123 micellization by disrupting the water solvation around PEO-PPO-PEO molecules. While the interactions between EAN and water can be affected by temperature, it is interesting to see how EAN addition affects Pluronic micellization at a higher temperature. We titrated micellar Pluronic P123 in EAN and water mixture into the same EAN and water mixture at 30 °C. The enthalpy change during titration shows that the addition of EAN at 30 °C further lowers the micellization enthalpy compared to the cases of either only increasing temperature or adding EAN to water (Figure 7). At the same concentration of EAN, the micelle formation region shifted to a lower Pluronic P123 concentration at 30 °C. For example, the CMC of Pluronic P123 with 0.5 M EAN addition is 0.12 mM at 20 °C, which shifted to a concentration much lower than 0.1 mM at 30 °C. This is in agreement with the literature suggesting that the hydrogen bonds between IL and hydrophilic block weakened as temperature increased and lead to enhancement of solvatophobic interactions and consequently decrease CMC [53]. The enthalpy of Pluronic P123 micellization in EAN aqueous solution can be determined by the extrapolation method described in Section 2. However, the very small micellization enthalpy (less than 30 kJ/mol) determined in this way may not be accurate due to the inability to determine the enthalpy of micelle dissociation as discussed in Section 3.1.

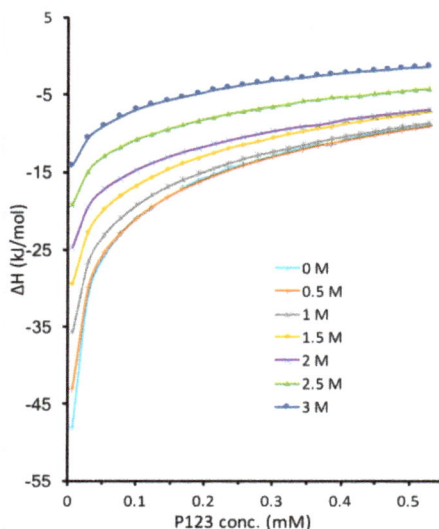

Figure 7. The enthalpy change of titrating 3.2 mM Pluronic P123 in EAN aqueous solution into EAN aqueous solution with the same content at 30 °C. The EAN concentration in each experiment is shown in the legend.

The polarity change in amphiphile solutions can reflect the solvent conditions as they affect the process of micelle formation. In order to measure the polarity of aqueous EAN Pluronic solution during micellization [12], pyrene fluorescence spectroscopy experiments were conducted for various concentrations of Pluronic P123 in EAN aqueous solutions. The higher the I_1/I_3 intensity ratio, the stronger the polarity of the medium is. We noticed that the I_1/I_3 intensity ratio of the solvent decreased linearly as with the EAN concentration, due to the less polar nature of EAN compared to water. The I_1/I_3 intensity ratio rapidly decreased with the Pluronic P123 concentration increase, caused by pyrene preferentially locating in the hydrophobic micelle core (Figure 8). The micropolarity in the micelle core did not change with EAN addition, suggesting that EAN does not partition in the PEO-PPO-PEO micelle core. CMC was determined by the breaking point of I_1/I_3 ratio from dramatically decreasing to reaching a plateau at higher Pluronic P123 concentration. CMC values obtained from pyrene fluorescence are consistent with the CMC values obtained from the analysis of ITC presented in Figure 6. To our best knowledge, there is no report available on pyrene fluorescence in EAN aqueous solutions.

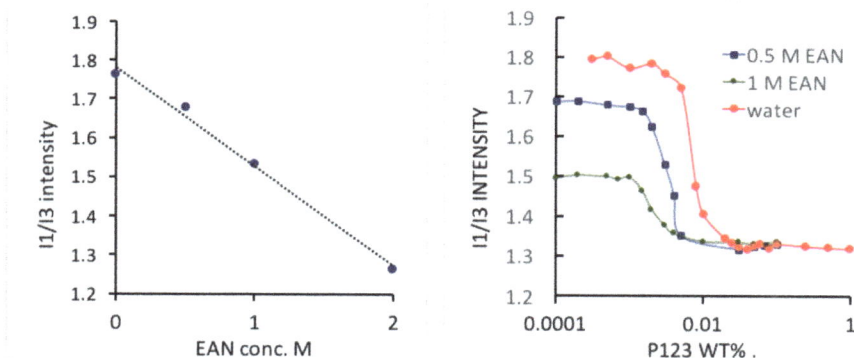

Figure 8. (**Left**) Pyrene fluorescence emission I_1/I_3 intensity ratio of EAN aqueous solutions; (**right**) pyrene fluorescence emission intensity I_1/I_3 ratio plotted against Pluronic P123 block copolymer concentration in aqueous solutions having different concentration of EAN.

3.3. Interactions between PEO-PPO-PEO Block Copolymer and EAN during Micellization

We evaluated in Section 3.2 the change of thermodynamic parameters of PEO-PPO-PEO block copolymer micellization in aqueous solution with EAN addition. We found that the micellization entropy and enthalpy increase at lower EAN concentrations and then decrease at higher EAN concentrations, which indicated that the interactions within this system may change as the EAN concentration increases. This three-component system comprises water, Pluronic and EAN; within this system each two components can interact and compete with the other component. It is reported that EAN behaves differently in water at varied concentrations [63]; the intermolecular interactions between EAN and PEO-PPO-PEO molecules were assessed during Pluronic micellization in pure EAN [37]. However, the interactions operating between EAN and Pluronic in aqueous solutions during micellization are still not established.

To assess the intermolecular interactions between PEO-PPO-PEO and EAN, we designed titration experiments as follows. First, we successively titrated aliquots (1.5 µL) 3.2 mM micellar Pluronic P123 in water into EAN aqueous solution (first titration), for which the observed enthalpy includes the enthalpy of EAN dilution in water and that of Pluronic interaction with EAN and water. Then, we kept all the other conditions the same and titrated the same amount (1.5 µL per injection) of water into EAN aqueous solution (second titration) in order to capture the enthalpy caused by EAN dilution. By subtracting the enthalpy of EAN dilution (second titration) from the total observed enthalpy from the first titration, the enthalpy caused by IL dilution in water can be excluded. By comparing the adjusted micellization enthalpy with that from titrating micellar Pluronic P123 into water (obtained in Section 3.1), we have a way to assess the enthalpy of EAN and Pluronic interactions [65,66] (Figure 9). Other researchers have used a similar method to probe intermolecular interactions between two components within a three- or more component system [67], for example, the effect of surfactant addition on Pluronic micellization [68].

In this analysis, we assume that the total observed enthalpy change (ΔH_{obs}) of titrating (Pluronic + water) into (water + EAN) equals the summation of the enthalpy of mixing between each of two components:

$$\Delta H_{obs} = \Delta H \text{ [EAN-water]} + \Delta H \text{ [EAN-P123]} + \Delta H \text{ [water-P123]}$$

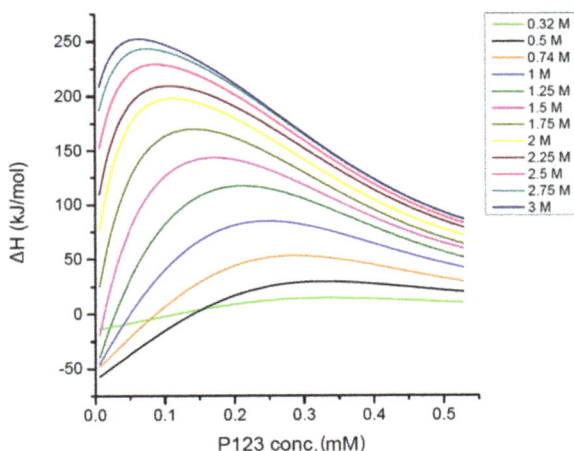

Figure 9. Enthalpy of Pluronic P123 and EAN interaction during micellization plotted as a function of Pluronic concentration. The interaction enthalpy is obtained by subtracting the EAN dilution enthalpy and the enthalpy of Pluronic in water from the observed enthalpy during titrating micellar Pluronic in water into EAN aqueous solution. The titrations were conducted at varied EAN concentrations as indicated in the legend.

However, during the actual PEO-PPO-PEO micellization in EAN aqueous solution, the three components interact simultaneously. The neglect of three-component interactions in the above-mentioned assumption might raise issues for accurate calculation of the Pluronic-EAN interaction enthalpy. Still, the trend of interactions between EAN and PEO-PPO-PEO during micellization at varied EAN concentration can be established in this way.

Below the CMC, EAN interacts with PEO-PPO-PEO unimers in the sample cell, and the interaction enthalpy increases with Pluronic P123 concentration. Above the CMC, EAN interacts with both unimers and micelles, the enthalpy of EAN-Pluronic interactions increases to a lesser extent and is further weakened by the dilution of EAN during titration. The maximum of the interaction enthalpy shifts to lower P123 concentration with increased EAN presence, which is the same trend as that of CMC. In this titration method, the EAN concentration in the sample cell is not constant during the titration; instead, EAN was diluted as more P123 aqueous solution was titrated into the sample cell. At the end of the titration, EAN was significantly diluted, which explains the decrease of the interaction enthalpy at the last several injections.

Different types of interactions in colloidal systems are not all additive. In colloidal systems, when we discuss interactions at distances smaller than tens of nanometers, the sizes of the colloidal particles are comparable to such small distances, and the discreteness of the material matters [69]. However, we can still obtain valuable insights of the interactions affecting the micellization process from the above discussed enthalpy change during P123 micellization. The exothermic process of Pluronic P123 micelle dilution and micelle dissociation in aqueous solution suggests a formation of water-Pluronic hydrogen bonds during the solvation of the Pluronic micelles and unimers. The presence of EAN affects Pluronic P123 micellization through electrostatic interactions and hydrogen bonding with water and with PEO-PPO-PEO molecules. It is known that the enthalpy of EAN mixing with water is endothermic [63], which indicates bond-breaking of the cation-anion interaction and of the water molecule network during the formation of water-EAN complexes. At lower EAN concentrations below CMC, the negative enthalpy of EAN-Pluronic interaction indicates the formation of hydrogen bonds between EAN and Pluronic unimers. It was reported that EAN is able to form a network of hydrogen bonds with PEO, which promotes segregation between PPO and PEO blocks [70].

The positive enthalpy (endothermic) of EAN-Pluronic P123 interactions during micelle formation (at higher Pluronic concentrations) indicates an enthalpy-unfavorable process, which may be due to a break of EAN-Pluronic hydrogen bonds and release of solvent into the bulk solution. The measured positive enthalpy of EAN-Pluronic interactions at higher EAN concentrations suggests that the addition of EAN disrupted the Pluronic-water interaction, resulting in a decrease of the fraction of hydrated methyl groups and consequently lowering the CMC. The dehydration effect of EAN on Pluronic aqueous solution is similar to a classic salt co-solute [66]. As the EAN concentration increases, the enthalpy of Pluronic-EAN interactions increases significantly (up to around 250 kJ/mol), which can explain the increase of the total observed enthalpy (from -375 to -110 kJ/mol for the first injection, shown in Figure 5) upon EAN addition.

4. Conclusions

The micellization thermodynamics of a representative Pluronic PEO-PPO-PEO amphiphilic block copolymer are evaluated here in the presence of the common protic ionic liquid ethylammonium nitrate (EAN). Isothermal titration calorimetry (ITC) experiments have been designed to obtain the thermodynamic parameters (including enthalpy, free energy and entropy) of Pluronic P123 ($EO_{20}PO_{70}EO_{20}$) micellization in EAN aqueous solution, from which the interactions between each component in this system can be assessed. The thermodynamic parameters of Pluronic P123 micellization in water at varied temperature are also examined as a reference system.

The presence of EAN promotes PEO-PPO-PEO block copolymer micellization in aqueous solution (with CMC for Pluronic P123 decreasing from 0.24 to 0.016 mM upon addition to water of 1.75 M EAN at 20 °C) through an increase in the entropy and/or decrease in enthalpy of the system, resulting in lower Gibbs free energy. The addition of EAN has a similar effect on micellization with increasing temperature. It is noted that at EAN concentrations lower than 1 M, the enthalpy and entropy of Pluronic P123 micellization increased with EAN concentration, whereas the enthalpy and entropy both decreased at EAN concentrations higher than 1 M, which can be attributed to EAN forming semi-ordered nano-domains in water.

The change of interactions between EAN and PEO-PPO-PEO during micellization at varying EAN concentrations was accessed through subtracting the EAN dilution enthalpy from the observed micellization enthalpy. We consider EAN as an additive and as a co-solvent, as it interacts with water and PEO-PPO-PEO molecules at the same time, which reflects the unique properties of ionic liquids during the micellization process. The observed enthalpy of Pluronic-EAN interactions suggests bond formation and breaking between EAN, Pluronic and water. EAN can also be viewed as a cosolvent that performs similar to glycerol in worsening the solvent condition and promoting micellization when added to water.

Our findings on the effects of EAN on PEO-PPO-PEO micellization in aqueous solution share similarities with the findings on the effect of salt addition. It would be interesting to explore further the similarities and differences between ionic liquids and classic salts on Pluronic micellization.

In this study, ionic liquid effects on amphiphilic block copolymer micellization thermodynamics in aqueous solution have been assessed through directly-measured enthalpy changes during this micellization process. The enthalpy of PEO-PPO-PEO–EAN interactions is reported for the first time. This improved understanding benefits applications concerning self-assembled or colloidal systems. The intermolecular interactions within self-assembly can guide the design of materials with improved properties. The tuning of micellization thermodynamics can benefit drug delivery with the selection of appropriate additives and solvents in use for nano-carrier assembly and disassembly. This work contributes to assessing the intermolecular interactions in ionic liquid-polymer solutions, which can be applied in biomass processing, polymer electrolytes [71] and membrane separations [72].

Acknowledgments: We thank Yingzhen Ma for assisting with ITC experiments and Yiyan Sui for her assistance with fluorescence experiments. We acknowledge the U.S. National Science Foundation for supporting research in our laboratory in the area of ionic liquids (Grants CBET 1033878 and 1159981).

Author Contributions: Zhiqi He and Paschalis Alexandridis conceived of and designed the experiments. Zhiqi He performed the experiments. Zhiqi He analyzed the data. Zhiqi He and Paschalis Alexandridis wrote the paper.

Conflicts of Interest: The authors declare no conflict of interest.

References

1. Pitto-Barry, A.; Barry, N.P.E. Pluronic block-copolymers in medicine: From chemical and biological versatility to rationalization and clinical advances. *Polym. Chem.* **2014**, *5*, 3291–3297. [CrossRef]
2. Alexandridis, P. Poly(ethylene oxide) poly(propylene oxide) block copolymer surfactants. *Curr. Opin. Colloid Interface Sci.* **1997**, *2*, 478–489. [CrossRef]
3. Alexandridis, P.; Lindman, B. *Amphiphilic Block Copolymers: Self-Assembly and Applications*; Elsevier Science B.V.: Amsterdam, The Netherlands, 2000.
4. Zhang, Y.M.; Song, W.T.; Geng, J.M.; Chitgupi, U.; Unsal, H.; Federizon, J.; Rzayev, J.; Sukumaran, D.K.; Alexandridis, P.; Lovell, J.F. Therapeutic surfactant-stripped frozen micelles. *Nat. Commun.* **2016**, *7*, 11649. [CrossRef] [PubMed]
5. Alexandridis, P.; Hatton, T.A. Poly(Ethylene Oxide)-Poly(Propylene Oxide)-Poly(Ethylene Oxide) Block-Copolymer Surfactants in Aqueous-Solutions and at Interfaces—Thermodynamics, Structure, Dynamics, and Modeling. *Colloids Surf. A* **1995**, *96*, 1–46. [CrossRef]
6. Kaizu, K.; Alexandridis, P. Effect of surfactant phase behavior on emulsification. *J. Colloid Interface Sci.* **2016**, *466*, 138–149. [CrossRef] [PubMed]
7. Alexandridis, P.; Yang, L. Micellization of polyoxyalkylene block copolymers in formamide. *Macromolecules* **2000**, *33*, 3382–3391. [CrossRef]
8. Sarkar, B.; Alexandridis, P. Block copolymer-nanoparticle composites: Structure, functional properties, and processing. *Prog. Polym. Sci.* **2015**, *40*, 33–62. [CrossRef]
9. Bodratti, A.M.; Sarkar, B.; Alexandridis, P. Adsorption of poly(ethylene oxide)-containing amphiphilic polymers on solid-liquid interfaces: Fundamentals and applications. *Adv. Colloid Interface Sci.* **2017**, *244*, 132–163. [CrossRef] [PubMed]
10. Linse, P.; Malmsten, M. Temperature-Dependent Micellization in Aqueous Block Copolymer Solutions. *Macromolecules* **1992**, *25*, 5434–5439. [CrossRef]
11. Sturcova, A.; Schmidt, P.; Dybal, J. Role of hydration and water coordination in micellization of Pluronic block copolymers. *J. Colloid Interface Sci.* **2010**, *352*, 415–423. [CrossRef] [PubMed]
12. Alexandridis, P.; Nivaggioli, T.; Hatton, T.A. Temperature Effects on Structural-Properties of Pluronic P104 and F108 PEO-PPO-PEO Block-Copolymer Solutions. *Langmuir* **1995**, *11*, 1468–1476. [CrossRef]
13. Alexandridis, P.; Holzwarth, J.F.; Hatton, T.A. Micellization of Poly(Ethylene Oxide)-Poly(Propylene Oxide)-Poly(Ethylene Oxide) Triblock Copolymers in Aqueous-Solutions—Thermodynamics of Copolymer Association. *Macromolecules* **1994**, *27*, 2414–2425. [CrossRef]
14. Verrna, P.; Nath, S.; Singh, P.K.; Kurnbhakar, M.; Pal, H. Effects of block size of pluronic polymers on the water structure in the corona region and its effect on the electron transfer reactions. *J. Phys. Chem. B* **2008**, *112*, 6363–6372. [CrossRef] [PubMed]
15. Mortensen, K. Structural studies of aqueous solutions of PEO-PPO-PEO triblock copolymers, their micellar aggregates and mesophases; a small-angle neutron scattering study. *J. Phys. Condens. Matter* **1996**, *8*, A103–A124. [CrossRef]
16. Alexandridis, P.; Yang, L. SANS investigation of polyether block copolymer micelle structure in mixed solvents of water and formamide, ethanol, or glycerol. *Macromolecules* **2000**, *33*, 5574–5587. [CrossRef]
17. Alexandridis, P.; Ivanova, R.; Lindman, B. Effect of glycols on the self-assembly of amphiphilic block copolymers in water. 2. Glycol location in the microstructure. *Langmuir* **2000**, *16*, 3676–3689. [CrossRef]
18. Sarkar, B.; Ravi, V.; Alexandridis, P. Micellization of amphiphilic block copolymers in binary and ternary solvent mixtures. *J. Colloid Interface Sci.* **2013**, *390*, 137–146. [CrossRef] [PubMed]
19. Kaizu, K.; Alexandridis, P. Micellization of polyoxyethylene-polyoxypropylene block copolymers in aqueous polyol solutions. *J. Mol. Liq.* **2015**, *210*, 20–28. [CrossRef]
20. Alexandridis, P.; Holzwarth, J.F. Differential scanning calorimetry investigation of the effect of salts on aqueous solution properties of an amphiphilic block copolymer (Poloxamer). *Langmuir* **1997**, *13*, 6074–6082. [CrossRef]

21. Kaizu, K.; Alexandridis, P. Glucose-induced sphere to ellipsoid transition of polyoxyethylene-polyoxypropylene block copolymer micelles in aqueous solutions. *Colloid Surf. A* **2015**, *480*, 203–213. [CrossRef]

22. Lin, Y.N.; Alexandridis, P. Cosolvent effects on the micellization of an amphiphilic siloxane graft copolymer in aqueous solutions. *Langmuir* **2002**, *18*, 4220–4231. [CrossRef]

23. He, Z.; Alexandridis, P. Ionic liquid and nanoparticle hybrid systems: Emerging applications. *Adv. Colloid Interface Sci.* **2017**, *244*, 54–70. [CrossRef] [PubMed]

24. Ghasemi, M.; Alexandridis, P.; Tsianou, M. Cellulose dissolution: Insights on the contributions of solvent-induced decrystallization and chain disentanglement. *Cellulose* **2017**, *24*, 571–590. [CrossRef]

25. Sivapragasam, M.; Moniruzzaman, M.; Goto, M. Recent advances in exploiting ionic liquids for biomolecules: Solubility, stability and applications. *Biotechnol. J.* **2016**, *11*, 1000–1013. [CrossRef] [PubMed]

26. Egorova, K.S.; Gordeev, E.G.; Ananikov, V.P. Biological Activity of Ionic Liquids and Their Application in Pharmaceutics and Medicine. *Chem. Rev.* **2017**, *117*, 7132–7189. [CrossRef] [PubMed]

27. Adawiyah, N.; Moniruzzaman, M.; Hawatulailaa, S.; Goto, M. Ionic liquids as a potential tool for drug delivery systems. *MedChemComm* **2016**, *7*, 1881–1897. [CrossRef]

28. Pereiro, A.B.; Araujo, J.M.M.; Esperanca, J.M.S.S.; Marrucho, I.M.; Rebelo, L.P.N. Ionic liquids in separations of azeotropic systems—A review. *J. Chem. Thermodyn.* **2012**, *46*, 2–28. [CrossRef]

29. Ventura, S.P.M.; Silva, F.A.E.; Quental, M.V.; Mondal, D.; Freire, M.G.; Coutinho, J.A.P. Ionic-Liquid-Mediated Extraction and Separation Processes for Bioactive Compounds: Past, Present, and Future Trends. *Chem. Rev.* **2017**, *117*, 6984–7052. [CrossRef] [PubMed]

30. Zhao, Y.; Bostrom, T. Application of Ionic Liquids in Solar Cells and Batteries: A Review. *Curr. Org. Chem.* **2015**, *19*, 556–566. [CrossRef]

31. He, Z.; Alexandridis, P. Nanoparticles in ionic liquids: Interactions and organization. *Phys. Chem. Chem. Phys.* **2015**, *17*, 18238–18261. [CrossRef] [PubMed]

32. Greaves, T.L.; Drummond, C.J. Solvent nanostructure, the solvophobic effect and amphiphile self-assembly in ionic liquids. *Chem. Soc. Rev.* **2013**, *42*, 1096–1120. [CrossRef] [PubMed]

33. Lopez-Barron, C.R.; Li, D.C.; Wagner, N.J.; Caplan, J.L. Triblock Copolymer Self-Assembly in Ionic Liquids: Effect of PEO Block Length on the Self-Assembly of PEO-PPO-PEO in Ethylammonium Nitrate. *Macromolecules* **2014**, *47*, 7484–7495. [CrossRef]

34. Tsoutsoura, A.; Alexandridis, P. Block copolymer ordered phases in ionic liquid solvents. *Polym. Mater. Sci. Eng.* **2010**, *102*, 168–169.

35. Fumino, K.; Wulf, A.; Ludwig, R. Hydrogen Bonding in Protic Ionic Liquids: Reminiscent of Water. *Angew. Chem. Int. Ed.* **2009**, *48*, 3184–3186. [CrossRef] [PubMed]

36. Werzer, O.; Warr, G.G.; Atkin, R. Conformation of Poly(ethylene oxide) Dissolved in Ethylammonium Nitrate. *J. Phys. Chem. B* **2011**, *115*, 648–652. [CrossRef] [PubMed]

37. Chen, Z.F.; Greaves, T.L.; Caruso, R.A.; Drummond, C.J. Amphiphile Micelle Structures in the Protic Ionic Liquid Ethylammonium Nitrate and Water. *J. Phys. Chem. B* **2015**, *119*, 179–191. [CrossRef] [PubMed]

38. Xie, L.L. Thermodynamics of AOT Micelle Formation in Ethylammonium Nitrate. *J. Dispers. Sci. Technol.* **2009**, *30*, 100–103. [CrossRef]

39. Fajalia, A.I.; Antoniou, E.; Alexandridis, P.; Tsianou, M. Self-assembly of sodium bis(2-ethylhexyl) sulfosuccinate in aqueous solutions: Modulation of micelle structure and interactions by cyclodextrins investigated by small-angle neutron scattering. *J. Mol. Liq.* **2015**, *210*, 125–135. [CrossRef]

40. Madhusudhana Reddy, P.; Venkatesu, P. Influence of ionic liquids on the critical micellization temperature of a tri-block co-polymer in aqueous media. *J. Colloid Interface Sci.* **2014**, *420*, 166–173. [CrossRef] [PubMed]

41. Khan, I.; Umapathi, R.; Neves, M.C.; Coutinho, J.A.P.; Venkatesu, P. Structural insights into the effect of cholinium-based ionic liquids on the critical micellization temperature of aqueous triblock copolymers. *Phys. Chem. Chem. Phys.* **2016**, *18*, 8342–8351. [CrossRef] [PubMed]

42. Benlhima, N.; Lemordant, D.; Letellier, P. Acidity scales in water-ethylammonium nitrate mixtures at 298 K. Solubility of organic compounds. *J. Chim. Phys. Phys. Chim. Biol.* **1989**, *86*, 1919–1939. [CrossRef]

43. Chen, Z.; Greaves, T.L.; Caruso, R.A.; Drummond, C.J. Effect of cosolvents on the self-assembly of a non-ionic polyethylene oxide-polypropylene oxide-polyethylene oxide block copolymer in the protic ionic liquid ethylammonium nitrate. *J. Colloid Interface Sci.* **2015**, *441*, 46–51. [CrossRef] [PubMed]

44. Zhang, G.D.; Chen, X.; Zhao, Y.R.; Ma, F.M.; Jing, B.; Qiu, H.Y. Lyotropic liquid-crystalline phases formed by Pluronic P123 in ethylammonium nitrate. *J. Phys. Chem. B* **2008**, *112*, 6578–6584. [CrossRef] [PubMed]

45. Atkin, R.; De Fina, L.M.; Kiederling, U.; Warr, G.G. Structure and Self Assembly of Pluronic Amphiphiles in Ethylammonium Nitrate and at the Silica Surface. *J. Phys. Chem. B* **2009**, *113*, 12201–12213. [CrossRef] [PubMed]

46. Abe, H.; Nakama, K.; Hayashi, R.; Aono, M.; Takekiyo, T.; Yoshimura, Y.; Saihara, K.; Shimizu, A. Electrochemical anomalies of protic ionic liquid—Water systems: A case study using ethylammonium nitrate—Water system. *Chem. Phys.* **2016**, *475*, 119–125. [CrossRef]

47. Prameela, G.K.S.; Phani Kumar, B.V.N.; Pan, A.; Aswal, V.K.; Subramanian, J.; Mandal, A.B.; Moulik, S.P. Physicochemical perspectives (aggregation, structure and dynamics) of interaction between Pluronic (L31) and surfactant (SDS). *Phys. Chem. Chem. Phys.* **2015**, *17*, 30560–30569. [CrossRef] [PubMed]

48. Loef, D.; Niemiec, A.; Schillen, K.; Loh, W.; Olofsson, G. A Calorimetry and Light Scattering Study of the Formation and Shape Transition of Mixed Micelles of $EO_{20}PO_{68}EO_{20}$ Triblock Copolymer (P123) and Nonionic Surfactant ($C_{12}EO_6$). *J. Phys. Chem. B* **2007**, *111*, 5911–5920. [CrossRef] [PubMed]

49. Thurn, T.; Couderc, S.; Sidhu, J.; Bloor, D.M.; Penfold, J.; Holzwarth, J.F.; Wyn-Jones, E. Study of Mixed Micelles and Interaction Parameters for ABA Triblock Copolymers of the Type EOm-POn-EOm and Ionic Surfactants: Equilibrium and Structure. *Langmuir* **2002**, *18*, 9267–9275. [CrossRef]

50. Naskar, B.; Ghosh, S.; Moulik, S.P. Solution Behavior of Normal and Reverse Triblock Copolymers (Pluronic L44 and 10R5) Individually and in Binary Mixture. *Langmuir* **2012**, *28*, 7134–7146. [CrossRef] [PubMed]

51. Porcedda, S.; Marongiu, B.; Schirru, M.; Falconieri, D.; Piras, A. Excess enthalpy and excess volume for binary systems of two ionic liquids plus water. *J. Therm. Anal. Calorim.* **2011**, *103*, 29–33. [CrossRef]

52. Bouchemal, K.; Agnely, F.; Koffi, A.; Ponchel, G. A concise analysis of the effect of temperature and propanediol-1, 2 on Pluronic F127 micellization using isothermal titration microcalorimetry. *J. Colloid Interface Sci.* **2009**, *338*, 169–176. [CrossRef] [PubMed]

53. Zhang, S.H.; Li, N.; Zheng, L.Q.; Li, X.W.; Gao, Y.A.; Yu, L. Aggregation behavior of pluronic triblock copolymer in 1-butyl-3-methylimidazolium type ionic liquids. *J. Phys. Chem. B* **2008**, *112*, 10228–10233. [CrossRef] [PubMed]

54. Merino-Garcia, D.; Andersen, S.I. Calorimetric evidence about the application of the concept of CMC to asphaltene self-association. *J. Dispers. Sci. Technol.* **2005**, *26*, 217–225. [CrossRef]

55. Liu, J.; Zhao, M.W.; Zhang, Q.; Sun, D.Z.; Wei, X.L.; Zheng, L.Q. Interaction between two homologues of cationic surface active ionic liquids and the PEO-PPO-PEO triblock copolymers in aqueous solutions. *Colloid Polym. Sci.* **2011**, *289*, 1711–1718. [CrossRef]

56. Evans, D.F.; Wennerström, H. *The Colloidal Domain: Where Physics, Chemistry, Biology, and Technology Meet*, 2nd ed.; Wiley-VCH: New York, NY, USA, 1999.

57. Tsui, H.W.; Hsu, Y.H.; Wang, J.H.; Chen, L.J. Novel Behavior of Heat of Micellization of Pluronics F68 and F88 in Aqueous Solutions. *Langmuir* **2008**, *24*, 13858–13862. [CrossRef] [PubMed]

58. Tsui, H.W.; Wang, J.H.; Hsu, Y.H.; Chen, L.J. Study of heat of micellization and phase separation for Pluronic aqueous solutions by using a high sensitivity differential scanning calorimetry. *Colloid Polym. Sci.* **2010**, *288*, 1687–1696. [CrossRef]

59. Su, Y.-L.; Liu, H.-Z.; Wang, J.; Chen, J.-Y. Study of Salt Effects on the Micellization of PEO-PPO-PEO Block Copolymer in Aqueous Solution by FTIR Spectroscopy. *Langmuir* **2002**, *18*, 865–871. [CrossRef]

60. Bedo, Z.; Berecz, E.; Lakatos, I. Enthalpy-Entropy Compensation of Micellization of Ethoxylated Nonyl-Phenols. *Colloid Polym. Sci.* **1992**, *270*, 799–805. [CrossRef]

61. Horn, R.G.; Evans, D.F.; Ninham, B.W. Double-layer and solvation forces measured in a molten salt and its mixtures with water. *J. Phys. Chem.* **1988**, *92*, 3531–3537. [CrossRef]

62. Jiang, W.; Wang, Y.; Voth, G.A. Molecular Dynamics Simulation of Nanostructural Organization in Ionic Liquid/Water Mixtures. *J. Phys. Chem. B* **2007**, *111*, 4812–4818. [CrossRef] [PubMed]

63. Zarrougui, R.; Dhahbi, M.; Lemordant, D. Transport and Thermodynamic Properties of Ethylammonium Nitrate-Water Binary Mixtures: Effect of Temperature and Composition. *J. Solut. Chem.* **2015**, *44*, 686–702. [CrossRef]

64. Velasco, S.B.; Turmine, M.; Di Caprio, D.; Letellier, P. Micelle formation in ethyl-ammonium nitrate (an ionic liquid). *Colloids Surf. A Physicochem. Eng. Asp.* **2006**, *275*, 50–54. [CrossRef]

65. Li, X.F.; Wettig, S.D.; Verrall, R.E. Isothermal titration calorimetry and dynamic light scattering studies of interactions between gemini surfactants of different structure and pluronic block copolymers. *Adv. Colloid Interface Sci.* **2005**, *282*, 466–477. [CrossRef] [PubMed]

66. Barbosa, A.M.; Santos, I.J.B.; Ferreira, G.M.D.; da Silva, M.D.H.; Teixeira, A.V.N.D.; da Silva, L.H.M. Microcalorimetric and SAXS Determination of PEO-SDS Interactions: The Effect of Cosolutes Formed by Ions. *J. Phys. Chem. B* **2010**, *114*, 11967–11974. [CrossRef] [PubMed]

67. Gianni, P.; Bernazzani, L.; Guido, C.A.; Mollica, V. Calorimetric investigation of the aggregation of lithium perfluorooctanoate on poly(ethyleneglycol) oligomers in water. *Thermochim. Acta* **2006**, *451*, 73–79. [CrossRef]

68. Wang, R.; Tang, Y.; Wang, Y. Effects of Cationic Ammonium Gemini Surfactant on Micellization of PEO-PPO-PEO Triblock Copolymers in Aqueous Solution. *Langmuir* **2014**, *30*, 1957–1968. [CrossRef] [PubMed]

69. Batista, C.A.S.; Larson, R.G.; Kotov, N.A. Nonadditivity of nanoparticle interactions. *Science* **2015**, *350*, 1242477. [CrossRef] [PubMed]

70. Chen, Z.; FitzGerald, P.A.; Kobayashi, Y.; Ueno, K.; Watanabe, M.; Warr, G.G.; Atkin, R. Micelle Structure of Novel Diblock Polyethers in Water and Two Protic Ionic Liquids (EAN and PAN). *Macromolecules* **2015**, *48*, 1843–1851. [CrossRef]

71. Wang, W.; Alexandridis, P. Composite Polymer Electrolytes: Nanoparticles Affect Structure and Properties. *Polymers* **2016**, *8*, 387. [CrossRef]

72. Lam, B.; Wei, M.; Zhu, L.; Luo, S.; Guo, R.; Morisato, A.; Alexandridis, P.; Lin, H.Q. Cellulose triacetate doped with ionic liquids for membrane gas separation. *Polymer* **2016**, *89*, 1–11. [CrossRef]

polymers

MDPI

Review

Self-Assembly of Block and Graft Copolymers in Organic Solvents: An Overview of Recent Advances

Leonard Ionut Atanase [1,2,*] and Gerard Riess [3]

1   Faculty of Dental Medicine, "Apollonia" University, 700399 Iasi, Romania
2   Research Institute "Academician Ioan Haulica", 700399 Iasi, Romania
3   University of Haute Alsace, Ecole Nationale Supérieure de Chimie de Mulhouse,
     Laboratoire de Photochimie et d'Ingénierie Macromoléculaires, 68093 Mulhouse CEDEX, France;
     gerard.riess@uha.fr
*   Correspondence: leonard.atanase@yahoo.com; Tel.: +40-741-686-687

Received: 20 November 2017; Accepted: 6 January 2018; Published: 11 January 2018

Abstract: This review is an attempt to update the recent advances in the self-assembly of amphiphilic block and graft copolymers. Their micellization behavior is highlighted for linear AB, ABC triblock terpolymers, and graft structures in non-aqueous selective polar and non-polar solvents, including solvent mixtures and ionic liquids. The micellar characteristics, such as particle size, aggregation number, and morphology, are examined as a function of the copolymers' architecture and molecular characteristics.

Keywords: self-assembly; micelle; organic solvents; block copolymers; graft copolymers; triblock terpolymers

1. Introduction

The self-assembly and micellization of block and graft copolymers with the formation of structured nanoparticles have attracted a major interest over the last decades. Since the pioneering studies of Tuzar and Kratochvil [1] on micellar systems in organic solvents, one of the major research trends afterward was to develop aqueous-based systems. These systems are of considerable interest in biomedical applications. This field has been quite extensively reviewed by different research groups [2–6]. During the same time period, the research activity on non-aqueous micellar systems was ongoing, but somehow to a minor extent, as outlined by the review articles published at the beginning of this century [7–9].

Therefore, the aim of the present review article is to update and highlight the recent advances related to the self-assembly of block and graft copolymers in non-aqueous media. This topic is one of interest not only from a theoretical point of view, but also due to its widespread application possibilities in membrane technology, surface modification of pigments and fillers, non-aqueous dispersion, lubricant additives, etc. In addition to these specific applications, non-aqueous micellar systems may have certain advantages with respect to the aqueous micellar dispersions. In fact, it has to be recalled that the micelle's characteristics generated by a polyA-*b*-polyB copolymer in the presence of selective aqueous or organic solvent S are depending to a large extent to the Flory-Huggins interaction parameter, such as χ_{AB}, χ_{AS} and χ_{BS}, which are the polymer/polymer and the polymer/solvent interaction parameters, respectively. These parameters are furthermore directly correlated with the solubility parameter δ of the compounds. In the common practice, with a homologues series of organic solvents, such as alcanes, esters, or alcohols, it will therefore be easier than for pure water to adjust the χ values of the solvent to χ values of the copolymer sequences. Finally, it is quite obvious that the self-assembly of hydrophobic/hydrophobic copolymers, which correspond to a large part of those synthesized up to now, may be achieved only in organic media.

In view of this background, the present literature survey from the last decade, including the authors' contributions, is structured as follows: (i) a brief section is provided in order to review the experimental techniques; (ii) the self-assembly of linear AB and ABA block copolymers in polar and non-polar solvents, as well as in specific solvent mixtures and in ionic liquids (ILs), are outlined; (iii) a major section is then devoted to ABC linear triblock terpolymers and their specific multi-compartment micellar morphologies; and, (iiii) the micellization aspects of graft copolymers and their possibility of forming unimolecular micelles will be described in the last section.

Out of scope of this review are the following topics: bottle-brush micelles; rod-coil copolymers; interpolymer complexes of block and graft copolymers; and, micellization in super critical carbon dioxide (CO_2).

2. Experimental Section

2.1. Synthesis and Molecular Characteristics

It is now accepted that well-defined block and graft copolymers can be obtained by living sequential and controlled radical polymerization techniques [10–16]. However, one has to bear in mind that even the so-called *well-defined* copolymers may present a non-negligible *polydispersity* in their composition and molecular weight for linear AB, ABA, and ABC block copolymers. The same type of limitation not only affects block copolymers, but graft copolymers as well. In addition, a *polydispersity* in graft density has to be considered for graft copolymers.

From a practical point of view, it is worth noting that AB block copolymers can also be synthesized by the so-called Polymerization Induced Self-Association (PISA) technique, which is carried out essentially in organic solvents. In-situ micellization is directly involved in this synthesis technique, as outlined recently by Armes and coworkers [17–21]. The determination of the copolymers' molecular characteristics, such as molecular weight, composition, and end-group functionality, is well-documented in recent review articles [10–12].

2.2. Solvents

It is well-established that the solubility of a given polymer is directly defined by the Flory-Huggins solvent/polymer interaction parameter (χ). However, this parameter is not always available. Therefore, for the selection of a selective solvent, which is required in order to induce the self-assembly of a block or graft copolymer, it is of common practice to use the concept of *"solubility parameter"* (δ) of solvents and polymers. Among the various solubility scales, the Hildebrand δ value is the most frequently used. The δ_t value involves the dispersive δ_d and polar δ_p, as well as the hydrogen bonding δ_h contributions, such as: $\delta_t = (\delta_d^2 + \delta_p^2 + \delta_h^2)^{1/2}$. For non-ionic solvents, and particularly for aliphatic hydrocarbon compounds, δ_p and δ_h are negligible with respect to δ_d. The δ_t values are therefore in the range of 15 to 18 $MPa^{1/2}$. This approach is advantageous for solvent mixtures, often used in micellization studies, because defining an *"average solubility parameter"* is possible. The usual non-aqueous solvents used in the micellization experiments are listed in Table 1, with an indication of their solubility parameter [22].

Table 1. Solubility parameters (δ) of the most usual non-polar and polar solvents.

Non-Polar Solvents	δ ($MPa^{1/2}$)	Polar Solvents	δ ($MPa^{1/2}$)
n-Hexane	14.9	Ethyl acetate	18.1
n-Heptane	15.2	Dichloromethane	19.8
n-Decane	15.8	Acetone	19.9
Hexadecane	16.3	THF	19.5–20.2
Cyclohexane	16.7	Isopropanol	23.6
Toluene	18.2	*n*-Propanol	24.4

Table 1. *Cont.*

Non-Polar Solvents	δ (MPa$^{1/2}$)	Polar Solvents	δ (MPa$^{1/2}$)
Chloroform	18.7	Acetonitrile	24.4
1,4-Dioxane	19.9–20.5	DMF	24.9
		DMSO	26.5
		Ethanol	26.5
		Methanol	29.7
		Water	47.9
		Ionic liquids	24.0–32.0

2.3. Self-Assembly Techniques

The self-assembly techniques and the micellar characterization methods used for amphiphilic block and graft copolymers have been described in several review articles [7–9,23]. The simplest method for the preparation of micelles is the direct dissolution of copolymer samples in a selective solvent for one of the sequences. It has to be kept in mind that this procedure is recommended only for copolymers with low molecular weights and short insoluble sequences. In another technique, the micellization occurs by the addition of a non-solvent to a common solvent, in which the copolymer is initially molecularly dispersed. However, this technique may induce aggregate formation. In order to avoid the formation of agglomerates, the dialysis technique, starting from a common solvent, is the recommended method for the micelle preparation. Notably, the nano-rings, an elaborate morphology, became available by surface induced self-assembly, by forming a thin film on specific surfaces, such as mica and silicone wafers. This procedure occurs through the solvent evaporation of the copolymer solution on the surface, which induces a *"frozen-in"* structure that is maintained after re-dissolution of the thin film [24]. From these studies, it turns out that, for a given copolymer sample, the self-association procedure has a major influence on its micellar characteristics, such as morphology, size, and size distribution.

It has to be noticed that some of these techniques may lead to kinetically *"frozen-in"* situations instead of a thermodynamic equilibrium between unimers and micelles, which are usually only observed when the core-forming polymer sequence has a low glass transition temperature (T_g). As a general remark, it appears from the present literature survey that a strict distinction is not always made between kinetic *"frozen-in"* self-assembled systems and micelles in thermodynamic equilibrium with their unimers.

Finally, by the "crystallization-driven self-assembly" (CDSA) method, intensively studied by the groups of Winnik and Manners for poly(ferrocenylsilanes)-based copolymers, especially cylindrical micelles can be obtained from crystallizable block copolymers having a crystalline core block that is much smaller than the corona sequence [25–27].

2.4. Characterization Techniques

The critical micellar concentration (CMC), as well as the critical micellization temperature (CMT), are fundamental characteristics of copolymers solutions in a unimer/micelle thermodynamic equilibrium. CMC values are usually accessible by different fluorescence techniques [28,29], and to a minor extent, by surface tension measurements. Moreover, it has to be noted that the CMC values determined for *"frozen-in"* systems might be questionable. The micellar hydrodynamic diameter, D_h, is determined by dynamic light scattering (DLS), transmission electron microscopy (TEM), and SAXS [30]. The micellar morphologies can be observed with different electron microscopy techniques, such as TEM and scanning electron microscopy (SEM). In particular, cryo-TEM and tomography has been applied by different authors [31–33] for the direct visualization of micellar nano-structures. In addition to these well-described size and morphology characterization techniques, other elaborate methods are available for the study of block copolymers self-assembly in organic media, such as super-resolution fluorescence microscopy technique, which is especially used for the visualization of cylindrical micelles obtained from crystallizable copolymers by the CDSA method [34].

3. AB and ABA Linear Block Copolymers in Organic Solvents

Within this overview of block copolymer self-assembly in organic solvents, the logical approach was to first consider the simplest molecular architectures of AB and ABA copolymers. This section provides a review of the micellar characteristics during the self-assembly of AB and ABA copolymers in non-polar and polar solvents, including their mixtures and ionic liquids (ILs), as well as in biocompatible solvents. This self-assembly is further described for AB copolymers, including crystallizable sequences.

The basic micellar morphologies correspond to spheres, cylinders, and vesicles, depending on the molecular characteristics of the copolymers, the selective solvent, and the micellization conditions. Figure 1 is an illustration of these basic morphologies.

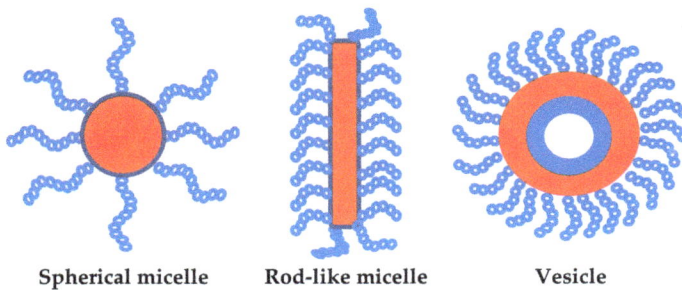

Spherical micelle   Rod-like micelle   Vesicle

Figure 1. Basic micellar morphologies for AB copolymers.

It has to be recalled that a given amphiphilic AB copolymer in a selective solvent of either the A or B block leads to two types of morphologies having either an A or B micellar core, respectively. This micellar structure inversion can also be created by thermal treatment, as demonstrated by Li et al. [35] for the self-assembly of poly(tert-butylmethacrylate)-*b*-poly[*N*-(4-vinylbenzyl)-*N*,*N*-diethylamine] (PtBMA-*b*-PVEA) copolymers in methanol. This possibility is schematically outlined in Figure 2:

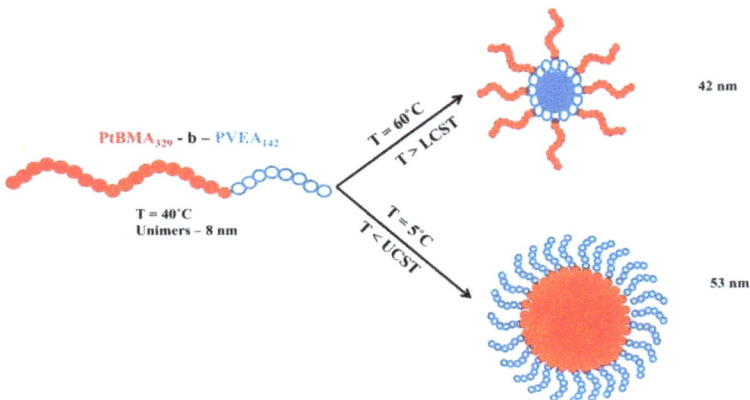

Figure 2. Micellization characteristics of PtBMA$_{329}$-*b*-PVEA$_{142}$ copolymer in methanol as a function of temperature (lower critical solution temperature (LCST) = 53 °C; upper critical solution temperature (UCST) = 32 °C). Adapted from Li et al. [35].

3.1. Self-Assembly of AB and ABA Block Copolymers in Non-Polar Selective Solvents

Micellization of well-defined AB block copolymers, such as poly(butadiene)-*b*-poly(styrene) (PB-*b*-PS) and poly(isoprene)-*b*-poly(styrene) (PI-*b*-PS), has been studied quite extensively in the second half of the 20th century [1,7,36–39]. From the relatively few papers published on this topic since then, the objectives of the authors were mainly to examine specific aspects of the self-assembly of AB diblock copolymers in non-aqueous media. An interesting example of this type of study was that of Sotiriou et al. [37]. For a series of PS-*b*-PI diblock copolymers, tagged with a ω-lithium sulfonate (SO$_3$Li) end group on either the PS or PI sequence, these authors examined the micellization behavior in *n*-decane as a function of the localization of the polar end-group. A micellar solution was obtained by direct dissolution and heating the block copolymer samples in *n*-decane. The principal micellar characteristics of these systems at 25 °C are provided in Table 2.

Table 2. Micellar characteristics for the PS-*b*-PI diblock copolymers in *n*-decane at 25 °C as determined by Sotiriou et al. [37].

Sample	Nagg	Rh (nm)
PS$_{60}$-*b*-PI$_{203}$	10.2	10.4
(SO$_3$Li)-PS$_{60}$-*b*-PI$_{203}$	23.4	14.2
PS$_{55}$-*b*-PI$_{151}$-(SO$_3$Li)	17.0 11.0 [b]	660.0 [a] 75.0 [b]

[a] large *polydispersity*. [b] values at 30 °C.

These results prove that the position of the SO$_3$Li end-group has a considerable influence on the micellar characteristics. The SO$_3$Li end-group fixed on the PS sequence, which forms the micellar core, leads to an increase in both N_{agg} and R_h, with respect to the unlabeled sample. This effect may be attributed to the dipolar interaction of the SO$_3$Li groups in a non-aqueous medium. For the SO$_3$Li end-groups on the PI sequence, this interaction occurs in the solvent phase between the micelles as a result of the formation of polydisperse and interconnected SO$_3$Li large aggregates with the PS cores.

Later on, Cheng et al. [40] studied the morphological changes of non-functionalized PS-*b*-PI copolymer in *n*-decane as a function of temperature and pressure using small-angle neutron spectroscopy (SANS). According to the authors, the increase in pressure from 200 to 16,000 psi, at room temperature, had no effect on the micellar characteristics (N_{agg}, R_g, and R_{core}), but led to the formation of micellar agglomerates. At 60 °C and high pressure, the micelles underwent a macro-phase separation with the formation of sheet-like aggregates. The authors indicated that these morphological changes are attributed to the decrease of the "*n-decane quality*" for the PI sequences.

Growney et al. [41] examined the self-assembly of poly(styrene)-*b*-poly(ethylene propylene) (PS-*b*-PEP) diblock copolymers that are obtained by the hydrogenation of the PI sequence of PS-*b*-PI copolymers. As selective non-polar solvents of PEP, either *n*-heptane (δ = 15.2 MPa$^{1/2}$) or *n*-dodecane (δ = 16.0 MPa$^{1/2}$) was used, and star-like micelles with an R_h of around 40 nm were obtained in both of the solvents. The self-association of another type of PI-based diblock copolymer, such as poly(ethylene oxide)-*b*-poly(isoprene) (PEO-*b*-PI), was investigated by Bartels et al. [42] in *n*-decane. In this case, the micellar solutions were obtained using a precipitation method starting from a homogeneous copolymer solution in tetrahydrofuran (THF).

Wang et al. [43] observed by DLS a bimodal distribution attributed to unimers and spherical micelles for the self-assembly of a series of poly(styrene)-*b*-poly(2-vinyl pyridine) (PS-*b*-P2VP) diblock copolymers in the presence of a pure non-polar solvent, such as toluene. Moreover, these authors investigated the micellar aggregate formation by mixing different architectures of this type of copolymers, such as linear diblocks and triblocks, as well as branched star-like copolymers.

Finally, Arai et al. [44] studied the self-aggregation behavior of PS-based diblock copolymers in chloroform and 1,2-dichloroethane (DCE), using DLS and SLS. The micellar characteristics a series of poly(styrene)-*b*-poly[(ar-vinylbenzyl)trimethylammonium chloride] (PS-*b*-PV) diblock copolymers, obtained by RAFT polymerization, were directly correlated with the number average polymerization degree (DP_n) ratio of the two blocks. For example, star-like micelles were observed for a ratio $DP_n(PS)/DP_n(PV) > 6$, whereas brush-like micelles were obtained if this ratio was smaller than 6.

3.2. Self-Assembly of AB and ABA Block Copolymers in Polar Selective Solvents

In this section, the recent micellization studies completed for AB and ABA block copolymers in different pure polar solvents are summarized (Table 3).

Table 3. Micellization studies of AB block copolymers in pure polar solvents.

AB Block Copolymers	Selective Solvent for A Block	Selective Solvent for B Block	Micellar Characteristics	Ref.
PS$_{48}$-*b*-PNVP$_{99}$	-	CH$_3$–OH	R_h = 16 nm R_g/R_h = 0.65 C.M.C. = 0.13 mg·mL^{-1} N_{agg} = 10	[45]
POSS-PMMA$_{144}$-*b*-P(MA-POSS)$_{2.6;9.6;11.3}$	THF	-	R_h = 85; 148; 80 nm Core-shell micelles	[46]
PMMA$_{340;400}$-*b*-PtBMA$_{134}$	-	2-ethylhexanol	R_h = 19.4 ÷ 28.2 nm Spheres and cylinders	[47]
PSAMA$_{15}$-*b*-P(Boc-Phe-HEMA)$_{7;17;37;75}$	-	DMF; DMSO; ACN	$R_{h(DMF)}$ = 119 ÷ 318 nm $R_{h(DMSO)}$ = 37 ÷ 90 nm $R_{h(ACN)}$ = 24 ÷ 48 nm Spheres	[48]
PEtOx$_{10}$-*b*-PNBA$_{7;17;31;48}$	-	CH$_2$Cl$_2$	R_h = 45 ÷ 60 nm Spheres	[49]
PVAc$_{57}$-*b*-(PFHE-stat-PVAc)$_{95}$	CH$_3$–OH	-	R_h = 10 ÷ 40 nm Spheres	[50]
PMMA$_{41;54;73}$-*b*-PsfMA$_{59;46;27}$	THF; ACN; CHCl$_3$	-	$N_{agg(THF)}$ = 8 $N_{agg(CHCl3)}$ = 26 $N_{agg(ACN)}$ = 410 $R_{h(THF)}$ = 60–70 nm Spheres	[51]

The copolymers are designated by PX$_{m,n}$-PY$_{m',n'}$, where m, n, m' and n' are the DP_n values. PNVP: poly(*N*-vinylpyrrolidone); POSS-PMMA: polyhedral oligomeric silsesquioxane-poly(methyl methacrylate); PSAMA: poly(2-(methacryloyloxy)ethyl stearate); P(Boc-Phe-HEMA): poly(tert-butyloxycarbonyl phenylalanine methacryloyloxyethyl ester); PEtOx: poly(2-ethyl-2-oxazoline); PNBA: poly(2-nitrobenzyl acrylate); PVAc: poly(vinyl acetate); PFHE: poly(perfluorohexylethylene); PsfMA: poly(1H,1H,2H,2H-perfluorodecyl methacrylate).

From Table 3, it appears that the interest of the authors was focused on the (meth)acrylic- and vinyl ester-based copolymers that might have practical industrial application possibilities. Furthermore, it can be noticed that solvents having a solubility parameter in the range of 18 to 30 MPa$^{1/2}$ (see Table 1), and in particular methanol, are suitable selective solvents.

3.3. Self-Assembly of AB and ABA Block Copolymers in Organic Solvent Mixtures

An alternative to pure organic solvents is provided by the mixtures of two solvents, which may allow for a gradual variation of the *"solvent quality"*. Sophisticated morphologies of micellar nano-particles become accessible using this typical approach.

The influence of the *"solvent quality"* on micellar characteristics was studied by Cho et al. [52] for a poly(styrene)-*b*-P4VP:poly(4-vinyl pyridine) (PS$_{400}$-*b*-P4VP$_{167}$) diblock copolymer sample in pure THF, THF/water, and THF/ethanol mixtures, respectively. The R_h of the micelles, with a P4VP core and PS corona, was around 23 nm in pure THF, significantly decreased to 14.1 nm for a solvent mixture composition of 95/5 *v/v*% THF/EtOH. Moreover, in the presence of ethanol, which is a *"good solvent"* for the P4VP core, micellar size *polydispersity* increased, whereas the aggregation number decreased.

Zhou et al. [53] investigated the self-assembly of poly(styrene)-*b*-PHFBMA:poly(2,2,3,3,4,4, 4-heptafluorobutyl methacrylate) (PS-*b*-PHFBMA) copolymers in a mixture of THF and ethyl acetate (EtOAc) at different volume ratios, such as 5:0; 4:1; 3:2; and, 0:5. For these mixtures, the authors calculated the solubility parameters in order to study their influence on the copolymer's self-aggregation behavior. As observed with TEM, the micellar morphology changed from spheres to vesicles with the increase in EtOAc content. By DLS, it turns out that the average size of the micelles increased from 140 to 190 nm, 235 and 267 nm, respectively, as a function of EtOAc volume fraction. Moreover, it was demonstrated by these authors that the morphology and the size of the micelles was highly influenced by the temperature.

Wang et al. [54] studied the self-aggregation of a PS_{64}-*b*-PEO_{827} diblock copolymer in a mixture of cyclohexane/1,4-dioxane (80/20 wt %), using DLS and TEM. For the preparation of the micellar solution, the copolymer was at first directly dissolved in 1,4-dioxane and then the cyclohexane was added very slowly. Spherical micelles with a D_h of 50 nm were obtained from this procedure. A micellar morphological transformation was observed, from spheres to cylinders and vesicles, by decreasing the temperature from 25 to 0 °C and then to −10 °C. This transformation from spheres to vesicles, attributed to the increasing interfacial energy between the solvent and the PEO core, was highly temperature dependent and accompanied by an increase in the D_h from 50 to 1680 nm.

In addition to DLS and TEM techniques, the micellar morphological modifications were confirmed with SAXS. In this context, Rao et al. [55] investigated the self-aggregation of a PS_{481}-*b*-$P2VP_{157}$ diblock copolymer in a mixture of dimethylformamide (DMF) and methanol in order to prepare micelle-functionalized silica particles. Core-shell spherical micelles, with a R_h value of 22.5 nm, were obtained by direct dissolution of the copolymer in DMF, as a common solvent, followed by a slow addition of methanol.

Choi et al. [56] studied the self-assembly of PS_{404}-*b*-PEP_{886} diblock copolymers in a mixture of 1-phenyldodecane ($\delta = 17.4$ MPa$^{1/2}$)/squalane ($\delta = 16.6$ MPa$^{1/2}$). At 110 °C and a 50/50 ratio of 1-phenyldodecane/squalane, the micellar R_h was equal to 280 nm, which is much higher than the value of 40 nm determined by Growney et al. [41] for the PS_{315}-*b*-PEP_{1203} micelles in *n*-heptane, at the same temperature. Due to this comparison, solvents, such as heptanes and *n*-decane with δ values of 15 to 16 MPa$^{1/2}$ appear to be more selective for PEP than the 1-phenyldodecane/squalane mixture, with an average δ value of around 17 MPa$^{1/2}$.

3.4. Self-Assembly of AB and ABA Block Copolymers in Ionic Liquids

Ionic liquids (ILs), which are a special class of polar solvents, are interesting for environmental reasons when compared to common volatile organic solvents. In addition, excellent chemical and thermal stability, wide liquid temperature ranges, and low toxicity are the most important typical properties of ionic liquids. Due to these specific properties, ILs have become efficient solvents for the synthesis of block copolymers using the PISA technique, as outlined in a recent review by Derry et al. [57]. Moreover, the self-assembly of block copolymers in ILs has led to the development of original micellar structures. In connection with this topic, the recent relevant publications are listed in Table 4.

Table 4. Self-assembly studies of AB block copolymers in pure ionic liquids.

AB Block Copolymers	Selective Solvent for A Block	Selective Solvents for B Block	Micellar Characteristics	Ref.
PEO_{432}-*b*-$PNIPAM_{106}$	[BMIM][BF4] $\delta = 24$ MPa$^{1/2}$	-	$R_h = 25$ nm L.C.M.T = 200 °C U.C.M.T = 60 °C	[58]
$PS_{529;548;981}$-*b*-$P2VP_{543;543;923}$	-	[BMIM][CF3SO3] $\delta = 25$ MPa$^{1/2}$	Spherical micelles	[59]
PEO_{341}-*b*-P(AzoMA-*r*-NIPAM)$_{177}$	[C4MIM][PF6] $\delta = 23$ MPa$^{1/2}$	-	$R_h \sim 120$ nm	[60]
PEO_{432}-*b*-$PnBMA_{99;183}$	[BMIM][TFSI] $\delta = 26$ MPa$^{1/2}$ [EMIM][TFSI] $\delta = 27$ MPa$^{1/2}$	-	C.M.T = 120–150 °C	[61]

Table 4. *Cont.*

AB Block Copolymers	Selective Solvent for A Block	Selective Solvents for B Block	Micellar Characteristics	Ref.
$PEGE_{109;113;104}$-b-$PEO_{54;115;178}$ $PGPrE_{98}$-b-PEO_{260}	PAN; EAN $\delta = 25\text{–}26$ MPa$^{1/2}$	-	spherical micelles for PEO_{178}-b-$PEGE_{104}$ Disk-shape micelles for PEO_{54}-b-$PEGE_{109}$	[62]
$PEGE_{104}$-b-PEO_{178}	[C4MIM][PF6] $\delta = 23$ MPa$^{1/2}$	-	$R_h = 13$ nm	[63]
$PMMA_{250}$-b-$PnBMA_{92;169;218;246;310;373;549}$	[EMIM][TFSI] [BMIM][TFSI]	-	$R_h = 17.8 \div 34.6$ nm $R_{core} = 5.8 \div 23.2$ nm $N_{agg} = 41 \div 432$	[64,65]

L.C.M.T and U.C.M.T: lower and upper critical micellization temperature; PNIPAM: poly(*N*-isopropylacrylamide); P(AzoMA): poly(4-phenylazophenyl methacrylate); PEGE: poly(ethyl glycidyl ether); PGPrE: poly(glycidyl propyl ether); [BMIM][CF3SO3]: 1-butyl-3-methylimidazolium trifluoromethanesulfonate; [C4MIM][PF6]: 1-butyl-3-methylimidazolium hexafluorophosphate; EAN: ethylammonium nitrate; PAN: propylammonium nitrate.

In addition to Table 4, typical and detailed examples are given to highlight the correlation between the micellar characteristics and the major system parameters, such as the relative sequence length of the copolymers, and, in particular, the respective solvent/copolymer solubility parameter.

Simone and Lodge [66] studied, by cryo-TEM and DLS, the self-assembly of a series of three PS-*b*-PMMA block copolymers with different compositions in 1-butyl-3-methylimidazolium hexafluorophosphate [BMIM][PF6]. This ionic liquid, with a δ value of 30 MPa$^{1/2}$, is a selective solvent for the PMMA block. The reduction of the PMMA content leads to a morphological transition from spherical to cylindrical micelles. More recently, for the same type of hydrophobic/hydrophobic PS-*b*-PMMA block copolymer, Mok et al. [67] investigated the effect of the composition on the C.M.C in 1-ethyl-3-methylimidazolium bis(trifluoromethylsulfonyl)imide [EMIM][TFSI], an IL with a δ value of 27 MPa$^{1/2}$. A decrease in the C.M.C values from 0.40 to 0.078 wt % was observed when the DP_n of the PS-core increased from 29 to 106.

The micellization in [BMIM][PF6] ionic liquid of a series of hydrophobic/hydrophilic PB-*b*-PEO diblock copolymers, with fixed PB and increasing PEO sequence lengths, was investigated by He et al. [68]. By increasing the molar fraction of the PEO, micellar morphology evolved from worm-like micelles and bilayered vesicles to spheres. An illustration of these morphologies determined by cryo-TEM is provided in Figure 3.

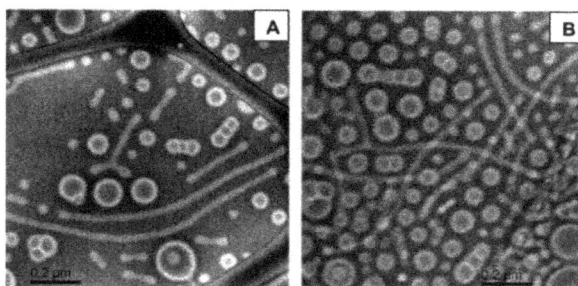

Figure 3. Cryo-TEM images of PB-*b*-PEO copolymer, with 0.25 mol % PEO, at a concentration of 1 wt % in [BMIM][PF6]. (**A**) Worm-like micelles with occasional Y-junctions and (**B**) micellar overlap. "Reprinted with permission from He, Y.; Li, Z.; Simone, P.; Lodge, T.P. Self-assembly of block copolymer micelles in an ionic liquid. *J. Am. Chem. Soc.* **2006**, *128*, 2745–2750. Copyright 2017 Americal Chemical Society".

More recently, the micellization of a quite similar PB-*b*-PEO copolymer series was investigated by Meli et al. [69] in [EMIM][TFSI] ionic liquid. These authors demonstrated that the direct dissolution of the copolymer led to the formation of large aggregates. However, a thermal treatment at 170 °C induced the formation of spherical micelles with R_h values of around 29 nm. For a given copolymer sample, it was of interest to compare the influence of the solubility parameter of both the PB-core and the solvent on the micellar characteristics. For this purpose, the authors studied the self-aggregation in a different IL, such as 1-butyl-3-methylimidazolium bis-(trifluoromethyl sulfonyl)imide [BMIM][TFSI], which has a slightly smaller δ value (26.7 MPa$^{1/2}$ when compared to 27.6 MPa$^{1/2}$ for [EMIM][TFSI]). Similar micellar characteristics were obtained for these two ILs. This experiment, performed with two different ILs with similar δ values, confirms that the δ value of the solvent is the key parameter in the macromolecular self-association process.

The self-assembly of ABA copolymers in ILs was predominantly investigated for PEO-*b*-PPO-*b*-PEO copolymers, also designated as Pluronics. Zhang et al. [70] determined the C.M.C values of three Pluronics with a constant PPO sequence length, such as L61 (PEO$_3$-PPO$_{30}$-PEO$_3$), L64 (PEO$_{13}$-PPO$_{30}$-PEO$_{13}$) and F68 (PEO$_{79}$-PPO$_{30}$-PEO$_{79}$), in both 1-butyl-3-methylimidazolium tetrafluoroborate [BMIM][BF4] (δ = 24 MPa$^{1/2}$) and [BMIM][PF6] (δ = 30 MPa$^{1/2}$) ionic liquids. These authors found that the critical micellar concentrations increased as expected with the PEO sequence length. As an extension of these results, Lopes-Barron et al. [71] studied the self-association in deuterated ethylammonium nitrate (dEAN) of a similar series of Pluronics, such as F127 (PEO$_{106}$-PO$_{70}$-PEO$_{106}$), P123 (PEO$_{20}$-PPO$_{70}$-PEO$_{20}$) and L121 (PEO$_5$-PPO$_{70}$-PEO$_5$). Pluronic samples with higher PEO/PPO molar ratios (F127 and P123) promoted the formation of spherical micelles, whereas small PEO/PPO ratios (L121) favor the formation of vesicles.

3.5. Self-Assembly of AB and ABA Block Copolymers in Biocompatible Organic Solvents

Among the above-mentioned non-aqueous block copolymer micellar systems, a large number may be considered as biocompatible. This especially occurs with block copolymer micelles in saturated aliphatic hydrocarbon solvents, such as *n*-decane, dodecane, etc. Another typical example of a non-aqueous self-assembly study was reported by Miller et al. [72] for poly(caprolactone)-*b*-poly(2-vinylpyrridine) (PCL-*b*-P2VP) copolymers in oleic acid, a biocompatible natural fatty acid. Spherical micelles, with a PCL core and an average size of 144 nm, were observed from the cryo-TEM images. Moreover, these authors investigated the loading of two model proteins into this micellar system.

Our research group studied the micellization of P2VP-*b*-PEO diblock copolymers in several biocompatible solvents, such as PEG400 (δ = 21.3 MPa$^{1/2}$), paraffin oil (δ = 15.3 MPa$^{1/2}$), and Miglyol 812 (δ = 17.3 MPa$^{1/2}$) [73–75]. Spherical micelles, having a P2VP core and a R_h in the range of 23 to 25 nm were obtained by the self-assembly of P2VP$_{37}$-*b*-PB$_{189}$ copolymer sample in paraffin oil and Miglyol 812, which is a glycerine ester. However, micelles with a higher R_h of around 60 nm and a PB core, were formed in PEG400. The driving force for the self-assembly of these diblock copolymers is the polymer/solvent interaction parameter χ. Notably, block copolymer micellar systems based on natural oils, such as Miglyol 812, may be used for biomedical or cosmetic applications [73].

3.6. Crystallization-Induced Self-Assembly of AB and ABA Block Copolymers in Organic Solvents

Block copolymers that include a crystallizable sequence may lead to micelles having a partially crystallized core. This so-called "crystallizable-driven self-assembly" (CDSA) method was reviewed by different authors [76,77], and more recently, by Tritschler et al. [27]. This process, involving phase separation above the melting temperature (T_m) and crystallization upon cooling, leads to a partially crystallized micellar core that is stabilized by the soluble sequence of the copolymer. By using this method, the groups of Winnik and Manners have obtained sophisticated and precise rod-coil micellar structures in organic solvents with well-controlled dimensions [27–29]. The width and the shape of these non-spherical micelles could be modified by varying the DP_n of the crystalline micellar core, the composition of the corona or the experimental conditions.

An example of crystallizable PB$_{54}$-*b*-PEO$_{61}$ block copolymer was studied by Mihut et al. [78] in *n*-heptane, which is a *"good solvent"* for PB. At 70 °C micelles are formed with a PEO core and PB corona, with an R_h of 12 nm. Upon cooling to 20 °C, a size increase to 140 nm was observed by DLS as a direct consequence of the PEO core crystallization. A similar example of crystallization-induced self-aggregation concerns the micellar solutions of poly(methylene)-*b*-poly(acrylic acid) (PM-*b*-PAA) block copolymer in DMF, which was reported by Wang et al. [79]. At 80 °C, the copolymer is molecularly dispersed in DMF. On cooling, the PM block crystallizes and self-aggregates into well-defined disk-like structures. In a further recent example, another research group [80] has studied the self-assembly of P2VP-*b*-PEO in a *n*-amyl acetate ($\delta = 17.4$ MPa$^{1/2}$)/*n*-butanol ($\delta = 23.1$ MPa$^{1/2}$) mixture. At 35 °C, these authors observed the formation of spherical micellar morphologies having a partially crystallizable PEO corona with a size of around 200 nm. With a fast temperature decrease, well-defined single crystals were obtained by crystallization of the PEO block. Recently, the groups of Winnik and Manners [81] studied the self-assembly of two amphiphilic crystalline-coil polyferrocenyldimethylsilane-*b*-poly(*N*-isopropylacrylamide) (PFS-*b*-PNIPAM) diblock copolymers in methanol, ethanol, and 2-propanol. Spherical micelles were formed in methanol and ethanol for the PFS$_{56}$-*b*-PNIPAM$_{190}$ sample whereas a mixture of spherical and cylindrical structures were noticed for this sample in 2-propanol. This behavior, which was also observed for PFS-*b*-P2VP copolymers, was probably due to the different solubility of the PFS in these solvents.

3.7. Concluding Remarks

To make conclusions about the literature survey concerning the self-assembly of polyA-*b*-polyB diblock copolymers, the aim of the present section is to provide a critical analysis of the published results. Moreover, our intention is to highlight the recent advances and trends that were observed in different steps of the self-assembly process of diblock copolymers in organic solvents.

For AB and ABA diblock copolymers, the recent investigations were focused on the synthesis of *"well-defined"* samples, with low *polydispersity* indices in composition and molecular weight. Although sequential anionic polymerization remains a favorite and well adapted synthesis technique, the so-called *"controlled free radical"* methods (RAFT, ATRP, NMP …) are now used to a large extent. In fact, these recent techniques have the advantage to provide access to a broader range of block copolymer types, in particular to those based on polar monomers. Among the recent synthesis trends, mention has to be made for the PISA process, which involves, in the polymerization step, the formation of a micellar system in organic solvent. A further synthesis trends concerns the preparation of end-functionalized and fluorescent labeled AB diblock copolymers. With respect to the determination of the *molecular characteristics*, NMR and SEC remain the *"classical"* analytic techniques. Unfortunately, quite a number of studies are published with indication of the *"equivalent PS"* number and weight average molecular weights of their products. Precise and actual molecular weight values of the copolymer sample would be accessible by SEC and simultaneous determination of the intrinsic viscosity [η]. Such multi-detector SEC devices with so-called *"universal calibration"* are presently standard equipments. In addition to SEC technique, diffusion ordered spectroscopy (DOSY), recently reviewed by Groves [82], could be a valuable tool to detect polyA and/or polyB homopolymer *"impurities"* in a synthesized polyA-*b*-polyB block copolymer.

As already previously mentioned in Section 2, most of the published micellar systems may be considered from a thermodynamic point of view as non-equilibrium systems between unimers and micelles. The characteristics, such as size and morphology of these *"frozen-in"* nanoparticles will depend to a large extend on their preparation procedure. The experimental conditions may in fact vary from a simple dissolution of the sample to an elaborated precipitation procedure in a selective solvent of either the polyA or polyB block of the copolymer. It has to be recalled that the most usual procedure consists in the solubilization of the copolymer in a common solvent. To this molecular dispersion is then added drop-wise the selective solvent of either the A or B block and the common solvent is then eliminated, in general, by dialysis, in order to end up with micelles dispersed in a pure selective solvent.

Among the recent trends in *self-assembly techniques*, mention has to be made of the preparation procedure involving *solvents mixtures*, either mixtures of two organic solvents or of an organic solvent and water. This type of approach has the advantage that the solubility parameter and the solvent/polymer interaction parameter χ may be triggered step by step as a function of the volume fraction of the two solvents in presence. From the studies concerning the *self-assembly in organic solvents mixtures*, mentioned in Section 3.3, it could be noticed that these mixtures have a strong influence on particle size and morphology. A first type of approach, illustrated by Zhou et al. [53], is to generate a solvent mixture, having a solubility parameter δ_{mix}, adapted as a selective solvent for either the A of B block of the AB copolymer. In a first approximation, it may be assumed that $\delta_{mix} = \emptyset_1\delta_1 + \emptyset_2\delta_2$, with $\emptyset_1$ and $\emptyset_2$ the volume fractions and δ_1, δ_2 the solubility parameters of the two solvents. The major limitation of this concept is that both solvents have to be selective solvents of either the A of the B block. If this requirement is not met, then a partition of the solvents may occur, which leads to a swelling of the micellar core by either one or both solvents in presence. A second possibility to use organic solvent mixtures in the self-assembly process of polyA-*b*-polyB copolymers was investigated by different authors [54–56]. Their approach consists in the solubilization of the copolymer in a common solvent S, a "*good*" solvent for both polyA and polyB sequences. To this molecularly dispersed copolymer is ther added a given amount of selective solvent, such as for instance a solvent S_A selective for the polyA block. This procedure leads to the precipitation of the polyB block and to the formation of micelles having a polyB core more or less swelled by the common solvent S, as schematically illustrated in Figure 4. These micelles are stabilized by the polyA sequence solubilized in the $S + S_A$ solvent mixture. The presence of the common solvent S in the final micellar system represents the major difference with respect to the "*classical*" systems, where the common solvent is eliminated, in general, by dialysis, in the final preparation stage.

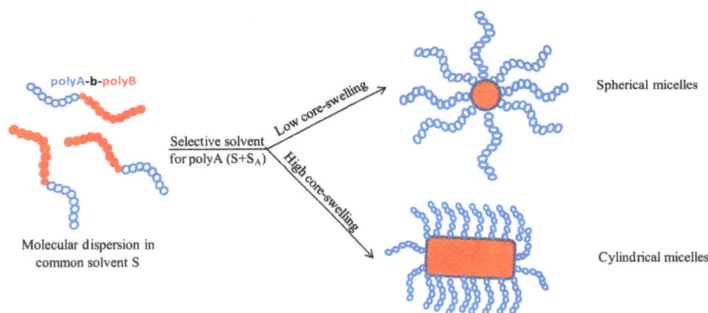

Figure 4. Schematization of the polyA-polyB diblock copolymer self-assembly process in solvents mixture.

For polyA-*b*-polyB micellar systems in a selective solvent it is now well established that the spherical, cylindrical, and vesicular morphology is mainly determined by the relative volume fractions of the core and corona. Spherical morphologies, for instance, are in general generated when DP_A, the polymerization degree of the polyA sequence, is low with respect to DP_B (the corona forming block). In the case of solvent mixtures, the relative volume fractions of core and corona are mainly determined by their swelling characteristics as a function of the solvent compositions and temperature. At a given temperature, spherical micelles have a tendency to be formed at a lower swelling degree of the micellar core. At this point, there is still a lack of information concerning: (i) the onset of self-assembly as a function of the molecular characteristics of the copolymer; (ii) the evaluation of the copolymer conformation with increasing selective solvent concentration; (iii) the partition of common and selective solvents in the micellar core and corona; and, (iiii) the evolution of the swelling degree of the core as a function of the volume fraction of the solvents. Regarding the *self-assembly in solvents mixtures*, it is established that this process opens interesting perspectives for the development of new morphologies and it could also provide an important insight in the *"classical"* self-assembly process itself.

In our opinion, the scattering and electron microscopy for *micellar characterization techniques* are well documented in the literature. As a minor point, it might be suggested that the DLS characterizations should be carried out as a function of concentration, with extrapolation to zero concentration. The determination of the *interphase*, the dimensions of the transition zone between core and corona, could be of interest for specific applications, such as those where micellar systems are used as nanoreactors. Finally, it is quite surprising that in contrast to water-based micellar systems, only very few results were published concerning the CMC and CMT values of block copolymers in organic solvents. A similar remark can be made for the determination of the aggregation number N_{agg}, the average number of polymer chains per micelles.

The *"crystallization-induced self-assembly"* topic, including rod-coil and other micellar structures, was very recently reviewed by Tritschler et al. [27]. This self-assembly method offers the possibility to prepare non-spherical micellar morphologies with controlled dimensions having promising applications in various domains.

4. Self-Assembly of Linear ABC Triblock Terpolymers in Organic Solvents

Linear ABC triblock terpolymers have opened an extensive research area for the development of sophisticated micellar structures over the last decade. Up to now, several excellent review papers have been published on this topic [2,7,83]. For the self-assembly of ABC copolymers it turns out once again that a majority of the studies were focused on aqueous-based micellar systems and only relatively few of them on the micellization in organic media. Wyman and Liu [83], for instance, showed the fascinating morphologies that are becoming available for a given ABC copolymer by precise control of the micellization conditions, including the solvent mixtures. More recently, Gröschel and Müller [84] extended this review by providing a detailed insight of the multi-compartment nanostructures that are accessible, as well with AB and ABC triblock terpolymers.

The objective of this chapter is to highlight the recent developments, including the present author's contributions, of ABC triblock terpolymer micellar systems generated in organic solvents. Polar and non-polar solvents will be taken into consideration, as well as mixtures of organic solvents. According to Wyman and Liu [83], the basic micellar spherical morphologies of ABC triblock terpolymers are outlined in Figure 5, as a function of the solvent selectivity.

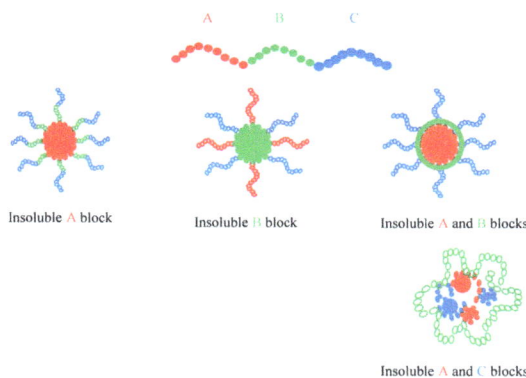

Figure 5. Basic micellar morphologies for linear ABC triblock terpolymers.

The sequence organization, as schematically displayed in Figure 5 for spherical micelles, is similar for cylindrical (worm-like) and vesicle micellar morphologies.

For ABC-type linear triblock terpolymers, the sequence arrangement is an additional parameter that must be considered with respect to diblock copolymers. In fact, for a given ABC sample in selective solvent conditions, the micellar characteristics are different from the corresponding ACB and BAC structures. Up to now, the effect of the sequence arrangement was demonstrated for the self-assembly of triblock terpolymers in aqueous media [85,86] or in water/organic solvent mixtures [87]. Marsat et al. [85] examined the micellar formation in aqueous medium of an ABC, ACB, and BAC triblock terpolymer comprising a hydrophilic, a hydrophobic, and a fluorophilic sequence, such as poly[oligo(ethylene oxide) monomethyl ether acrylate]-poly(benzyl acrylate)-poly(1H,1H-perfluorobutyl acrylate) (POEGA-PBzA-PFA). The chemical structure of this ABC copolymer is shown in Figure 6.

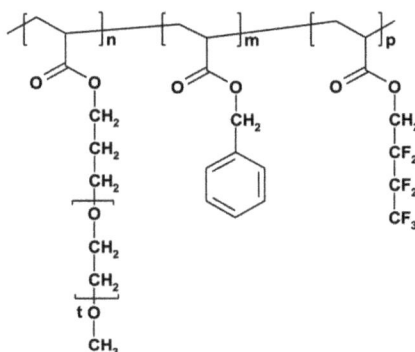

Figure 6. Chemical structure of the ABC triblock terpolymer POEGA-PBzA-PFA [85].

Analogous to the previous chapter, the present one will be organized in a similar way by considering the self-assembly of ABC triblock terpolymers in non-polar, polar, and solvent mixtures. This means that a given ABC copolymer may appear in different subsections.

4.1. Self-Assembly of Linear ABC Triblock Terpolymers in Non-Polar Selective Solvents

A typical example of self-assembly in non-polar solvent has been studied in our research group for a series of PB-P2VP-PEO triblock terpolymers in *n*-heptane, a selective solvent of the PB sequence, as well as in aqueous medium [7,88,89].

This type of ABC copolymers has been selected as the solubility parameter of the sequences is well differentiated, such as 17, 21, and 20.8 MPa$^{1/2}$ for the PB, P2VP, and PEO, respectively. Moreover, the P2VP sequence is easily protonated or quaternized, yielding a pH- and electrolyte-sensitive water-soluble block. Figure 7 illustrates the variation of the micellar particle size as a function of the total DP_n.

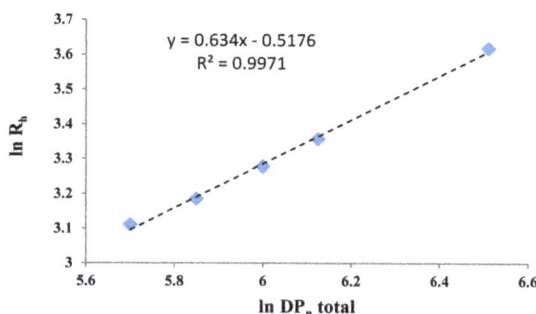

Figure 7. Logarithmic variation of the R_h as a function of DP_n total of PB-P2VP-PEO triblock terpolymers in *n*-heptane [88].

From Figure 7, it can be noticed that R_h scales as $DP_n^{0.64}$ in reasonable agreement with the exponent of 0.68 predicted by Noolandi and Hong [90] for diblock copolymers. In a simplified approach, the R_h values could further be correlated with the individual DP_n values of the copolymers [88]. From a practical point of view, it was demonstrated that PB-P2VP-PEO copolymers with a P2VP middle block are efficient dispersing and stabilizing agents for TiO$_2$ pigments in non-aqueous as well as in aqueous media [89]. Another example of the self-assembly studies of P2VP-based copolymers investigated in our research group was the micellization of poly(2-vinyl pyridine)-poly(tert-butyl methacrylate)-poly(cyclohexyl methacrylate) (P2VP-PtBMA-PCHMA) triblock terpolymers in methylcyclohexane, which are of practical interest as "viscosity improvers in motor-oil formulations" [7].

The self-assembly of P2VP-based triblock terpolymers, such as PS-P2VP-PMMA, was also studied by Tsitsilianis and Sfika [91] in toluene, a selective solvent for the PS and PMMA sequences. These authors found that both the N_{agg} and R_h of spherical core-shell micelles, having a P2VP-core and a corona of PS and PMMA, depend on the DP_n of the insoluble P2VP sequence. Core-shell micelles, based on amino (meth)acrylates, were also obtained by Bütün et al. [92] for poly[2-(diisopropylamino)ethyl methacrylate]-poly[2-(dimethylamino)ethyl methacrylate]-poly[2-(N-morpholino)ethyl methacrylate] (PDPA-PDMA-PMEMA) triblock terpolymers in hexane. In this case, the PDMA and PMEMA sequences are in the core of the micelles, which have a R_h of 46 nm, whereas the PDPA block formed the solvated corona.

4.2. Self-Assembly of Linear ABC Triblock Terpolymers in Polar Selective Solvents

The recent publications concerning the self-assembly of linear ABC triblock terpolymers in polar solvents are summarized in Table 5.

Table 5. Self-assembly of linear ABC triblock terpolymers in polar solvents.

ABC Triblock Terpolymers	Selective Solvent for A Block	Selective Solvent for B Block	Selective Solvent for C Block	Micellar Characteristics	Ref.
PtBA$_{107}$-b-PCEMA$_{193}$-b-PGMA$_{115}$	CH$_3$–OH	-	CH$_3$–OH	Vesicles and cylinders	[93]
PB$_{800}$-b-P2VP$_{190}$-b-PtBMA$_{380;550}$	Acetone	-	-	R_h = 43; 44 nm R_g = 34; 36 nm N_{agg} = 203; 174	[94]
PtBA$_{110}$-b-PCEMA$_{195}$-b-PSGMA$_{115}$	Propanol	-	-	Cylinders	[95]
PnBA$_{28}$-b-PS$_{37}$-b-P2VP$_{73}$	-	-	CH$_3$–OH	R_h = 27 nm Spheres	[96]
PS$_{306;510;516}$-b-PB$_{151;258;140}$-b-PMMA$_{340;260;76}$	DMAc	-	DMAc	Spheres	[97]
PS$_{385}$-b-PI$_{485}$-b-P2VP$_{829}$	-	-	C$_2$H$_5$–OH	Ellipsoid	[98]

PtBA: poly(tert-butyl acrylate); PCEMA: poly(2-cinnamoyloxyethyl methacrylate); PGMA: poly(glyceryl monomethacrylate); PSGMA: poly(sucinnated glyceryl monomethacrylate); PnBA: poly(*n*-butyl acrylate); DMAc: dimethylacetamide.

From Table 5, it may be noticed that the recent investigations concerning the self-assembly in polar solvents were mainly focused on alcohols, such as methanol, which is a selective solvent of the end-blocks. This implies that the middle block preferentially forms the micellar core.

4.3. Self-Assembly of Linear ABC Triblock Terpolymers in Organic Solvent Mixtures

The self-assembly of ABC terpolymers in organic solvent mixtures was performed essentially through the precipitation method starting from a homogeneous solution in a common solvent for all three blocks, followed by the addition of a non-solvent. As previously described, Bütün et al. [92] studied the self-assembly of two PDPA-PDMA-PMEMA triblock terpolymers, not only in *n*-hexane, but also in a mixture of $CHCl_3/n$-hexane, starting from a unimer's solution in $CHCl_3$. In this case, core-shell-corona micelles were obtained with a PMEMA core and a R_h value of 15 nm.

The morphology of core-shell-corona structures, formed by the self-assembly of PS-P4VP-PEC triblock terpolymers in DMF/ethanol mixtures, was reported by Wang et al. [99] as a function of temperature. A further interesting study concerning the morphological behavior as a function of the "*solvent quality*" was performed by Löbling et al. [97,100] for a series of PS-PB-PMMA and PS-PB-PtBMA triblock terpolymers in acetone/isopropanol mixtures. The micellar morphology modifications determined by cryo-TEM as a function of the isopropanol content are illustrated in Figure 8.

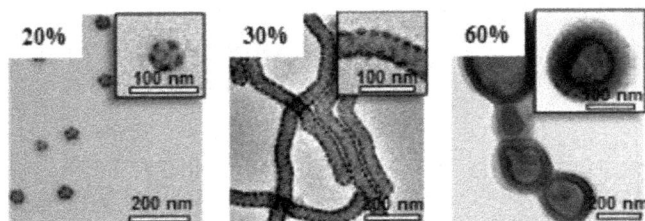

Figure 8. Micellar morphological changes of PS-PB-PMMA triblock terpolymers in acetone/isopropanol mixtures as a function of isopropanol content. "Reprinted with permission from Löbling, T.I.; Ikkala, O.; Gröschel, A.H.; Müller, A.H.E. Controlling multicompartment morphologies using solvent conditions and chemical modification. *ACS Macro Lett.* **2016**, *5*, 1044–1048. Copyright 2017 Americal Chemical Society".

More recently, Cong et al. [98] completed a similar study for PI-PS-P2VP triblock terpolymers in THF/ethanol mixtures.

4.4. Miscellaneous Self-Assembly Studies of Linear ABC Triblock Terpolymers

Given this context, and in analogy to AB diblock copolymers, studies have been completed on the behavior of ABC self-assembly in ionic liquids, the influence of partially crystallized moieties, and the development of biocompatible or thermo-sensitive systems.

Self-assembly of AB diblock copolymers in ILs has been studied extensively, as shown in Section 3. However, it seems that up to now ABC triblock terpolymers have attracted less attention. Mention could be made only of the publication of Kitazawa et al. [101]. These authors investigated the morphological behavior of a poly(benzyl methacrylate)-poly(methyl methacrylate)-poly(2-phenylethyl methacrylate) (PBnMA-PMMA-PPhEtMA) triblock terpolymer, as a function of temperature in 1-ethyl-3-methylimidazolium bis(trifluoromethanesulfonyl)amide [C2MIM][NTF2] ($\delta \sim 26$ $MPa^{1/2}$) at high copolymer concentrations, such as 10 and 20 wt %.

The groups of Winnik and Manners have extended the CDSA method from AB diblock copolymers to ABC triblock terpolymers by studying the morphology of a series of well-defined coil-crystalline-coil poly(ferrocenylphenylphosphine)-poly(ferrocenyldimethylsilane)-poly (dimethylsiloxane) (PFP-*b*-PFS-*b*-PDMS) terpolymers in hexane, which is a selective solvent for PDMS [102]. Cylindrical micelles were obtained for the sample having the DP_n(PFP) <6, whereas spherical micelles were observed if the DP_n(PFP) = 11. The longer PFP chains hinder the crystallization of the PFS sequence leading thus to the formation of spherical morphologies. More recently, the crystallization-induced self-assembly of poly(styrene)-poly(ethylene)-poly(methyl methacrylate) (PS-PE-PMMA) triblock terpolymer in toluene and dioxane, a good and bad solvent for the semicrystalline PE, respectively, was investigated by Schmelz et al. [103]. As a function of the *"solvent quality"*, either spherical or worm-like crystalline-core micelles were observed by these authors.

The self-assembly of ABC triblock terpolymers in biocompatible non-aqueous solvents was investigated in our research group for a PB-P2VP-PEO sample. At low concentrations in PEG400, this triblock terpolymer forms spherical micelles with a PB-core and P2VP/PEO corona. An interesting feature appeared at concentrations higher than 3 wt %, at which a thermo-reversible gel, due to the formation of H–bonds, was noticed for these micellar systems [104].

4.5. Concluding Remarks

It is undeniable that ABC triblock terpolymers have opened an important research area, in particular for the development of sophisticated micellar morphologies. This topic having recently be reviewed in detail, the aim of the present section is to highlight more specific aspects related to the synthesis and the self-assembly of ABC linear triblock terpolymers. A very large range of this type of copolymers has been prepared up to now. In our opinion, it would be of interest for further developments of the field to synthesize homologous series of ABC, BCA, and BAC structures in order to complete the demonstration that the sequence distribution has a major influence on the micellar characteristics. For this type of copolymers, it would further worthwhile to check in different selective solvents the CMC, CMT values, as well as the N_{agg}, which are scarcely reported up to now.

In the same manner as for AB diblock copolymers, the self-assembly of ABC copolymers in *solvent mixtures* could contribute to the understanding of the micellization process. This type of studies may be carried-out in specific solvent mixtures, as already demonstrated by different authors [85,90,97]. In order to complete the results reported by Zhang et al. [87] and by Bethausen et al. [105], it could be of interest to examine in detail the self-assembly of ABC terpolymers in solvent/water mixtures. Of further interest will be the self-assembly studies of ABC terpolymers in ionic liquids and biocompatible organic solvents, which are only very scarcely discussed in the literature.

Crystallization of ABC terpolymers in bulk, thin films and on surface, such as two-dimensional (2D) has been examined in detail over the last years. However, in comparison with AB diblock copolymers only a few studies are available concerning the CDSA method for ABC terpolymers.

5. Self-Assembly of AB and ABC Graft Copolymers in Organic Solvents

Analogous to AB block copolymers, the AB graft copolymers may develop the basic sphere, cylinder, and vesicle micellar morphologies, as a function of molecular characteristics, selective solvent type, copolymer concentration, and self-aggregation process. Micellization of synthetic- and polysaccharide-based graft copolymers in aqueous media has been reviewed very recently by the present authors [23]. Therefore, it was necessary to complete this topic by providing an overview on the self-assembly of various types of graft copolymers in organic solvents. After recalling some basic graft copolymer characteristics, this section focus on AB graft copolymers, including the concept of unimolecular micelle formation, followed by the morphologies of ABC graft structures.

As schematically shown in Figure 9, a graft copolymer is essentially defined by the DP_A and DP_B of the backbone and graft chain, respectively. A major molecular characteristic is further the graft density, which represents the number of side chains per 100 backbone monomer units. Moreover, assuming that the graft chains are randomly distributed along the backbone, the average distance ΔP between two grafting sites is thus accessible.

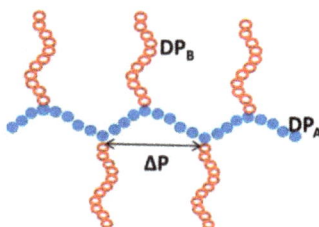

Figure 9. Schematically representation of an AB graft copolymer.

Already at the end of the last century, Kikuchi and Nose [106] clearly demonstrated that graft copolymers have a greater tendency than the corresponding block copolymers to form unimolecular micelles by intramolecular association of the backbone sequences. This tendency was well illustrated for PMMA-*g*-PS in organic solvents, such as iso-amyl acetate or in the presence of a mixture of acetonitrile and acetoacid ether. It appeared that the tendency of forming unimolecular micelles increases with an increasing graft density and molecular weight of the grafted chains. At high graft densities, unimolecular micelles are formed by intramolecular collapse of the backbone chain. In a theoretical approach, Borisov and Zhulina [107] admitted that the collapsing of the backbone chain occurs with formation of a *"pearl neck-lace"* stabilized by steric repulsion of the soluble side chains.

From the literature survey, as already mentioned, it turned out that the number of publications related to self-assembly of graft copolymers in aqueous media increased over the last two decades due to the predominant applications possibilities of these aqueous systems. Purely organic solvent-based systems have become rather scarce. In the following section, an overview of the recent investigations concerning the self-assembly of synthetic and natural-based graft copolymers in pure organic solvents, are reviewed.

5.1. Self-Assembly of AB Graft Copolymers

For the self-assembly studies of polysaccharide-based graft copolymers in organic solvents, a special mention could be made of the investigation of Liu et al. [108]. These authors reported the synthesis of ethyl cellulose-*g*-poly(acrylic acid) (EC-*g*-PAA), which leads to unimolecular micelles in DMF, methanol, and their mixtures with water by increasing the graft density. Francis et al. [109] described the synthesis of chitosan-*g*-PS and prepared micellar dispersions in DMF as these systems are of interest as metal complexing agents.

Analogous to the peptide/PtBMA water and organo-soluble composite investigations by Saha et al. [110], Bose et al. [111] reported peptide-based graft copolymers, such as tyrosine-*g*-polyoxazoline. The micellar structures of this type of copolymer were examined in aqueous and non-aqueous media. Concerning the recent example of a purely synthetic graft copolymer, Stepánek et al. [112] investigated the micellization in methanol and water of poly(4-methylstyrene)-*g*-poly(methacrylic acid) (P4MS-*g*-PMAA). In methanol, a tendency to form unimolecular micelles was demonstrated.

PtBMA-based graft copolymers, such as PS-*g*-PtBMA, were reported by Gromadzki et al. [113] These authors examined the conformation of this type of copolymers in a non-selective solvent, such as a THF, as well as in n-amylalcohol, a selective solvent of the PtBMA graft chains. By increasing the graft density, a stretching of the PS backbone was noticed for the unimolecular micelles.

A typical example of fluorinated graft copolymers was studied by Koda et al. [114]. This type of thermo-sensitive copolymer leads to micelle formation in both fluorinated solvents and in water. As a final and recent example from this category of graft copolymers, mention has to be made of poly(phenyl carbodiimide)-*g*-poly(styrene) (PCD-*g*-PS) self-assembly in methanol described by Kurilov et al. [24]. The nano-ring morphologies obtained for this rod-coil copolymer were generated by surface-induced self-assembly in thin films, prepared by the controlled evaporation of THF/methanol copolymer solutions.

5.2. Self-Assembly of ABC Graft Copolymers

Various structures are available for ABC three-component graft copolymers and their micellization characteristics in aqueous media have been recently reviewed [23]. Concerning the micelle formation in organic media, a typical example that is related to the self-assembly of ABC graft copolymers, such as poly(1-dodecene-co-pMS)-g-PEO, was published by Liu et al. [115]. These authors performed an extensive study of their micellization behavior in *n*-octanol. As a function of the molecular weight of the PEG, 350, 750 and 2000 g·mol^{-1}, respectively, the C.M.C increased from 1.28×10^{-4} to 1.95×10^{-4} g·mL^{-1}. For the reversed spherical micelles, the size distributions were relatively large with typical diameters in the range of 71 to 186 nm.

A detailed micellization study by small angle neutron scattering (SANS) was carried out by Alexander et al. [116] for PI-g-Pluronics. These authors investigated the micellization behavior of these graft copolymers in different solvents, such as THF, a common solvent for PI and Pluronics, and hexane and ethanol, which are selective solvents for PI and Pluronics, respectively. As expected, *"flower-like"* micelles were obtained in hexane, the selective solvent for the PI backbone chain, whereas *"star-like"* micelles were formed in ethanol. These morphologies are illustrated in Figure 10. This study was completed by the determination of the micellization onset as a function of the copolymer concentration in THF/hexane mixtures, ranging from pure THF to pure hexane.

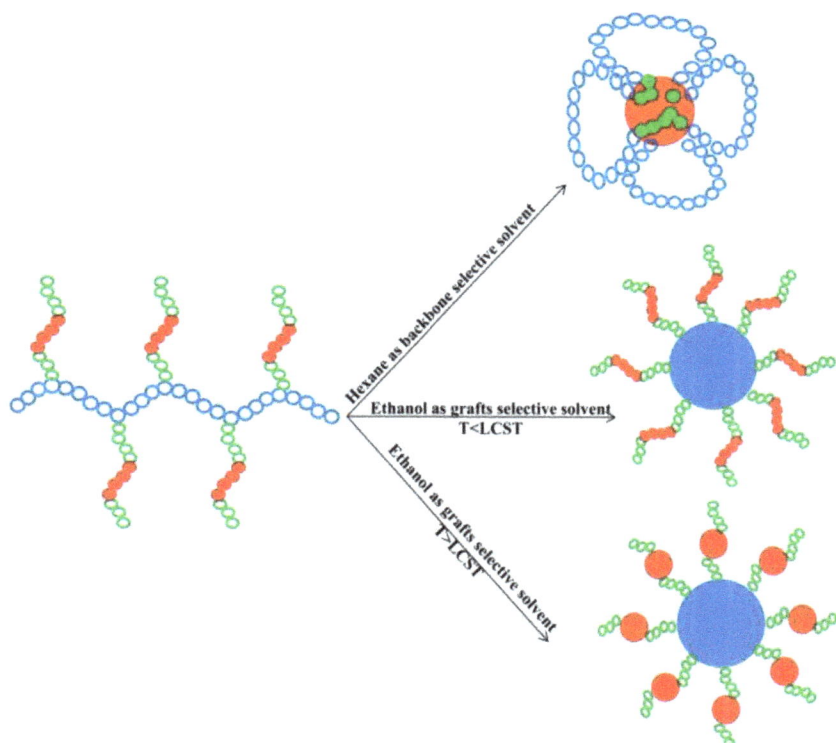

Figure 10. Micellar conformations of PI-g-Pluronic copolymers in hexane (backbone selective solvent) and in ethanol (grafts selective solvent). Adapted from Alexander et al. [116].

Mo et al. [117] extended the self-assembly concept to multigraft copolymers. Using the *"click chemistry"* preparation technique, they synthesized a series of PGMA-*g*-(PCEMA-PtBA-MPEG) having the same P(GMA-N$_3$) backbone, and three different type of side chains. In DMF, a common *"good solvent"* for both the backbone and side-chains, they observed by AFM a stretched *"worm-like"* conformation. With three different selective solvents (CH$_2$Cl$_2$, CH$_3$OH, and H$_2$O), the morphologies of the unimolecular micelles varied from a *"neck-lace"* to *"worm-like"* and to multi-component spherical structures, as schematically illustrated in Figure 11.

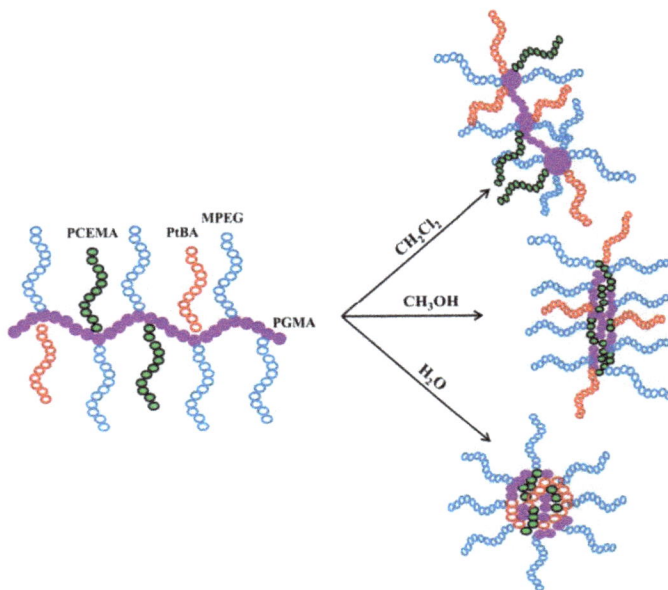

Figure 11. Micellar conformation of PGMA-*g*-(PCEMA-PtBA-MPEG) copolymers in CH$_2$Cl$_2$, CH$_3$OH and H$_2$O selective solvents, respectively. Adapted from Mo et al. [117].

5.3. Concluding Remarks

With the present synthesis methods, outlined in our recent review article [23], a wide range of AB and ABC graft copolymers with various architectures are available in order to examine their behavior in organic solvents. In fact, over the last decade, this topic has attracted less research interest than self-assembly studies in aqueous media. With respect to block copolymers, graft copolymers have the advantage to form *unimolecular micelles* by adjusting the graft density. Moreover, they provide also access to natural-based products, such as polysaccharides comprising structures. By taking into account these specific features, original developments may be expected in this research area.

6. General Conclusions and Perspectives

The self-assembly of block and graft copolymers in aqueous media is undeniably at the present a major research trend given their biomedical application possibilities. Nevertheless, from our overview it turns out that micellar systems in non-aqueous solvents are of ongoing theoretical and practical interest.

For AB and ABA block copolymers, which are considered as *"model molecular architectures"*, the recent studies were mainly focused on their self-assembly in polar and non-polar solvents, as well as in solvent mixtures. This last approach has the advantage that the micellar characteristics, such as size and morphology, can be triggered in a continuous way by adjusting the solvent mixture composition,

Polymers **2018**, *10*, 62

and thus the *"solvent quality"*. Over these last years, a major interest was further devoted to the micellization in ionic liquids as selective solvents. As this research area has reached its maturity, further developments may be expected for polyA-*b*-poly(B-*co*-C) copolymers with a random or gradient B and C blocks. This approach is in fact an alternative to adjust the solubility parameter.

ABC triblock terpolymers have developed into an extensive research area for the study of sophisticated micellar morphologies. Substantial advances were reported, in particular, for aqueous and non-aqueous systems, yielding multi-compartment micellar structures. For these multi-compartment nanoparticles, it was demonstrated that the sequence arrangement of the ABC triblock terpolymers is a major structural parameter. Different morphologies are obtained in a given selective solvent for ABC, BAC, and BCA of the same composition and molecular weight. Even if the challenge remains for the polymer chemist, this concept should be estended to ABC, BAC, and BCA copolymers to self-aggregation in organic solvents, including the biocompatible ILs.

Despite their wide-spread application possibilities, AB and ABC graft copolymers have attracted up-to-now relatively minor interest concerning the micellization in organic solvents. With respect to block copolymers, graft copolymers have the advantage that by adjusting the graft density brush-like structures, and unimolecular micelles can be obtained. Graft copolymers, based on natural polymeric precursors, may create opportunities for self-assemblies in organic solvents.

Finally, as well for block copolymers as for graft copolymers, an efficient control of the molecular architectures and the self-assembly parameters are still a major challenge for further investigations in this research area.

Author Contributions: Leonard Ionut Atanase and Gerard Riess conceived, designed and wrote this review.

Conflicts of Interest: The authors declare no conflict of interest.

References

1. Tuzar, Z.; Kratochvil, P. Block and graft copolymer micelles in solution. *Adv. Colloid Interface Sci.* **1976**, *6*, 201–232. [CrossRef]
2. Nakashima, K.; Bahadur, P. Aggregation of water-soluble block copolymers in aqueous solutions: Recent trends. *Adv. Colloid Interface Sci.* **2006**, *123–126*, 75–96. [CrossRef] [PubMed]
3. Cho, H.K.; Cheong, I.W.; Lee, J.M.; Kim, J.H. Polymeric nanoparticles, micelles and polymersomes from amphiphilic block copolymer. *Korean J. Chem. Eng.* **2010**, *27*, 731–740. [CrossRef]
4. Kulthe, S.S.; Choudhari, Y.M.; Inamdar, N.N.; Mourya, V. Polymeric micelles: Authoritative aspects for drug delivery. *Des. Monomers Polym.* **2012**, *15*, 465–521. [CrossRef]
5. Xu, W.; Ling, P.; Zhang, T. Polymeric micelles, a promising drug delivery system to enhance bioavailability of poorly water-soluble drugs. *J. Drug Deliv.* **2013**. [CrossRef] [PubMed]
6. Ahmad, Z.; Shah, A.; Siddiq, M.; Kraatz, H.B. Polymeric micelles as drug delivery vehicles. *RSC Adv.* **2014**, *4*, 17028–17038. [CrossRef]
7. Riess, G. Micellization of block copolymers. *Prog. Polym. Sci.* **2003**, *28*, 1107–1170. [CrossRef]
8. Gohy, J.F. Block copolymer micelles. *Adv. Polym. Sci.* **2005**, *190*, 65–136. [CrossRef]
9. Mai, Y.; Eisenberg, A. Self-assembly of block copolymers. *Chem. Soc. Rev.* **2012**, *41*, 5969–5985. [CrossRef] [PubMed]
10. Bhattacharya, A.; Misra, B.N. Grafting: A versatile means to modify polymers. Techniques, factors and applications. *Prog. Polym. Sci.* **2004**, *29*, 767–814. [CrossRef]
11. Uhrig, D.; Mays, J. Synthesis of well-defined multigraft copolymers. *Polym. Chem.* **2011**, *2*, 69–76. [CrossRef]
12. Feng, C.; Li, Y.; Yang, D.; Hu, J.; Zhang, X.; Huang, X. Well-defined graft copolymers: From controlled synthesis to multipurpose applications. *Chem. Soc. Rev.* **2011**, *40*, 1282–1295. [CrossRef] [PubMed]
13. Deng, Y.; Zhang, S.; Lu, G.; Huang, X. Constructing well-defined star graft copolymers. *Polym. Chem.* **2013**, *4*, 1289–1299. [CrossRef]
14. Matsuo, Y.; Konno, R.; Ishizone, T.; Goseki, R.; Hirao, A. Precise synthesis of block polymers composed of three or more blocks by specially designed linking methodologies in conjunction with living anionic polymerization system. *Polymers* **2013**, *5*, 1012–1040. [CrossRef]

15. Keddie, D.J. A guide to the synthesis of block copolymers using reversible-addition fragmentation chain transfer (RAFT) polymerization. *Chem. Soc. Rev.* **2014**, *43*, 496–505. [CrossRef] [PubMed]

16. Jennings, J.; He, G.; Howdle, S.M.; Zetterlund, P.B. Block copolymer synthesis by controlled/living radical polymerization in heterogeneous systems. *Chem. Soc. Rev.* **2016**, *45*, 5055–5084. [CrossRef] [PubMed]

17. Lopez-Oliva, A.P.; Warren, N.J.; Rajkumar, A.; Mykhaylyk, O.O.; Derry, M.J.; Doncom, K.E.B.; Rymaruk, M.J.; Armes, S.P. Polydimethylsiloxane-based diblock copolymer nano-objects prepared in nonpolar media via RAFT-mediated polymerization-induced self-assembly. *Macromolecules* **2015**, *48*, 3547–3555. [CrossRef]

18. Jones, E.R.; Semsarilar, M.; Wyman, P.; Boerakker, M.; Armes, S.P. Addition of water to an alcoholic RAFT PISA formulation leads to faster kinetics but limits the evolution of copolymer morphology. *Polym. Chem.* **2016**, *7*, 851–859. [CrossRef]

19. Ratcliffe, L.P.D.; McKenzie, B.E.; Le Bouedec, G.M.D.; Williams, C.N.; Brown, S.L.; Armes, S.P. Polymerization-induced self-assembly of all-acrylic diblock copolymers via RAFT dispersion polymerization in alkanes. *Macromolecules* **2015**, *48*, 8594–8607. [CrossRef]

20. Canning, S.L.; Smith, G.N.; Armes, S.P. A critical appraisal of RAFT-mediated polymerization-induced self-assembly. *Macromolecules* **2016**, *49*, 1985–2001. [CrossRef] [PubMed]

21. Canning, S.L.; Cunningham, V.J.; Ratcliffe, L.P.D.; Armes, S.P. Phenyl acrylate is a versatile monomer for the synthesis of acrylic diblock copolymer nano-objects via polymerization-induced self-assembly. *Polym. Chem.* **2017**, *8*, 4811–4821. [CrossRef]

22. Van Krevelen, D.W. *Properties of Polymers*, 3rd ed.; Elsevier Sci.: Amsterdam, The Netherlands, 2003; pp. 189–224. ISBN 044482877X.

23. Atanase, L.I.; Desbrieres, J.; Riess, R. Micellization of synthetic and polysaccharides-based graft copolymers in aqueous media. *Prog. Polym. Sci.* **2017**, *73*, 32–60. [CrossRef]

24. Kulikov, O.V.; Siriwardane, D.A.; McCandless, G.T.; Mahmood, S.F.; Novak, B.M. Self-assembly studies on triazolepolycarbodiimide-g-polystyrene copolymers. *Polymer* **2016**, *92*, 94–101. [CrossRef]

25. Hudson, Z.M.; Boott, C.E.; Robinson, M.E.; Rupar, P.A.; Winnik, M.A.; Manners, I. Tailored hierarchical micelle architectures using living crystallization-driven self-assembly in two dimensions. *Nat. Chem.* **2014**, *6*, 893–898. [CrossRef] [PubMed]

26. Hailes, R.L.N.; Oliver, A.M.; Gwyther, J.; Whittell, G.R.; Manners, I. Polyferrocenylsilanes: Synthesis, properties and applications. *Chem. Soc. Rev.* **2016**, *45*, 5358–5407. [CrossRef] [PubMed]

27. Tritschler, U.; Pearce, S.; Gwyther, J.; Whittell, G.R.; Manners, I. Functional nanoparticles from the solution self-assembly of block copolymers. *Macromolecules* **2017**, *50*, 3439–3463. [CrossRef]

28. Winnik, F.M.; Regismond, S.T.A.; Anghel, D.F. Fluorescent labels: Versatile tools for studying the association of amphiphilic polymers in water. In *Associative Polymers in Aqueous Media*; Glass, J.E., Ed.; American Chemical Society: Washington, DC, USA, 2010; Volume 17, pp. 286–302. ISBN 9780841236592.

29. Štěpánek, M. Fluorescence spectroscopy studies of amphiphilic block copolymer micelles in aqueous solutions. In *Fluoresce Studies of Polymer Containing Systems*; Procházka, K., Ed.; Springer: Cham, Switzerland, 2016; Volume 16, pp. 203–215. ISBN 978-3-319-26786-9.

30. Walker, L.M. Scattering from polymer-like micelles. *Curr. Opin. Colloid Interface Sci.* **2009**, *14*, 451–454. [CrossRef]

31. Chavda, S.; Yusa, S.; Inoue, M.; Abezgauz, L.; Kesselman, E.; Danino, D.; Bahadur, P. Synthesis of stimuli responsive PEG$_{47}$-*b*-PAA$_{126}$-*b*-PSt$_{32}$ triblock copolymer and its self-assembly in aqueous solutions. *Eur. Polym. J.* **2013**, *49*, 209–216. [CrossRef]

32. Löbling, T.I.; Haataja, J.S.; Synatschke, C.V.; Schacher, F.H.; Müller, M.; Hanisch, A.; Gröschel, A.H.; Müller, A.H.E. Hidden structural features of multicompartment micelles revealed by cryogenic transmission electron tomography. *ACS Nano* **2014**, *8*, 11330–11340. [CrossRef] [PubMed]

33. Franken, L.E.; Boekema, E.J.; Stuart, M.C.A. Transmission electron microscopy as a tool for the characterization of soft materials: Application and interpretation. *Adv. Sci.* **2017**, *4*, 1600476. [CrossRef] [PubMed]

34. Boott, C.E.; Laine, R.F.; Mahou, P.; Finnegan, J.R.; Leitao, E.M.; Webb, S.E.D.; Kaminski, C.F.; Manners, I. In situ visualization of block copolymer self-assembly in organic media by super-resolution fluorescence microscopy. *Chem. Eur. J.* **2015**, *21*, 18539–18542. [CrossRef] [PubMed]

35. Li, S.; Huo, F.; Li, Q.; Gao, C.; Su, Y.; Zhang, W. Synthesis of a doubly thermo-responsive schizophrenic diblock copolymer based on poly[*N*-(4-vinylbenzyl)-*N,N*-diethylamine] and its temperature-sensitive flip-flop micellization. *Polym. Chem.* **2014**, *5*, 3910–3918. [CrossRef]

36. Sotiriou, K.; Nannou, A.; Velis, G.; Pispas, S. Micellization behavior of PS(PI)$_3$ miktoarm star copolymers. *Macromolecules* **2002**, *35*, 4106–4112. [CrossRef]

37. Sotiriou, K.; Pispas, S.; Hadjichristidis, N. Controlling the colloidal behavior of styrene-isoprene diblock copolymers by selective end functionalization. *Colloids Surf. Part A Physicochem. Eng. Asp.* **2007**, *293*, 51–57. [CrossRef]

38. Di Cola, E.; Lefebvre, C.; Deffieux, A.; Narayanan, T.; Borsali, R. Micellar transformations of poly(styrene-*b*-isoprene) block copolymers in selective solvents. *Soft Matter* **2009**, *5*, 1081–1090. [CrossRef]

39. Lund, R.; Willner, L.; Lindner, P.; Richter, D. Structural properties of weakly segregated PS-PB block copolymer micelles in *n*-alkanes: Solvent entropy effects. *Macromolecules* **2009**, *42*, 2686–2695. [CrossRef]

40. Cheng, G.; Hammouda, B.; Perahia, D. Polystyrene-block-polyisoprene diblock-copolymer micelles: Coupled pressure and temperature effects. *Macromol. Chem. Phys.* **2014**, *215*, 776–782. [CrossRef]

41. Growney, D.J.; Mykhaylyk, O.O.; Armes, S.P. Micellization and adsorption behavior of a near-monodisperse polystyrene-based diblock copolymer in nonpolar media. *Langmuir* **2014**, *30*, 6047–6056. [CrossRef] [PubMed]

42. Bartels, J.W.; Cauet, S.I.; Billings, P.L.; Lin, L.Y.; Zhu, J.; Fidge, C.; Pochan, D.J.; Wooley, K.L. Evaluation of isoprene chain extension from PEO macromolecular chain transfer agents for the preparation of dual, invertible block copolymer nanoassemblies. *Macromolecules* **2010**, *43*, 7128–7138. [CrossRef] [PubMed]

43. Wang, X.; Davis, J.L.; Hinestrosa, J.P.; Mays, J.W.; Kilbey, S.M. Control of self-assembled structure through architecturally and compositionally complex block copolymer surfactant mixtures. *Macromolecules* **2014**, *47*, 7138–7150. [CrossRef]

44. Arai, T.; Masaoka, M.; Michitaka, T.; Watanabe, Y.; Hashidzume, A.; Sato, T. Aggregation behavior of polystyrene-based amphiphilic copolymers in organic media. *Polym. J.* **2014**, *46*, 189–194. [CrossRef]

45. Hussain, H.; Tan, B.H.; Gudipati, C.S.; He, C.B.; Liu, Y.; Davis, T.P. Micelle formation of amphiphilic polystyrene-*b*-poly(*N*-vinylpyrrolidone) diblock copolymer in methanol and water-methanol binary mixtures. *Langmuir* **2009**, *25*, 5557–5564. [CrossRef] [PubMed]

46. Shao, Y.; Aizhao, P.; Ling, He. POSS end-capped diblock copolymers: Synthesis, micelle self-assembly and properties. *J. Colloid Interface Sci.* **2014**, *425*, 5–11. [CrossRef] [PubMed]

47. Liaw, C.Y.; Henderson, K.J.; Burghardt, W.R.; Wang, J.; Shull, K.R. Micellar morphologies of block copolymer solutions near the sphere/cylinder transition. *Macromolecules* **2015**, *48*, 173–183. [CrossRef]

48. Jena, S.S.; Roy, S.G.; Azmeera, V.; De, P. Solvent-dependent self-assembly behavior of block copolymers having side-chain amino acid and fatty acid block segments. *React. Funct. Polym.* **2016**, *99*, 26–34. [CrossRef]

49. Jana, S.; Saha, A.; Paira, T.K.; Mandal, T.K. Synthesis and self-aggregation of poly(2-ethyl-2-oxazoline)-based photo-cleavable block copolymer: Micelle, compound micelle, reverse micelle, and dye encapsulation/release. *J. Phys. Chem. B* **2016**, *120*, 813–824. [CrossRef] [PubMed]

50. Demarteau, J.; Ameduri, B.; Ladmiral, V.; Mees, M.A.; Hoogenboom, R.; Debuigne, A.; Detrembleur, C. Controlled synthesis of fluorinated copolymers via cobalt-mediated radical copolymerization of perfluorohexylethylene and vinyl acetate. *Macromolecules* **2017**, *50*, 3750–3760. [CrossRef]

51. Pospiech, D.; Jehnichen, D.; Eckstein, K.; Scheibe, P.; Komber, H.; Sahre, K.; Janke, A.; Reuter, U.; Häußler, L.; Schellkopf, L.; et al. Semifluorinated PMMA block copolymers: Synthesis, nanostructure, and thin film properties. *Macromol. Chem. Phys.* **2017**, *218*, 1600599. [CrossRef]

52. Cho, H.; Park, H.; Park, S.; Choi, H.; Huang, H.; Chang, T. Development of various PS-b-P4VP micellar morphologies: Fabrication of inorganic nanostructures from micellar templates. *J. Colloid Interface Sci.* **2011**, *356*, 1–7. [CrossRef] [PubMed]

53. Zhou, Y.N.; Cheng, H.; Luo, Z.H. Fluorinated AB diblock copolymers and their aggregates in organic solvents. *J. Polym. Sci. Part A Polym. Chem.* **2011**, *49*, 3647–3657. [CrossRef]

54. Wang, L.; Yu, X.; Yang, S.; Zheng, J.X.; Van Horn, R.M.; Zhang, W.B.; Xu, J.; Cheng, S.Z.D. Polystyrene-block-poly(ethylene oxide) reverse micelles and their temperature-driven morphological transitions in organic solvents. *Macromolecules* **2012**, *45*, 3634–3638. [CrossRef]

55. Rao, J.; Zhang, H.; Gaan, S.; Salentinig, S. Self-assembly of polystyrene-b-poly(2-vinylpyridine) micelles: From solutions to silica particles surfaces. *Macromolecules* **2016**, *49*, 5978–5984. [CrossRef]

56. Choi, S.H.; Lee, W.B.; Lodge, T.P.; Bates, F.S. Structure of poly(styrene-*b*-ethylene-alt-propylene) diblock copolymer micelles in binary solvent mixtures. *J. Polym. Sci. Part A Polym. Phys.* **2016**, *54*, 22–31. [CrossRef]
57. Derry, M.J.; Fielding, L.A.; Armes, S.P. Polymerization-induced self-assembly of block copolymer nanoparticles via RAFT non-aqueous dispersion polymerization. *Prog. Polym. Sci.* **2016**, *52*, 1–18. [CrossRef]
58. Lee, H.N.; Bai, Z.; Newell, N.; Lodge, T.P. Micelle/inverse micelle self-assembly of a PEO-PNIPAm block copolymer in ionic liquids with double thermoresponsivity. *Macromolecules* **2010**, *43*, 9522–9528. [CrossRef]
59. Lu, H.; Akgun, B.; Wei, X.; Li, L.; Satija, S.K.; Russel, T.P. Temperature-triggered micellization of block copolymers on an ionic liquid surface. *Langmuir* **2011**, *27*, 12443–12450. [CrossRef] [PubMed]
60. Ueki, T.; Nakamura, Y.; Lodge, T.P.; Watanabe, M. Light-controlled reversible micellization of a diblock copolymer in an ionic liquid. *Macromolecules* **2012**, *45*, 7566–7573. [CrossRef]
61. Hoarfrost, M.L.; Lodge, T.P. Effects of solvent quality and degree of polymerization on the critical micelle temperature of poly(ethylene oxide-*b*-*n*-butyl methacrylate) in ionic liquids. *Macromolecules* **2014**, *47*, 1455–1461. [CrossRef]
62. Chen, Z.; FitzGerald, P.A.; Kobayashi, Y.; Ueno, K.; Watanabe, M.; Warr, G.G.; Atkin, R. Micelle structure of novel diblock polyethers in water and two protic ionic liquids (EAN and PAN). *Macromolecules* **2015**, *48*, 1843–1851. [CrossRef]
63. Kobayashi, Y.; Kitazawa, Y.; Komori, T.; Ueno, K.; Kokubo, H.; Watanabe, M. Self-assembly of polyether diblock copolymers in water and ionic liquids. *Macromol. Rapid Commun.* **2016**, *37*, 1207–1211. [CrossRef] [PubMed]
64. Ma, Y.; Lodge, T.P. Chain exchange kinetics in diblock copolymer micelles in ionic liquids: The role of χ. *Macromolecules* **2016**, *49*, 9542–9552. [CrossRef]
65. Ma, Y.; Lodge, T.P. Poly(methyl methacrylate)-block-poly(*n*-butyl methacrylate) diblock copolymer micelles in an ionic liquid: Scalling of core and corona size with core block length. *Macromolecules* **2016**, *49*, 3639–3646. [CrossRef]
66. Simone, P.M.; Lodge, T.P. Micellization of PS-PMMA diblock copolymers in an ionic liquid. *Macromol. Chem. Phys.* **2007**, *208*, 339–348. [CrossRef]
67. Mok, M.M.; Thiagarajan, R.; Flores, M.; Morse, D.C.; Lodge, T.P. Apparent critical micelle concentrations in block copolymer/ionic liquid solutions: Remarkably weak dependence on solvophobic block molecular weight. *Macromolecules* **2012**, *45*, 4818–4829. [CrossRef]
68. He, Y.; Li, Z.; Simone, P.; Lodge, T.P. Self-assembly of block copolymer micelles in an ionic liquid. *J. Am. Chem. Soc.* **2006**, *128*, 2745–2750. [CrossRef] [PubMed]
69. Meli, L.; Santiago, J.M.; Lodge, T.P. Path-dependent morphology and relaxation kinetics of highly amphiphilic diblock copolymer micelles in ionic liquids. *Macromolecules* **2010**, *43*, 2018–2027. [CrossRef]
70. Zhang, S.; Li, N.; Zheng, L.; Li, X.; Gao, Y.; Yu, L. Aggregation behavior of pluronic triblock copolymer in 1-butyl-3-methylimidazolium type ionic liquids. *J. Phys. Chem. B* **2008**, *112*, 10228–10233. [CrossRef] [PubMed]
71. Lopez-Barron, C.R.; Li, D.; Wagner, N.J.; Caplan, J.L. Triblock copolymer self-assembly in ionic liquids Effect of PEO block length on the self-assembly of PEO-PPO-PEO in ethylammonium nitrate. *Macromolecules* **2014**, *47*, 7484–7495. [CrossRef]
72. Miller, A.C.; Bershteyn, A.; Tan, W.; Hammond, P.T.; Cohen, R.E.; Irvine, D.J. Block copolymer micelles as nanocontainers for controlled release of proteins from biocompatible oil phases. *Biomacromolecules* **2009**, *10*, 732–741. [CrossRef] [PubMed]
73. Atanase, L.I.; Riess, G. Block copolymer stabilized nonaqueous biocompatible sub-micron emulsions for topical applications. *Int. J. Pharm.* **2013**, *448*, 339–345. [CrossRef] [PubMed]
74. Atanase, L.I.; Riess, G. Stabilization of non-aqueous emulsions by poly(2-vinylpyridine)-*b*-poly(butadiene) block copolymers. *Colloids Surf. Part A Physicochem. Eng. Asp.* **2014**, *458*, 19–24. [CrossRef]
75. Atanase, L.I.; Riess, G. PEG 400/Paraffin oil non-aqueous emulsions stabilized by PBut-block-P2VP block copolymers. *J. Appl. Polym. Sci.* **2015**, *132*, 41390–41397. [CrossRef]
76. He, W.N.; Xu, J.T. Crystallization assisted self-assembly of semicrystalline block copolymers. *Prog. Polym. Sci.* **2012**, *37*, 1350–1400. [CrossRef]
77. Crassous, J.J.; Schurtenberger, P.; Ballauff, M.; Mihut, A.M. Design of block copolymer micelles via crystallization. *Polymer* **2015**, *62*, 1–13. [CrossRef]

78. Mihut, A.M.; Crassous, J.J.; Schmalz, H.; Ballauf, M. Crystallization-induced aggregation of block copolymer micelles: Influence of crystallization kinetics on morphology. *Colloid Polym. Sci.* **2010**, *288*, 573–578. [CrossRef]

79. Wang, H.; Wu, C.; Xia, G.; Ma, Z.; Mo, G.; Song, R. Semi-crystalline polymethylene-*b*-poly(acrylic acid) diblock copolymers: Aggregation behavior, confined crystallization and controlled growth of semycrystalline micelles from dilute DMF solution. *Soft Matter* **2015**, *11*, 1778–1787. [CrossRef] [PubMed]

80. Wang, T.; Zhang, X.; Li, X.; Gao, X.; Song, L. Crystallization and morphology transition of P2VP-b-PEO block copolymer micelles composed of an amorphous core and a crystallizable corona. *Polym. Bull.* **2015**, *73*, 773–789. [CrossRef]

81. Zhou, H.; Lu, Y.; Zhang, M.; Guerin, G.; Manners, I.; Winnik, M.A. PFS-b-PNIPAM: A first step toward polymeric nanofibrillar hydrogels based on uniform fiber-like micelles. *Macromolecules* **2016**, *49*, 4265–4276. [CrossRef]

82. Groves, P. Diffusion ordered spectroscopy (DOSY) as applied to polymers. *Polym. Chem.* **2017**, *8*, 6700–6708. [CrossRef]

83. Wyman, I.W.; Liu, G. Micellar structures of linear triblock terpolymers: Three blocks but many possibilities. *Polymer* **2013**, *54*, 1950–1978. [CrossRef]

84. Gröschel, A.H.; Müller, A.H.E. Sell-assembly concepts for multicompartment nanostructures. *Nanoscale* **2015**, *7*, 11841–11876. [CrossRef] [PubMed]

85. Marsat, J.N.; Heydenreich, M.; Kleinpeter, E.; Berlepsch, H.; Böttcher, C.; Laschewsky, A. Self-assembly into multicompartment micelles and selective solubilization by hydrophilic-lipophilic-flurophilic block copolymers. *Macromolecules* **2011**, *44*, 2092–2105. [CrossRef]

86. Kyeremateng, S.O.; Busse, K.; Kohlbrecher, J.; Kressler, J. Synthesis and self-organization of poly(propylene oxide)-based amphiphilic and triphilic block copolymers. *Macromolecules* **2011**, *44*, 583–593. [CrossRef]

87. Zhang, Y.; Lin, W.; Jing, R.; Huang, J. Effect of block sequence on the self-assembly of ABC terpolymers in selective solvent. *J. Phys. Chem. B* **2008**, *112*, 16455–16460. [CrossRef] [PubMed]

88. Lerch, J.P.; Atanase, L.I.; Purcar, V.; Riess, G. Self-aggregation of poly(butadiene)-*b*-poly(2-vinylpyridine0-*b*-poly(ethylene oxide) triblock copolymers in heptanes studied by viscometry and dynamic light scattering. *C. R. Chim.* **2017**, *20*, 724–729. [CrossRef]

89. Lerch, J.P.; Atanase, L.I.; Riess, G. Adsorption of non-ionic ABC triblock copolymers: Surface modification of TiO₂ suspensions in aqueous and non-aqueous medium. *Appl. Surf. Sci.* **2017**, *419*, 713–719. [CrossRef]

90. Noolandi, J.; Hong, K.M. Theory of block copolymer micelles in solution. *Macromolecules* **1983**, *16*, 1443–1448. [CrossRef]

91. Tsitsilianis, C.; Sfika, V. Heteroarm star-like micelles formed from polystyrene-block-poly(2-vinyl pyridine)-block-poly(methyl methacrylate) ABC triblock copolymers in toluene. *Macromol. Rapid Commun.* **2001**, *22*, 647–651. [CrossRef]

92. Bütün, V.; Taktak, F.F.; Tuncer, C. Tertiary amine methacrylate-based ABC triblock copolymers: Synthesis, characterization, and self-assembly in both aqueous and nonaqueous media. *Macromol. Chem. Phys.* **2011**, *212*, 1115–1128. [CrossRef]

93. Njikang, G.; Han, D.; Wang, J.; Liu, G. ABC triblock micelle-like aggregates in selective solvents for A and C. *Macromolecules* **2008**, *41*, 9727–9735. [CrossRef]

94. Schacher, F.; Walther, A.; Ruppel, M.; Drechsler, M.; Müller, A.H.E. Multicompartment core micelles of triblock terpolymers in organic media. *Macromolecules* **2009**, *42*, 3540–3548. [CrossRef]

95. Dupont, J.; Liu, G. ABC triblock copolymer hamburger-like micelles, segmented cylinders, and Janus particles. *Soft Matter* **2010**, *6*, 3654–3661. [CrossRef]

96. Muslim, A.; Shi, Y.; Yan, Y.; Yao, D.; Rexit, A.A. Preparation of cylindrical multi-compartment micelles by the hierarchical self-assembly of ABC triblock polymer in solution. *RSC Adv.* **2015**, *5*, 85446–85452. [CrossRef]

97. Löbling, T.I.; Ikkala, O.; Gröschel, A.H.; Müller, A.H.E. Controlling multicompartment morphologies using solvent conditions and chemical modification. *ACS Macro Lett.* **2016**, *5*, 1044–1048. [CrossRef]

98. Cong, Y.; Zhou, Q.; Fang, J.; Zhu, K. Morphology transformation of multicompartment self-assemblies of ABC triblock copolymers. *Polymer* **2017**, *116*, 173–177. [CrossRef]

99. Wang, L.; Huang, H.; He, T. ABC triblock terpolymer self-assembled core-shel-corona nanotubes with high aspect ratios. *Macromol. Rapid. Commun.* **2014**, *35*, 1387–1396. [CrossRef] [PubMed]

100. Löbling, T.I.; Borisov, O.; Haataja, J.S.; Ikkala, O.; Gröschel, A.H.; Müller, A.H.E. Rational design of ABC triblock terpolymer solution nanostructures with controlled patch morphology. *Nat. Commun.* **2016**, *7*, 12097. [CrossRef] [PubMed]

101. Kitazawa, Y.; Ueki, T.; McIntosh, L.D.; Tamura, S.; Niitsuma, K.; Imaizumi, S.; Lodge, T.P.; Watanabe, M. Hierarchical sol-gel transition induced by thermosensitive self-assembly of an ABC triblock polymer in an ionic liquid. *Macromolecules* **2016**, *49*, 1414–1423. [CrossRef]

102. Wang, X.S.; Winnik, M.A.; Manners, I. Synthesis and solution self-assembly of coil-crystalline-coil polyferrocenylphosphine-*b*-polyferrocenylsilane-b-polysiloxane triblock copolymers. *Macromolecules* **2002**, *35*, 9146–9150. [CrossRef]

103. Schmelz, J.; Karg, M.; Hellweg, T.; Schmalz, H. General pathway toward crystalline-core micelles with tunable morphology and corona segregation. *ACS Nano* **2011**, *5*, 9523–9534. [CrossRef] [PubMed]

104. Atanase, L.I.; Lerch, J.P.; Riess, G. Water dispersibility of non-aqueous emulsions stabilized and viscosified by a poly(butadiene)-poly(2-vinylpyridine)-poly(ethylene oxide) (PBut-P2VP-PEO) triblock copolymer. *Colloids Surf. Part A Physicochem. Eng. Asp.* **2015**, *464*, 89–95. [CrossRef]

105. Betthausen, E.; Hanske, C.; Müller, M.; Fery, A.; Schacher, F.H.; Müller, A.H.E.; Pochan, D.J. Self-assembly of amphiphilic triblock terpolymers mediated by multifunctional organic acids: Vesicles, toroids, and (undulated) ribbons. *Macromolecules* **2014**, *47*, 1672–1683. [CrossRef]

106. Kikuchi, A.; Nose, T. Unimolecular micelle formation of poly(methyl methacrylate)-graft-polystyrene in mixed selective solvents of acetonitrile/acetoacetic acid ethyl ether. *Macromolecules* **1996**, *29*, 6770–6777. [CrossRef]

107. Borisov, O.V.; Zhulina, E.B. Amphiphilic graft copolymer in a selective solvent: Intramolecular structures and conformational transitions. *Macromolecules* **2005**, *38*, 2506–2514. [CrossRef]

108. Liu, W.; Liu, Y.; Hao, X.; Zeng, G.; Wang, W.; Liu, R.; Huang, Y. Backbone-collapsed intra- and inter-molecular self-assembly of cellulose-based dense graft copolymer. *Carbohydr. Polym.* **2012**, *88*, 290–298. [CrossRef]

109. Francis, R.; Baby, D.K.; Gnanou, Y. Synthesis and self-assembly of chitosan-g-polystyrene copolymer: A new route for the preparation of heavy metal nanoparticles. *J. Colloid Interface Sci.* **2015**, *438*, 110–115. [CrossRef] [PubMed]

110. Saha, A.; Jana, S.; Mandal, T.K. Peptide-poly(*tert*-butyl methacrylate) conjugate into composite micelles in organic solvents versus peptide-poly(methacrylic acid) conjugate into spherical and worm-like micelles in water: Synthesis and self-assembly. *J. Polym. Sci. Part A Polym. Chem.* **2016**, *54*, 3019–3031. [CrossRef]

111. Bose, A.; Jana, S.; Saha, A.; Mandal, T.K. Amphiphilic polypeptide-polyoxazoline graft copolymer conjugate with tunable thermoresponsiveness: Synthesis and self-assembly into various micellar structures in aqueous and nonaqueous media. *Polymer* **2017**, *110*, 12–24. [CrossRef]

112. Stepánek, M.; Kosovan, P.; Procházka, K. Self-assembly of poly(4-methylstyrene)-*g*-poly(methacrylic acid) graft copolymer in selective solvents for grafts: Scattering and molecular dynamics simulation study. *Langmuir* **2010**, *26*, 9289–9296. [CrossRef] [PubMed]

113. Gromadzki, D.; Filippov, S.; Netopilik, M.; Makuska, R.; Jigounov, A.; Plestil, J.; Horsky, J.; Stepánek, P. Combination of "living" nitroxide-mediated and photoiniferter-induced "grafting from" free-radical polymerizations: From branched copolymers to unimolecular micelles and microgels. *Eur. Polym. J.* **2009**, *45*, 1748–1758. [CrossRef]

114. Koda, Y.; Terachima, T.; Sawamoto, M. Multimode self-folding polymers via reversible and thermoresponsive self-assembly of amphiphilic/fluorous random copolymers. *Macromolecules* **2016**, *49*, 4534–4543. [CrossRef]

115. Liu, M.; Fu, Z.; Wang, Q.; Xu, J.; Fan, Z. Study of amphiphilic poly(1-dodecene-*co*-para-methylstyrene)-graft-poly(ethylene glycol). Part II: Preparation and micellization behavior of the amphiphilic copolymers. *Eur. Polym. J.* **2008**, *44*, 4122–4128. [CrossRef]

116. Alexander, S.; Cosgrove, T.; de Vos, W.M.; Castle, T.C.; Prescott, S.W. Aggregation behavior of polyisoprene-pluronic graft copolymers in selective solvents. *Langmuir* **2014**, *30*, 5747–5754. [CrossRef] [PubMed]

117. Mo, Y.; Liu, G.; Tu, Y.; Lin, S.; Song, J.; Hu, J.; Liu, F. Morphological switching of unimolar micelles of ternary graft copolymers in different solvents. *J. Polym. Sci. Part A Polym. Chem.* **2017**, *55*, 1021–1030. [CrossRef]

![polymers logo] *polymers*

MDPI

Article

Theory of the Flower Micelle Formation of Amphiphilic Random and Periodic Copolymers in Solution

Takahiro Sato

Department of Macromolecular Science, Osaka University, 1-1 Machikaneyama-cho, Toyonaka, Osaka 560-0043, Japan; tsato@chem.sci.osaka-u.ac.jp; Tel.: +81-6-6850-5461

Received: 9 December 2017; Accepted: 11 January 2018; Published: 14 January 2018

Abstract: The mixing Gibbs energy Δg_m for the flower-micelle phase of amphiphilic random and periodic (including alternating) copolymers was formulated on the basis of the lattice model. The formulated Δg_m predicts (1) the inverse proportionality of the aggregation number to the degree of polymerization of the copolymer, (2) the increase of the critical micelle concentration with decreasing the hydrophobe content, and (3) the crossover from the micellization to the liquid–liquid phase separation as the hydrophobe content increases. The transition from the uni-core flower micelle to the multi-core flower necklace as the degree of polymerization increases was also implicitly indicated by the theory. These theoretical results were compared with experimental results for amphiphilic random and alternating copolymers reported so far.

Keywords: amphiphilic polymers; random copolymers; alternating copolymers; flower micelles; flower necklaces; vesicle

---

1. Introduction

Borisov and Halperin [1–5] proposed theoretical models of flower micelles, flower necklaces, and bouquets of polymer micelles formed by amphiphilic periodic copolymers composed of hydrophilic and hydrophobic monomer units in aqueous solutions. They assumed that the main chain of the periodic copolymer is perfectly flexible, and all hydrophobes in the copolymer chain are included in the hydrophobic core(s) of the micelle.

Afterward, experimental studies on amphiphilic random and periodic (including alternating) copolymers bearing hydrophobic side chains demonstrated the formation of flower micelles and flower necklaces in aqueous solutions [6–11]. However, experimental results indicated that not all hydrophobic side chains on the copolymer chain are included in the hydrophobic core(s) of the micelle, being different from the Borisov–Halperin model, and that the loop-chain size is determined by the main-chain stiffness rather than the content and sequence of the hydrophobic side chain on the copolymer chain. Thus, we need a new theory to discuss the micellization behavior of such flower micelles and flower necklaces.

The present paper proposes a lattice-model theory for dilute aqueous solutions of amphiphilic random and periodic copolymers bearing hydrophobic linear side chains, which can be regarded as graft copolymer chains bearing hydrophobic graft (side) chains to demonstrate the formation of the flower micelle. Recently, Sato and Takahashi [12] presented a similar lattice-model theory for amphiphilic block copolymer solutions to discuss the competition between micellization and liquid–liquid phase separation in the solutions. The present theory is the random and periodic copolymer version of this theory.

2. Theory

Let us consider the graft copolymer illustrated in Figure 1a. The main chain and graft chains consist of P_M units and P'_G units, respectively. The mole fraction of the branch units on the main chain is denoted as x, and the distribution of the branch units along the main chain is assumed to be random or periodic (not block-like). The total number of the graft-chain units per copolymer chain is $P_G = xP_M P'_G$, and the total degree of polymerization of the graft copolymer chain is $P = P_M + P_G = P_M(1 + xP'_G)$. It is assumed that the main-chain and graft-chain units as well as the solvent S molecule occupy lattice sites with a common size a.

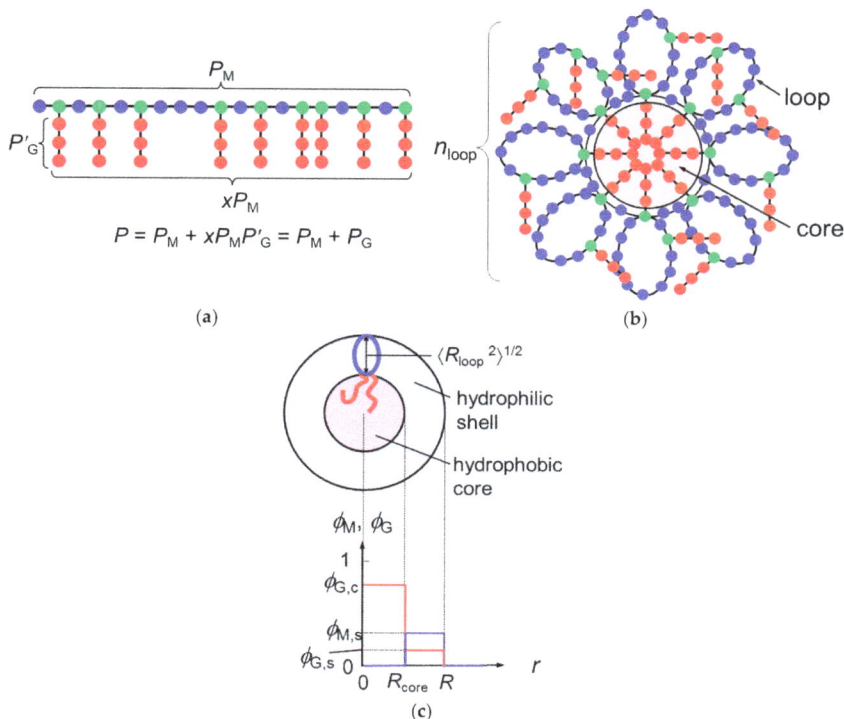

$$P = P_M + xP_M P'_G = P_M + P_G$$

(a)   (b)

(c)

Figure 1. Schematic diagrams of the graft copolymer chain (**a**), the flower micelle formed by the copolymer chain (**b**), and the radial concentration profiles of the main-chain and graft-chain units in the flower micelle (**c**). In Panel b, the flower micelle is constructed by m copolymer chains. In Panel a, blue and green circles are referred to as the main chain, and red circles as the graft chain. In Panels a and b, blue circles are called the A unit, and red and green circles as the B unit to discuss the mixing enthalpy (cf. Equations (9)–(11)).

If the graft-chain unit is sufficiently hydrophobic, graft chains of the copolymer tend to aggregate to form a hydrophobic core, and the main chain tends to form loop chains in aqueous medium. As a result, the m copolymer chains construct a flower micelle illustrated in Figure 1b; m is the copolymer-chain aggregation number of the micelle. Only graft chains attaching to roots of the loops can enter the hydrophobic core, and the remaining graft chains are outside the core.

According to the wormlike chain model [13–15], the ring closure probability of the chain rapidly diminishes to zero at the chain contour length reducing to ca. $1.6q$, where q is the persistence length. This means that the main chain portion shorter than $1.6q$ cannot form the loop because of the chain stiffness. In what follows, we consider the flower micelle consisting of loop chains with this "minimum loop size" [6]. The number of main-chain units per the minimum loop chain P_{loop}, and the number of loop chains per chain n_{loop} are calculated by

$$P_{loop} = \frac{1.6q}{a}, n_{loop} = \frac{P_M}{P_{loop}} \tag{-}$$

The numbers of graft chains included in the hydrophobic core and outside of the core, $x_c P_M$ and $x_s P_M$, respectively, are calculated by

$$x_c P_M = \lambda n_{loop}, x_s = x - x_c \tag{2}$$

where λ is the number of side chains included in the core at each root of the loop. (Figure 1b illustrates the case of $\lambda = 1$). It has been assumed in Equation (2) that n_{loop} is much larger than unity.

In the previous paper [12], we regarded the spherical micelle formed by di-block copolymer chains as a thermodynamic phase, assuming that the aggregation number of the micelle is sufficiently large. Similarly, the present study regards the flower micelle as a thermodynamic phase to demonstrate the micellization of the graft copolymer in a selective solvent. Furthermore, we use a simple model for the flower micellar phase, of which radial concentration profiles (volume fractions) of the main-chain and graft-chain units are given by

$$\phi_M = \begin{cases} 0, & 0 \leq r < R_{core} \\ \phi_{M,s}, & R_{core} \leq r < R \\ 0, & R \leq r \end{cases}, \phi_G = \begin{cases} \phi_{G,c}, & 0 \leq r < R_{core} \\ \phi_{G,s}, & R_{core} \leq r < R \\ 0, & R \leq r \end{cases} \tag{3}$$

(cf. Figure 1c). Here, R_{core} and R are the radii of the micelle core and the whole micelle, respectively, and the solvent volume fraction is given by $\phi_S = 1 - \phi_M - \phi_G$ at each radial distance r. Furthermore, using the wormlike chain model, R_{core}^2 and the mean square distance from the end to the midpoint of the loop $\langle R_{loop}^2 \rangle$ (cf. Figure 1c) are expressed in terms of the persistence lengths of the graft chain (q_G) and of the copolymer main chain (q), respectively, by [15]

$$(R_{core}/a)^2 = (2q_G/a)P'_G - 2(q_G/a)^2\left(1 - e^{-P'_G a/q_G}\right) \tag{4}$$

$$\frac{\langle R_{loop}^2 \rangle}{a^2} = \frac{P_{loop}^2}{\sqrt{42.5 - \frac{32}{3}\left(aP_{loop}/2q\right) + 16\left(aP_{loop}/2q\right)^2}} \tag{5}$$

(cf. Appendix A). The radius R of the whole micelle is calculated by

$$R = R_{core} + \langle R_{loop}^2 \rangle^{1/2} \tag{6}$$

The average volume fraction ϕ_P of the copolymer in the flower micelle phase is given by

$$\phi_P = \frac{3a^3 Pm}{4\pi R^3} \tag{7}$$

and the volume fractions $\phi_{M,s}$, $\phi_{G,s}$, and $\phi_{G,c}$ are related to ϕ_P by

$$\phi_{M,s} = \frac{R^3 P_M \phi_P}{(R^3 - R_{core}^3)P}, \phi_{G,s} = \frac{R^3 x_s P_M P'_G}{(R^3 - R_{core}^3)P}\phi_P, \phi_{G,c} = \frac{R^3 x_c P_M P'_G}{R_{core}^3 P}\phi_P \tag{8}$$

From the last equation for $\phi_{G,c}$ in Equation (8), it can be seen that ϕ_P must be equal to or less than $PR_{core}{}^3/x_c P_M P'_G R^3$, because $\phi_{G,c}$ does not exceed unity. Furthermore, since m must be larger than unity, P_M must be smaller than $4\pi R^3 \phi_P/3a^3(1 + xP'_G)$ from Equation (7).

For amphiphilic random or periodic copolymers, the ionizable group or hydrophilic side-chain group of each hydrophilic monomer unit is substituted by the hydrophobic graft chain. Thus, the branch unit in the main chain (green circles in Figure 1a,b) may be hydrophobic, having interaction parameters much different from those of the non-branch unit (i.e., the hydrophilic monomer unit) in the main chain but similar to those of the graft-chain unit. We refer to the non-branch unit in the main chain as the A unit and to the graft-chain unit as well as the branch unit in the main chain as the B unit, neglecting the difference in the interaction between the graft-chain unit and the branch unit in the main chain. The volume fractions of the A and B units in the shell and core phases are given by

$$\phi_{A,s} = \frac{R^3(1-x)P_M}{(R^3 - R_{core}{}^3)P}\phi_P, \quad \phi_{B,s} = \frac{R^3(x + x_s P'_G)P_M}{(R^3 - R_{core}{}^3)P}\phi_P, \quad \phi_{B,c} = \frac{R^3 x_c P'_G P_M}{R_{core}{}^3 P}\phi_P \qquad (9)$$

and the mole fractions of the A and B units in the copolymer chain are written as

$$x_A = (1-x)\frac{P_M}{P}, \quad x_{B,s} = (x + x_s P'_G)\frac{P_M}{P}, \quad x_{B,c} = x_c P'_G \frac{P_M}{P}, \quad x_A + x_{B,s} + x_{B,c} = 1 \qquad (10)$$

where $x_{B,s}$ and $x_{B,c}$ are the mole fractions of the B unit in the shell and core regions, respectively. The solvent volume fractions in the shell and core regions are given by $\phi_{S,s} = 1 - \phi_{A,s} - \phi_{B,s}$ and $\phi_{S,c} = 1 - \phi_{B,c}$, respectively.

We apply the Flory–Huggins theory [16] to the flower micelle phase to formulate the mixing Gibbs energy per lattice site Δg_m of the micelle phase, which consists of the mixing entropy ΔS, the mixing enthalpy ΔH, and the interfacial Gibbs energy $4\pi R_{core}{}^2 \gamma$ (γ: the interfacial tension between the core and shell regions of the micelle). The formulation method is described in Appendix B. The final result is written as

$$
\begin{aligned}
\frac{\Delta g_m}{k_B T} &= \left(\frac{-T\Delta S + \Delta H + 4\pi R_{core}{}^2 \gamma}{k_B T}\right) \Big/ \frac{4\pi R^3}{3a^3} \\
&= \frac{\phi_P}{P}\ln(\kappa \phi_P) + \frac{R^3 - R_c{}^3}{R^3}\phi_{S,s}\ln\phi_{S,s} + \frac{R_c{}^3}{R^3}\phi_{S,c}\ln\phi_{S,c} \\
&\quad + [x_A \phi_{S,s}\chi_{AS} + (x_{B,s}\phi_{S,s} + x_{B,c}\phi_{S,c})\chi_{BS} - x_A(x_{B,s} + x_{B,c} - \phi_{B,s})\chi_{AB}]\phi_P \\
&\quad + \frac{3(R_{core}/a)^2}{(R/a)^3}\frac{a^2}{k_B T}\gamma
\end{aligned} \qquad (11)
$$

where χ_{AS}, χ_{BS}, and χ_{AB} are the interaction parameters between S and A, between S and B, and between A and B, respectively, κ is defined by Equation (B11), and $(a^2/k_B T)\gamma$ is calculated by Equation (B13) with Equation (B14). The term $\ln \kappa$ includes the conformational entropy loss at the formation of the flower micelle.

When the graft copolymer solution is homogeneous, the mixing Gibbs energy per lattice site Δg_h is given by [16]

$$\frac{\Delta g_h}{k_B T} = (1 - \phi_P)\ln(1 - \phi_P) + \frac{\phi_P}{P}\ln\phi_P + \bar{\chi}(1 - \phi_P)\phi_P \qquad (12)$$

with the average interaction parameter $\bar{\chi}$ between the graft copolymer chain and solvent, defined by [17]

$$\bar{\chi} \equiv x_A \chi_{AS} + (1 - x_A)\chi_{BS} - x_A(1 - x_A)\chi_{AB}. \qquad (13)$$

3. Results and Discussion

Because we did not consider above the interaction among flower micellar phases in the solution, the following discussion is limited to dilute solutions of random and periodic copolymers. Ueda et al. [9] reported the molecular weight dependence of the micellization behavior for the amphiphilic alternating copolymer of sodium maleate and dodecyl vinyl ether, P(MAL/C12), in dilute aqueous solutions including 0.05 M NaCl. First, we examine theoretically the micellization behavior of an alternating copolymer mimicking P(MAL/C12).

In the lattice theory, the choice of the unit lattice site is rather arbitrary. Here, we assume the main-chain portion (the C_2 unit) of maleate or dodecyl vinyl ether monomer unit is chosen as the unit lattice site. Then, the hydrophobic dodecyl side chain is assumed to occupy six lattice sites, i.e., $P'_G = 6$. (The carboxy group and the ether oxygen atom in the maleate and dodecyl vinyl ether monomer units are not considered explicitly; they are assumed to be included in the main-chain portions). In aqueous solutions, a strong electrostatic repulsion acts among maleate units (the A unit), while a hydrophobic attraction acts among the C_2 units of the dodecyl group (the B unit). The strong electrostatic repulsion and hydrophobic attraction are expressed using a negative χ_{AS} and positive χ_{BS}, respectively. (To account for the long range electrostatic interaction, the unit lattice site may have to be larger than the C_2 unit, but the following results do not essentially change by the choice of the unit lattice site). Since we here focus on the amphiphilicity of the graft copolymer, we assume χ_{AB} to be zero, as in the previous study [12]. (The change of the χ_{AB} value may be compensated by adjusting values of χ_{AS} and χ_{BS}).

Figure 2 shows the copolymer concentration dependences of Δg_m (red curve) and Δg_h (black curve) calculated by Equations (11) and (12). We have chosen $P_M = 50$, $x = 0.5$, $\chi_{AS} = -15$, $\chi_{BS} = 3$, and $\chi_{AB} = 0$ ($\bar{\chi} = 0.75$). All remaining parameters included in Equation (11) can be calculated from $a = 0.25$ nm (the contour length per the main-chain monomer (C_2) units), and $q = 3$ nm, $q_G = 0.53$ nm, and $\lambda = 3$ determined previously [9]. We can draw a common tangent (the thin line) to the dilute side of the black curve and red curve. (It is seen that the black curve has a downward convex shape around $\phi_P = 0$, if it is enlarged). The copolymer volume fractions at the two points of contact of the common tangent, denoted as $\phi_{P,d}$ and $\phi_{P,m}$, are binodal concentrations of the coexisting dilute and micellar phases, respectively. The tangent line is below the common tangent line (the thin broken line) for the thick black curve for Δg_h, indicating that the micellization is thermodynamically more stable than the phase separation into two homogeneous phases.

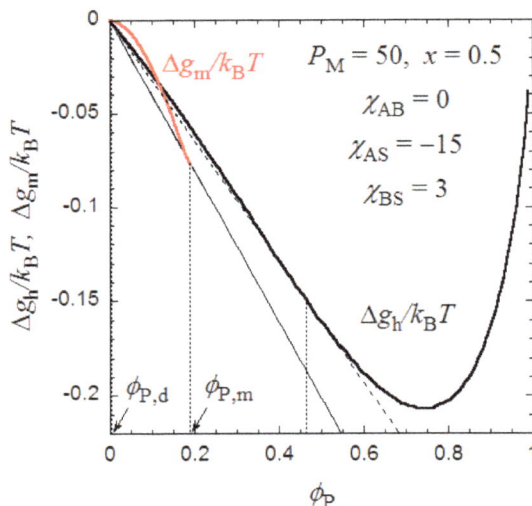

Figure 2. Concentration dependences of Δg_m and Δg_h at $x = 0.5$ calculated by Equations (9) and (10).

Similar curves for Δg_m and Δg_h were obtained for different P_M, and the volume fraction $\phi_{P,m}$ of the equilibrium micellar phase were determined by the above method. The aggregation number m of the copolymer chains per micelle can be calculated from Equation (7), i.e.,

$$m = \frac{4\pi R^3}{3a^3 P} \phi_{P,m} \tag{14}$$

Figure 3 shows the degree of polymerization P_M dependence of m such obtained as well as the product mP_M (the number of monomer units per micelle) at the interaction parameters identical to those in Figure 2. It is seen that m is inversely proportional to P_M, and the product mP_M is independent of P_M. (Because P is proportional to P_M and R is independent of P_M, the inverse proportionality of m to P_M comes from the P_M independence of $\phi_{P,m}$ calculated from the comparison between of the Δg_m and Δg_h curves). This relation was observed experimentally for P(MAL/C12) in 0.05 M aqueous NaCl solution [9] as well as for a random copolymer of poly(ethylene glycol) methyl ether methacrylate and dodecyl methacrylate, P(PEGMA/DMA), in water [18]; however, for P(PEGMA/DMA) with $x = 0.5$, the constant mP_M is slightly larger than 300. The value of mP_M changes by values of q, λ, and the interaction parameters. It is noted that the formulation of Δg_m in the previous section can apply both to periodic and random copolymers.

When P_M approaches 300 in Figure 3, m tends to unity, and $\phi_{P,d}$ corresponding to the critical micelle concentration (cmc) of the coexisting dilute phase becomes very low (not shown). That is, when P_M approaches 300, the flower micelle is formed by one copolymer chain (the unimer micelle), and the cmc tends to zero. This situation resembles the liquid–liquid phase separation in a homopolymer polymer solution with an infinitely high-molecular-weight polymer, where the polymer volume fraction at the critical point is predicted to be zero by the conventional Flory–Huggins theory [16].

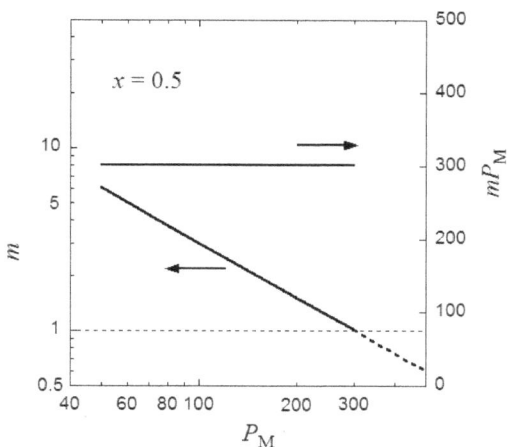

Figure 3. Degree of polymerization dependences of the aggregation number m and mP_M at the interaction parameters identical to those in Figure 2.

When the same calculation of m is performed where $P_M > 300$, the inverse proportionality of m to P_M still holds even if P_M exceeds 300, as indicated by the dashed line in Figure 3. However, because the aggregation number is less than unity, some portion of the main chain is not included in the flower micelle at $P_M > 300$. For example, at $P_M = 600$ where $m = 0.5$, half of the main chain is not included in the flower micelle. This half main-chain portion may form another flower micelle. As a result, the whole copolymer chain forms a double-core flower necklace. (Strictly speaking, the double-core flower necklace needs a bridge chain connecting two unit flowers, so that P_M must be slightly larger than 600 to form the double-core flower necklace). In fact, Ueda et al. [9] reported the transition from the flower micelle to the flower necklace at P_M exceeding 300.

The flower micelle is formed also by amphiphilic random copolymers with hydrophobic dodecyl side chains of $x < 0.5$ in aqueous solutions. Next, we examine the hydrophobic monomer content dependence of the micellization for an aqueous solution of an amphiphilic random copolymer, calculated in the same way from the Δg_m and Δg_h curves as in Figure 2. The number λ of side chains included in the core at each root of the loop appearing in Equation (2) may be dependent on the monomer content x. In the limit of $x = 1/P_{loop}$, each loop chain has only one hydrophobic side chain on average. Thus, $\lambda = 1$ at $x = 1/P_{loop}$. When x increases, λ may first increase from unity and approach an asymptotic value. For a given value of λ, $\phi_{P,d}$ and $\phi_{P,m}$ of the coexisting dilute and micellar phases can be calculated as functions of x from the curves of Δg_m and Δg_h as mentioned above.

Figure 4 shows $\phi_{P,d}$ and $\phi_{P,m}$ obtained for the amphiphilic random copolymer with a P_M of 50 and the same interaction parameters used in Figures 2 and 3, in the x-ϕ_P phase diagram. The x dependence of λ used is shown in the insert of Figure 4. When x is decreased from 0.5, $\phi_{P,d}$ (cmc) increases, and the copolymer in a dilute solution ($\phi_P < 0.08$) transforms from the flower micelle to the random coil at passing the bimodal curve for $\phi_{P,d}$ (cmc). At $x < 1/P_{loop}$, the loop size of the flower micelle should be larger than the minimum size given by Equation (1). We do not discuss such a loose flower micelle here.

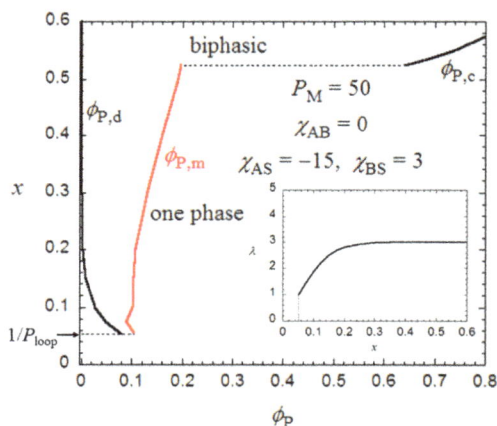

Figure 4. Monomer content-concentration phase diagram for an aqueous solution of a random copolymer with $P_M = 50$ and the same interaction parameters as those used in Figures 2 and 3.

On the other hand, when x increases from 0.5, the Δg_h – ϕ_P curve goes down relative to the Δg_m – ϕ_P curve, and as shown in Figure 5, at $x = 0.524$, we can draw a common tangent (the thin line) to the dilute and concentrated sides of the black curve (Δg_h) and the red curve (Δg_m). When $x > 0.524$, the phase separation into dilute and concentrated homogeneous phases with concentrations $\phi_{P,d}$ and $\phi_{P,c}$ becomes thermodynamically more stable than the micellization. As a result, the phase gap in the x-ϕ_P phase diagram is abruptly enlarged when $x > 0.524$, as shown in Figure 4. To the best of my knowledge, there have hitherto been no reports of the corresponding crossover from micellization to liquid–liquid phase separation as the hydrophobic content x increases.

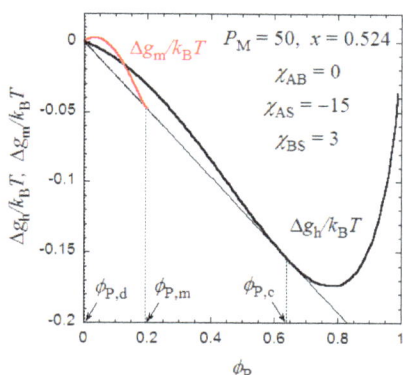

Figure 5. Concentration dependences of Δg_m and Δg_h at $x = 0.524$ calculated by Equations (9) and (10).

Eisenberg et al. [19] investigated the random copolymer of styrene and methacrylic acid with $x \sim 0.8$, which was first dissolved in dioxane or tetrahydrofuran (THF), followed by the addition of water, and observed "large compound micelles" and "bowl-shaped aggregates" by transmission electronic microscopy (TEM). Here, the "large compound micelle" is the large homogeneous polymer-rich spheres, corresponding to the droplet of the concentrated homogeneous phase formed by the liquid–liquid phase separation, predicted in Figure 4, and a "bowl-shaped aggregate" may be formed from the concentrated-phase droplet in which solvent bubbles are trapped [19]. Wang et al. [20] reported the formation of uniform colloidal spheres by an amphiphilic random copolymer, poly{2-[4-(phenylazo)phenoxy]ethyl acrylate-*co*-acrylic acid}, where $x = 0.5$ in THF–water mixtures with high water concentrations. This may be another example of the liquid–liquid phase separation of the amphiphilic random copolymer in solution. Zhang et al. [21] studied the self-association of amphiphilic graft (periodic) copolymers in a hypothetical solution of the two-dimensional space by the self-consistent field theory. Although they assumed perfect flexibility and comparable chain lengths of the main and graft chains, being different conditions from the present study, they observed a "large compound micelle" at higher graft density (i.e., higher hydrophobic content x) under weaker amphiphilicity (cf. Figure 8a in [21], where the graft chain number = 5).

Yusa et al. [22] found a transition from the unimer micelle to the single random coil chain of a random copolymer of hydrophilic sodium 2-(acrylamido)-2-methylpropanesulfonate and hydrophobic 11-acrylamidoundecanoic acid (AmU) where $x = 0.5$ in 0.1 M aqueous NaCl solution by changing pH. At pH = 3, where the carboxy group is not ionized, AmU was strongly hydrophobic, and the copolymer formed a unimer micelle with $m = 1$. The degree of polymerization P_M of the copolymer sample (= 475) was slightly larger than 300 (cf. Figure 3), maybe due to the difference in the parameters, e.g., P'_G and λ, from those used in Figure 3. On the other hand, at pH = 9, where the carboxy group of AmU is ionized, the copolymer was transformed to a random coil. In Figure 2, the Δg_m curve goes up and the Δg_h has no inflection point when χ_{BS} is decreased from 3, i.e., AmU becomes more hydrophilic. As a result, the random coil conformation in the homogeneous phase becomes more stable than the flower micelle, which is consistent with Yusa et al.'s finding. The transition from the unimer micelle to the single random coil chain by decreasing x, predicted in Figure 4, was reported by Fujimoto and Sato [11].

Recently, several authors have reported that amphiphilic random copolymers form vesicles in dilute solutions [23–26], which was not considered in the present study. Zhu and Liu [24] investigated vinyl polymers bearing L-glutamic acid moieties and dodecyl groups in the random sequence to find the vesicle in water at a high hydrophobic content x (>0.75). Their random copolymer samples possess low degrees of polymerization (<36). For these samples to form the flower micelle, n_{loop} should be less than 2 and one loop chain should bear many hydrophobes. The present theory may not be able to be applied to such random copolymers.

Tian et al. [25] observed vesicles as well as hollow tubes and wormlike rods formed by poly(hydroxyethyl methacrylate) (PHEMA) partially and randomly modified by the hydrophobic 2-diazo-1,2-naphthoquinone in solution. These copolymer samples were dissolved in dimethyl-formamide, followed by the addition of water, and finally dialyzed against water to form the vesicle. Because even PHEMA is insoluble in water, the vesicle formed must not be in the thermodynamically stable state, which cannot be treated in the present statistical thermodynamic theory.

Ghosh et al. [26] reported that an amphiphilic random copolymer of hydrophilic tri(oxyethylene) methacrylamide and hydrophobic n-octyl methacrylate exhibited a thermally induced vesicle to spherical micelle transition. However, it should be noted that the illustration of the spherical micelle by these authors (cf. Scheme 1 in [26]) was inconsistent with the experimental TEM observation of spherical aggregates (diameter in the range of 70–80 nm) at 60 °C. In the illustration, the hydrophilic and hydrophobic side chains were in the coronal and core regions of the micelle, respectively, and the whole copolymer main chain was confined to the corona-core interface. If this is the case, the diameter of the micelle must be equal to twice the sum of the hydrophilic and hydrophobic side chain lengths. Even if the side chains are fully extended, such an estimated diameter is as small as 6 nm, which is much smaller than the diameter of the spherical aggregate at 60 °C. Thus, the spherical aggregate at 60 °C may not be the spherical micelle indicated in their illustration, but the phase-separated concentrated phase droplet, because both kinds of side chains are hydrophobic at 60 °C above the lower critical solution temperature [26].

4. Conclusions

The flower micelle formed by amphiphilic random and periodic copolymers in solution was regarded as a thermodynamic phase to formulate the mixing Gibbs energy. The formulated mixing Gibbs energy of the micelle was compared with that of the homogeneous phase to calculate (1) the aggregation number m of the micelle as a function of the degree of polymerization P_M of the copolymer chain, (2) the cmc as a function of the hydrophobic content x, and (3) the crossover x from micellization to liquid–liquid phase separation.

The above theoretical results were compared with experimental results for amphiphilic random and alternating copolymers reported previously. Prediction (1) was confirmed experimentally [9,18], and the experimentally observed transition from the uni-core flower micelle to the multi-core flower necklace [9] was also consistent with the present theory. The "large compound micelle" previously observed for amphiphilic random copolymers [19,20] may correspond to the concentrated-phase droplets produced by liquid–liquid phase separation, which was predicted to occur in this theory. The transition from the unimer micelle to the single random coil chain [11,22] was also predicted by this theory.

The limitation of the present theory was also discussed. The present theory may not be able to be applied to amphiphilic random copolymers of low degrees of polymerization and high hydrophobic contents [24], nor to frozen micelles that are not in a thermodynamic equilibrium state [25].

Acknowledgments: I thank Dr. Daichi Ida at Kyoto University for calculating Equation (A4) in Appendix A.

Conflicts of Interest: The author declares no conflict of interest.

Appendix A. Mean Square Distance from the End to the Midpoint of the Loop

The mean square distance from the end to the midpoint of the loop $\langle R_{loop} \rangle$ (cf. Figure 1c) near the rod and coil limits is calculated using the wormlike chain model. Yamakawa and Stockmayer [13] formulated $\langle R_{loop}{}^2 \rangle$ for the wormlike chain near the rod limit. Their result is written as

$$\langle R_{loop}{}^2 \rangle = [2q(I_3/2I_2)N_K]^2 (N_K \ll 1) \tag{A1}$$

where q and N_K is the persistence length and the Kuhn statistical segment number, respectively, and I_2 and I_3 are calculated by

$$I_2 \equiv \int_{\theta/2}^{\pi} \frac{1}{\sqrt{C - \cos\omega}} d\omega, \, I_3 \equiv \frac{1}{2I_2} \int_{\theta/2}^{\pi} \frac{\sin\omega}{\sqrt{C - \cos\omega}} d\omega \tag{A2}$$

with the angle θ formed by the tangent vectors at both chain ends and a constant C determined by the equation

$$I_1 \equiv \int_{\theta/2}^{\pi} \frac{\cos\omega}{\sqrt{C - \cos\omega}} d\omega = 0 \tag{A3}$$

Near the rod limit, the energetically most stable loop conformation gives us the results, $\theta = 1.7208$, $C = 0.6522$, $I_2 = 3.29$, and $I_3 = 2.58$.

Using the first Daniels approximation, we can calculate $\langle R_{loop}^2 \rangle$ near the coil limit as [27]

$$\langle R_{loop}^2 \rangle = (2q)^2 \left(\frac{1}{4} N_K + \frac{1}{12} \right) (N_K \gg 1) \tag{A4}$$

Equation (5) in the text is the interpolation of $\langle R_{loop}^2 \rangle$ given by Equations (A1) and (A4) near rod and coil limits by use of the Padé approximation.

Appendix B. Mixing the Gibbs Energy of the Flower Micelle Phase

To calculate the mixing entropy of the flower micelle phase, we counted the number Ω of arrangements of m graft copolymer chains into the concentric spherical lattice with the inner and outer spherical radii R_{core} and R, respectively, shown in Figure 1c in the text. Each main chain of the graft copolymer may form loops, trains, and tails on the core–shell interface, but we assume that both hydrophilicity of the main chain and hydrophobicity of the graft-chain are so strong that both train and tail chains are negligibly short.

The first unit of the main chain in the first copolymer chain must be located at one of the lattice sites on the core–shell interface. The number of such lattice sites is given by $N_{intf} = 4\pi (R_{core}/a)^2$. The number ω'_i of lattice sites where the first unit of the i-th copolymer chain is given by

$$\omega'_i = N_{intf} f'_{i-1}, f'_{i-1} = 1 - \frac{n_{loop}}{N_{intf}} (i-1)(1 \le i \le m) \tag{B1}$$

where f'_{i-1} is the probability of the vacancy for the lattice site on the core–shell interface when first units of $i-1$ copolymer chains have been already arranged.

The flower micelle contains mn_{loop} loop chains and $mx_s P_M$ graft chains in the shell region. The first unit of the first copolymer chain is identical with the first unit of the first loop chain, and the last unit of the first loop chain must be located in the neighboring site of the first one of the same loop chain on the core–shell interface. Furthermore, the loop chain cannot be located in the core region, i.e., the core–shell interface acts as a reflecting barrier. The first loop chain possesses $x_s P_{loop}$ graft chains with the degree of polymerization P'_G. Thus, the number of arrangements $\omega_{s,1}$ of the first loop chain is given by [12,28]

$$\omega_{s,1} = \frac{(z-1)^{P_{loop}}}{(\sqrt{\pi}/3) P_{loop}} G(0; aP_{loop}/2q)(z-1)^{x_s P_{loop} P'_G} \tag{B2}$$

where $G(0, aP_{loop}/2q)$ is the ring closure probability. Shimada and Yamakawa [14] proposed an expression of the probability for the wormlike chain as

$$G(0; aP_{loop}/2q) = \frac{28.01}{\left(aP_{loop}/2q \right)^5} \exp\left[-\frac{7.027}{aP_{loop}/2q} + 0.492 \left(aP_{loop}/2q \right) \right] \tag{B3}$$

Similarly, the number of arrangements $\omega_{s,i}$ of the i-th loop chain is given by

$$\omega_{s,j} = \omega_{s,1} f_{s,j-1}^{P_{\text{loop}}(1 + x_s P'_G)}, f_{s,j-1} = 1 - \frac{\phi_{M,s} + \phi_{G,s}}{mn_{\text{loop}}}(j-1)\left(1 \leq j \leq mn_{\text{loop}}\right) \quad \text{(B4)}$$

Here, $f_{s,j-1}$ is the probability of the vacancy for the lattice site in the shell region when $j-1$ loop chains have been already arranged. The core region of the flower micelle consists of $mx_c P_M$ graft chains. Numbers of arrangements $\omega_{c,1}$ and $\omega_{c,i}$ of the first and i-th graft chains in the core region are written as

$$\omega_{c,1} = \frac{(z-1)^{P'_G - 1}}{(\sqrt{\pi}/3)P'_G}, \omega_{c,k} = \omega_{c,1} f_{c,k-1}^{P'_G}, f_{c,k-1} = 1 - \frac{\phi_{G,c}}{\lambda mn_{\text{loop}}}(k-1)\left(1 \leq k \leq \lambda mn_{\text{loop}}\right) \quad \text{(B5)}$$

where $f_{c,k-1}$ is the probability of the vacancy for the lattice site in the core region when $k-1$ graft chains have been already arranged.

Using the above results, the number of arrangements Ω of the total m graft copolymer chains on the concentric spherical lattice is calculated by

$$\Omega = \prod_{i=1}^{m} \omega'_i \prod_{j=1}^{mn_{\text{loop}}} \omega_{s,j} \prod_{k=1}^{\lambda mn_{\text{loop}}} \omega_{c,k} \quad \text{(B6)}$$

or

$$\ln \Omega = m \ln\left[N_{\text{intf}}\left(\omega_{s,1}\omega_{c,1}^{\lambda}\right)^{n_{\text{loop}}}\right] + \sum_{i=1}^{m} \ln f'_{i-1} + \sum_{j=1}^{mn_{\text{loop}}} \ln\left[f_{s,j-1}^{P_{\text{loop}}(1 + x_s P'_G)} f_{c,j-1}^{\lambda P'_G}\right] \quad \text{(B7)}$$

The numbers of arrangements of the uniform bulk copolymer (Ω_P) and the bulk solvent (Ω_S) are given respectively by [16]

$$\ln \Omega_P = m \ln(mP\omega_{P,1}^{n_{\text{loop}}}) + \sum_{i=1}^{m} \ln f'_{P,i-1} + \sum_{j=1}^{mn_{\text{loop}}} \ln f_{P,j-1}, \ln \Omega_S = 0 \quad \text{(B8)}$$

where

$$\omega_{P,1} = (z-1)^{P_{\text{loop}}(1 + xP'_G)}, f'_{P,i-1} = 1 - \frac{i-1}{m}, f_{P,j-1} = 1 - \frac{j-1}{mn_{\text{loop}}} \quad \text{(B9)}$$

Therefore, the entropy of mixing ΔS in the micellar phase is given by

$$\begin{aligned} \frac{\Delta S}{mk_B} &= \frac{1}{m}(\ln \Omega - \ln \Omega_P - \ln \Omega_S) \\ &= -\ln \kappa - \ln \phi_P - n_{\text{loop}}\left[P_{\text{loop}}(1 + x_s P'_G)\frac{\phi_{S,s}}{1-\phi_{S,s}} \ln \phi_{S,s} + \lambda P'_G \frac{\phi_{S,c}}{1-\phi_{S,c}} \ln \phi_{S,c}\right] \end{aligned} \quad \text{(B10)}$$

where $\phi_S = 1 - \phi_M - \phi_G$, k_B is the Boltzmann constant, and κ is defined by

$$\kappa \equiv \frac{(R/a)^3}{3(R_{\text{core}}/a)^2}\left[\frac{\pi P_{\text{loop}} P'_G{}^{\lambda}}{9G(0; \tilde{l}_{\text{loop}})}\right]^{n_{\text{loop}}} \quad \text{(B11)}$$

To obtain Equation (B10), we used Equation (5) in the text, approximated $f'_{i-1} \approx f'_{P,i-1}$, and replaced the summations with respect to j in Equations (B7) and (B8) by integrations on the assumption of $mn_{\text{loop}} \gg 1$. The parameter κ can be calculated from P'_G, P_{loop}, and $2q/a$ using Equations (3), (4) and (B3).

At calculating the mixing enthalpy ΔH of the micelle phase and the interfacial tension γ between the core and shell regions of the flower micelle, we assume that the branch units in the main chain have the same interaction parameters as those of the graft-chain units. In what follows, the non-branch unit in the main chain is referred to as the A unit, and the graft-chain unit as well as the branch unit in the main chain are referred to as the B unit. Under the mean-field approximation [16], ΔH is given by

$$\frac{\Delta H}{mPk_\mathrm{B}T} = x_\mathrm{A}\phi_{\mathrm{S,s}}\chi_{\mathrm{AS}} + (x_{\mathrm{B,s}}\phi_{\mathrm{S,s}} + x_{\mathrm{B,c}}\phi_{\mathrm{S,c}})\chi_{\mathrm{BS}} - x_\mathrm{A}(x_{\mathrm{B,s}} + x_{\mathrm{B,c}} - \phi_{\mathrm{B,s}})\chi_{\mathrm{AB}} \qquad (\mathrm{B}12)$$

where $\chi_{\alpha\beta}$ ($\alpha, \beta = \mathrm{S, A, B}$) is the interaction parameter between species α and β (S stands for the solvent; the definition of $\chi_{\alpha\beta}$ is slightly different from that of [16]), and x_A, $x_{\mathrm{B,s}}$, and $x_{\mathrm{B,c}}$ are the mole fractions of the A and B units (existing in the shell and core regions) defined by Equation (8).

Noolandi and Hong [17] formulated the interfacial tension γ between the core and shell regions of the spherical micelle. We may extend their result to γ for the flower micelle, where the asymptotic volume fractions of the A and B units are $\phi_{\mathrm{A,s}}$ and $\phi_{\mathrm{B,s}}$ in the shell region and 0 and $\phi_{\mathrm{B,c}}$ in the core region, respectively. The result is given by

$$\frac{a^2}{k_\mathrm{B}T}\gamma = \sqrt{\frac{\phi_{\mathrm{A,s}} + \phi_{\mathrm{B,s}} + \phi_{\mathrm{B,c}}}{3}}\{f[\tfrac{1}{2}\phi_{\mathrm{A,s}}, \tfrac{1}{2}(\phi_{\mathrm{B,s}} + \phi_{\mathrm{B,c}})] - \tfrac{1}{2}[f(\phi_{\mathrm{A,s}}, \phi_{\mathrm{B,s}}) + f(0, \phi_{\mathrm{B,c}})]\} \qquad (\mathrm{B}13)$$

where the function $f(x, y)$ is defined by

$$f(x, y) \equiv [\ln(1 - x - y) + \chi_{\mathrm{AS}}x + \chi_{\mathrm{BS}}y](1 - x - y) + \chi_{\mathrm{AB}}xy \qquad (\mathrm{B}14)$$

There is one more interface between the shell and solvent regions in Figure 1, but this interface is not so sharp that we did not consider its interfacial Gibbs energy.

Flory [29] extended the Flory–Huggins theory [16] to solutions of a semiflxible polymer with energetically unfavorable "bend" conformations and demonstrated that the equilibrium degree of the bending of the chain is independent of the polymer concentration. This result indicates that the chain stiffness does not contribute to the mixing Gibbs energy ΔG because of the cancelation of the bending energy in the solution and bulk states. Therefore, the above formulations of ΔS and ΔH can be applied to semiflexible polymer solutions.

References

1. Borisov, O.V.; Halperin, A. Micelles of polysoaps. *Langmuir* **1995**, *11*, 2911–2919. [CrossRef]
2. Borisov, O.V.; Halperin, A. Micelles of polysoaps: The role of bridging interactions. *Macromolecules* **1996**, *29*, 2612–2617. [CrossRef]
3. Borisov, O.V.; Halperin, A. Polysoaps: Extension and compression. *Macromolecules* **1997**, *30*, 4432–4444. [CrossRef]
4. Borisov, O.V.; Halperin, A. Self-assembly of polysoaps. *Curr. Opin. Colloid Interface Sci.* **1998**, *3*, 415–421. [CrossRef]
5. Halperin, A. *Supramolecular Polymers*; Ciferri, A., Ed.; Marcel Dekker Ltd.: New York, NY, USA; Basel, Switzerland, 2000; ISBN 9780824723316-CAT# DK3116.
6. Kawata, T.; Hashidzume, A.; Sato, T. Micellar Structure of Amphiphilic Statistical Copolymers Bearing Dodecyl Hydrophobes in Aqueous Media. *Macromolecules* **2007**, *40*, 1174–1180. [CrossRef]
7. Sato, T.; Kimura, T.; Hashidzume, A. Hierarchical Structures in Amphiphilic Random Copolymer Solutions. *Prog. Theor. Phys. Suppl.* **2008**, *175*, 54–63. [CrossRef]
8. Tominaga, Y.; Mizuse, M.; Hashidzume, A.; Morishima, Y.; Sato, T. Flower Micelle of Amphiphilic Random Copolymers in Aqueous Media. *J. Phys. Chem. B* **2010**, *114*, 11403–11408. [CrossRef] [PubMed]
9. Ueda, M.; Hashidzume, A.; Sato, T. Unicore-Multicore Transition of the Micelle Formed by an Amphiphilic Alternating Copolymer in Aqueous Media by Changing Molecular Weight. *Macromolecules* **2011**, *44*, 2970–2977. [CrossRef]
10. Uramoto, K.; Takahashi, R.; Terao, K.; Sato, T. Local and Global Conformations of Flower Micelles and Flower necklaces Formed by an Amphiphilic Alternating Copolymer in Aqueous Solution. *Polym. J.* **2016**, *48*, 863–867. [CrossRef]

11. Fujimoto, M.; Sato, T. Aggregation and Micellization Behavior of Amphiphilic Random Copolymers Bearing Various Hydrophobic Groups in Aqueous Solution. *Kobunshi Ronbunshu* **2016**, *73*, 547–555. [CrossRef]

12. Sato, T.; Takahashi, R. Competition between the micellization and the liquid–liquid phase separation in amphiphilic block copolymer solutions. *Polym. J.* **2017**, *49*, 273–277. [CrossRef]

13. Yamakawa, H.; Stackmayer, W.H. Statistical Mechanics of Wormlike Chains. II. Excluded Volume Effects. *J. Chem. Phys.* **1972**, *57*, 2843–2854. [CrossRef]

14. Shimada, J.; Yamakawa, H. Statistical mechanics of helical worm-like chains. XV. Excluded-volume effects. *J. Chem. Phys.* **1986**, *85*, 591–600. [CrossRef]

15. Yamakawa, H.; Yoshizaki, T. *Helical Wormlike Chains in Polymer Solutions*, 2nd ed.; Springer: Berlin/Heidelberg, Germany, 2016; ISBN 978-3-662–48714–3.

16. Flory, P.J. *Principles of Polymer Chemistry*; Cornell University Press: New York, NY, USA, 1953; ISBN 978-0-8014-0134-3.

17. Noolandi, J.; Hong, K.M. Theory of Block Copolymer Micelles in Solution. *Macromolecules* **1983**, *16*, 1443–1448. [CrossRef]

18. Hirai, Y.; Terashima, T.; Takenaka, M.; Sawamoto, M. Precision Self-Assembly of Amphiphilic Random Copolymers into Uniform and Self-Sorting Nanocompartments in Water. *Macromolecules* **2016**, *49*, 5084–5091 [CrossRef]

19. Liu, X.; Kim, J.-S.; Wu, J.; Eisenberg, A. Bowl-Shaped Aggregates from the Self-Assembly of an Amphiphilic Random Copolymer of Poly(styrene-*co*-methacrylic acid). *Macromolecules* **2005**, *38*, 6749–6751. [CrossRef]

20. Li, Y.; Deng, Y.; Tong, X.; Wang, X. Formation of Photoresponsive Uniform Colloidal Spheres from an Amphiphilic Azobenzene-Containing Random Copolymer. *Macromolecules* **2006**, *39*, 1108–1115. [CrossRef]

21. Zhang, L.; Lin, J.; Lin, S. Aggregate Morphologies of Amphiphilic Graft Copolymers in Dilute Solution Studied by Self-Consistent Field Theory. *J. Phys. Chem. B* **2007**, *111*, 9209–9217. [CrossRef] [PubMed]

22. Yusa, S.; Sakakibara, A.; Yamamoto, T.; Morishima, Y. Reversible pH-Induced Formation and Disruption of Unimolecular Micelles of an Amphiphilic Polyelectrolyte. *Macromolecules* **2002**, *35*, 5243–5249. [CrossRef]

23. Li, L.; Raghupathi, K.; Song, C.; Prasad, P.; Thayumanavan, S. Self-Assembly of Random Copolymers. *Chem. Commun.* **2014**, *50*, 13417–13432. [CrossRef] [PubMed]

24. Zhu, X.; Liu, M. Self-Assembly and Morphology Control of New L-Glutamic Acid-Based Amphiphilic Random Copolymers: Giant Vesicles, Vesicles, Spheres, and Honeycomb Film. *Langmuir* **2011**, *27*, 12844–12850. [CrossRef] [PubMed]

25. Tian, F.; Yu, Y.; Wang, C.; Yang, S. Consecutive Morphological Transitions in Nanoaggregates Assembled from Amphiphilic Random Copolymer via Water-Driven Micellization and Light-Triggered Dissociation. *Macromolecules* **2008**, *41*, 3385–3388. [CrossRef]

26. Dan, K.; Bose, N.; Ghosh, S. Vesicular Assembly and Thermo-Responsive Vesicle-to-Micelle Transition from an Amphiphilic Random Copolymer. *Chem. Commun.* **2011**, *47*, 12491–12493. [CrossRef] [PubMed]

27. Ida, D. Private communication, 2017.

28. Hoeve, C.A.J. Density Distribution of Polymer Segments in the Vicinity of an Adsorbing Interface. *J. Chem. Phys.* **1965**, *43*, 3007–3008. [CrossRef]

29. Flory, P.J. Statistical Thermodynamics of Semi-Flexible Chain Molecules. *Proc. Royal Soc. London A* **1956**, *234*, 60–89. [CrossRef]

polymers

MDPI

Article

A Comparative Study on Micellar and Solubilizing Behavior of Three EO-PO Based Star Block Copolymers Varying in Hydrophobicity and Their Application for the In Vitro Release of Anticancer Drugs

Bijal Vyas [1], Sadafara A. Pillai [1], Anita Bahadur [2] and Pratap Bahadur [1,*]

[1]   Department of Chemistry, Veer Narmad South Gujarat University, Surat 395007, India;
      vyasbijal80@gmail.com (B.V.); sadafarashirgar@gmail.com (S.A.P.)
[2]   Department of Zoology, PT Sarvajanik College of Science, Surat 395001, India; anita26p@gmail.com
*   Correspondence: pbahadur2002@gmail.com; Tel.: +91-987-913-2125

Received: 16 December 2017; Accepted: 13 January 2018; Published: 15 January 2018

Abstract: The temperature and pH dependent self-assembly of three star shaped ethylene oxide-propylene oxide (EO-PO) block copolymers (Tetronics® 304, 904 and 908) with widely different hydrophobicity was examined in aqueous solutions. Physico-chemical methods viz. viscosity, cloud point, solubilization along with thermal, scattering and spectral techniques shows strongly temperature and salt dependent solution behavior. T304 possessing low molecular weight did not form micelles; moderately hydrophilic T904 remained as micelles at ambient temperature and showed micellar growth while very hydrophilic T908 formed micelles at elevated temperatures. The surface activity/micellization/solubilization power was favored in the presence of salt. The copolymers turn more hydrophilic in acidic pH due to protonation of central ethylene diamine moiety that hinders micelle formation. The solubilization of a model insoluble azo dye 1-(o-Tolylazo)-2-naphthol (Orange OT) and hydrophobic drugs (quercetin and curcumin) for copolymer solutions in aqueous and salt solutions are also reported. Among the three copolymers, T904 showed maximum solubility of dye and drugs, hence the in vitro release of drugs from T904 micelles was estimated and the effect on cytotoxicity of loading the drugs in T904 micelles was compared with the cytotoxicity of free drugs on the CHO-K1 cells. The results from the present work provide a better insight in selection of Tetronics® for their application in different therapeutic applications.

Keywords: Tetronics; micelles; in vitro release; anticancer drugs; solubilization; drug delivery

1. Introduction

Block copolymers contain at least two incompatible blocks and thus show micro domain formation in solid state and self-assembly in selective solvents; these properties coupled with advances in polymerization techniques have made polymeric surfactants highly useful materials [1–3]. Pluronics® and Tetronics® are commercially available poly(ethylene oxide) (PEO)-poly(propylene oxide) (PPO) block copolymers with unique temperature dependent micellization/surface activity and reversible thermorheological behavior that make these substances widely useful in cosmetic/detergents/food/pharmaceutical industries. Due to their nontoxicity and low immune response, some of these copolymers are now Food and Drug Administration (FDA) approved. Their emerging applications in fabrication of mesoporous materials [4,5], synthesis of nanoparticles [6,7] and as nanocarriers for drug delivery systems [8–12] have generated more interest in researchers [13,14]. Strongly temperature dependent micellization and gelation of these amphiphilic copolymers have been thoroughly examined in the last few decades though mostly on linear PEO–PPO–PEO triblock copolymers and there exist several reviews [1,15,16]. However, only a few authors have reported on the aqueous solution behavior of their star shaped counterparts [17–30].

Tetronics® (also known as poloxamines) present an X-shaped structure made of an ethylenediamine central group bonded to four chains of PPO–PEO blocks. Tetronics® are synthesized by the sequential reaction of the acceptor ethylenediamine molecule first with propylene oxide (PO) and then with ethylene oxide (EO) precursors, resulting in a four-arm PEO-terminated molecular structure (Figure 1). The unique structure of Tetronics® provides them with multistimulus responsiveness. In this context, the two tertiary amine central groups play an essential role in conferring thermodynamic stability and pH sensitivity. The micelle formation of Tetronics® is slightly different from the Pluronics® as the former also show some pH responsiveness. Like Pluronics®, a slight increase in the temperature can induce surface activity/micellization/gelation due to the dehydration of PPO and PEO blocks. The central diamine unit in Tetronics® molecule is pH sensitive and can be protonated in acidic solution. Low pH and low temperature may thus hamper micellization. Albeit still limited, the studies on Tetronics® have revealed their potential in different fields. These cover broad area of applications including petroleum industries where these are used in comparatively higher concentrations either as de-emulsifiers or as antifoaming agents [31,32], as an important ingredient in contact lens washing solutions [33,34], in pharmaceutical and biomedical field as constituents of transdermal formulations [35], in nanoparticle engineering [36] and as tissue scaffoldings [37,38]. De Lisi et al. [39] studied the self-assembly and oil solubilizing behavior of T1107 as a function of temperature and pH and revealed improved solubilization and oil induced micellization, which can be tuned further by varying temperature and pH. Larrañeta et al. [40] synthesized different types of gels by the addition of α-cyclodextrin in the T904 solutions for sustained drug delivery applications. González-Gaitano et al. [41] explored the effect of different native and modified cyclodextrins on T904 micelles and revealed that most substituted cyclodextrins induced micellar breakdown while native cyclodextrins promoted the formation of inclusion complexes. Recently, Poellmann [42] based on his study on T1107 with denatured hen egg white lysozyme found a potential application of T1107 as a synthetic chaperone that enhances the molecular repair phenomenon in cells. Gonzalez-Lopez et al. [20] explored the micellization of different Tetronics® with varying structural features in acidic media. There are also contributions from our group concerning the effect of different additives on micellar behavior of Tetronics® [43–49].

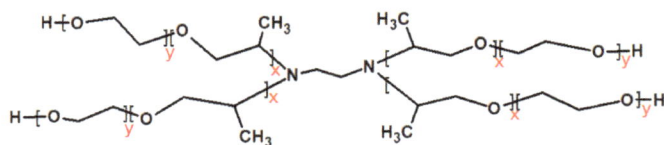

Figure 1. General structure of Tetronics®.

It can be understood from the above literature survey that reports on individual behavior of different Tetronics® in the presence of variety of additives have been published in literature. However, a systematic study comparing the self-assembly of different Tetronics® on the basis of their molecular architecture in aqueous and salt solutions is still missing, to the best of our knowledge. The knowledge attained from this work can be useful in anticipating the performance of the copolymer for the desired application. With this view point, we have tried to elucidate the aggregation behavior of three Tetronics®, mainly T304, T904 and T908 incorporated with different %EO (in case of T904 and T908, %PEO = 40 and 80) and with different molecular weight (in the case of T304 and T904, but with same %PEO = 40). This study was further extended by determining the solubility of a hydrophobic dye (orange-OT) and drugs (quercetin and curcumin) in micelles under different solution conditions. T904 with a moderately hydrophobic character showed a maximum solubility. Hence, in vitro release of the hydrophobic drugs from its micelles was measured and the influence on cytotoxicity of the drugs being loaded in micelles was compared with the free drugs. The findings of this work will be highly useful for the proper exploitation of copolymer micelles in several industrial and pharmaceutical applications.

2. Materials and Methods

Tetronics® 304, 904 and 908 were gift samples from the BASF Corporation (Parsippany, NJ, USA) and were used as received. The structural formula for Tetronics® block copolymers (Figure 1) and molecular/physico-chemical properties are shown in Table 1.

Table 1. Structural properties of different Tetronics®.

Tetronic®	M_w [a]	N_{EO}	N_{PO}	HLB [a]	CP [a] (°C)	pKa$_1$ [a]	pKa$_2$ [a]
T304	1650	3.7	4.3	12-18	72	4.3	8.1
T904	6700	15	17	12-18	78	4.0	7.8
T908	25,000	114	21	>24	>100	5.2	7.9

[a] Data is taken from BASF website and from references [20,25]. M_w—molecular weight, HLB—hydrophilic-lipophilic balance, CP—cloud point, N_{EO} = y and N_{PO} = x.

Sodium chloride (Merck, Mumbai, India, analytical grade), the drugs, quercetin and curcumin (Sigma Aldrich, Mumbai, India) and the dye, Orange-OT (TCI Chemicals, Chennai, Tamilnadu, India) were used as received. Solutions for dynamic light scattering (DLS) measurements were prepared in nano-pure water obtained from Millipore Milli-Q purification system (Mumbai, India). D_2O (99.9%) obtained from Sigma Chemical Company (Mumbai, India) was used for nuclear magnetic resonance (NMR) and small-angle neutron scattering (SANS) measurements. The chinese hamster ovary (CHO-K1) cell line for the toxicity assay was procured from American Type Culture Collection (ATCC).

2.1. Methods

2.1.1. Cloud Point (CP)

Cloud points were determined by visual observation of the turbidity of the solution (in 20 mL glass vials) immersed in a temperature controlled water bath. The solutions were stirred with a magnetic bar while being heated. All of the measured CP values were reproducible up to ±1.0 °C.

2.1.2. Surface Tension

The surface tension measurements were done with a KRUSS Easy Tensiometer from Kruss Gmbh (Hamburg, Germany) using the Wilhelmy plate method. The surface tension of double distilled water 71.8 mN·m^{-1} at 25.0 ± 0.1 °C was used to calibrate the instrument. The surface tension of each solution was measured by successive additions of the stock solutions in double distilled water after thorough mixing and equilibration. The series of measurements were repeated at least three times. The reproducibility of surface tension measurements is estimated to be within ±0.2 mN·m^{-1}.

2.1.3. Viscosity

The viscosities of solutions were measured using an Ubbelohde suspended level capillary viscometer. The viscometer was suspended vertically in a thermostat at ±0.1 °C. A clean and dry viscometer was used for each measurement. The flow time (usually exceeding 170 s) of a constant volume of the solution through the capillary was used to calculate the viscosity of the solution.

2.1.4. Nuclear Magnetic Resonance (NMR)

The ^{1}H-NMR spectra were recorded on a Bruker DMX Avance 600 spectrometer (Osaka, Japan) over a wide temperature range. The sample temperature was kept constant within ±0.1 °C by using a

Bruker BCU-05 temperature control unit. The samples were equilibrated at the desired temperature for at least 15 min prior to measurement.

2.1.5. High-Sensitivity Differential Scanning Calorimetry (HSDSC)

Calorimetric measurements were carried out using a Microcal MC-2 instrument (Microcal Inc., Amherst, MA, USA) and the DA-2 dedicated software package (provided by Microcal, Malvern, UK) for data acquisition. Samples were equilibrated in the HSDSC cells for a minimum of 60 min prior to each run, and scans performed at a scan rate of 60 Kh^{-1}.

2.1.6. Dynamic Light Scattering (DLS)

Dynamic light scattering (DLS) was used to determine the apparent hydrodynamic diameter (D_h) of the micelles. DLS measurements were carried out at 90° scattering angle on solutions using Autosizer 4800 (Malvern Instruments, Worcestershire, UK) equipped with 192 channel digital correlator (7132) and coherent (Innova, Santa Clara, CA, USA) Ar-ion laser at a wavelength of 514.5 nm. The average diffusion coefficients and hence the hydrodynamic size was obtained by the method of cumulants.

2.1.7. Small Angle Neutron Scattering (SANS)

The SANS experiments were performed using a SANS diffractometer at the Dhruva reactor, Bhabha atomic research centre, Trombay. For SANS, copolymer solutions in D_2O at different concentrations and temperatures were measured. The solutions were held in a quartz cell of 5 mm thickness with tight-fitting Teflon stoppers. The data were recorded in the Q range of 0.017–0.35 Å. All the measured SANS distributions were corrected for the background and solvent contributions. The data were normalized to the cross-sectional unit using standard procedures [50].

The copolymer micelles consist of a hydrophobic core of PPO surrounded by a hydrated shell of PEO. There is very good contrast between the hydrophobic core and the solvent. However, because of a large amount of D_2O (water of hydration) being present in the outer PEO corona, the scattering contrast between the hydrated corona and the solvent is expected to be poor. In view of this, we assume that the form factor F(Q) depends only on the hydrophobic core radius. The structure factor S(Q) of the spherical micelles in Equation (1) is calculated using the Percus–Yevick approximation for the case of hard sphere potential in the Ornstein–Zernike equation [51]:

$$\frac{d\Sigma}{d\Omega}(Q) = nV^2(\rho_P - \rho_S)^2 P(Q)S(Q) + B. \tag{1}$$

The mean core radius (R_c), hard sphere radius (R_{hs}) and volume fraction (Φ) of the micelles have been determined as the fitting parameters from the analysis. The aggregation number is calculated by the relation $N = 4\pi a^3/3v$, where v is the volume of the surfactant monomer.

2.1.8. Solubilization

Drug/dye solubilization measurements were carried out on Shimadzu (UV-2450) UV-Visible double beam spectrophotometers (Tokyo, Japan) with a matched pair of stoppered fused silica cells of 1 cm optical path length. Saturated drug/dye loaded solutions were prepared in glass vessels by mixing excess powdered drug/dye with copolymer solution and stirring at constant temperature at 200 rpm for 2 days. The solutions were filtered (Millipore, 0.45 μm) to remove insolubilized drug/dye. Blank experiments, without copolymer, were done to determine the solubility of the drug/dye in water. The amount of drug/dye solubilized was determined by measuring absorbance at 255/262/470 nm. Calibration with dilute solutions of the drug/dye dissolved in methanol gave satisfactory Beer–Lambert plots. In a solubilization experiment, the filtered solution was diluted thirty times with methanol, the amount of water after dilution being low enough to allow direct use of the calibration plot.

2.1.9. In Vitro Release Studies

The in vitro release of the poorly water soluble anticancer drugs, quercetin and curcumin, from the micelles was investigated using a pre swelled dialysis bag (M_w cut-off 12,000–14,000 Da). In brief, 10 mL of the drug formulations containing approximately 5 mgs of drugs was transferred into respective dialysis bags and immersed into 100 mL of phosphate buffer saline (PBS), which was placed in a shaking water bath at 37 °C. In addition, 3 mL sample aliquots were taken from the release medium at scheduled time intervals and the same volume of fresh buffer was refilled to maintain the volume. The concentration of drugs released into PBS was quantified based on their absorbance at 470 nm (λ_{max} for orange OT), 255 nm (λ_{max} for quercetin) and 262 nm (λ_{max} for curcumin), respectively using a Shimadzu 160 spectrophotometer on a UV-Vis curve, to further conclude the rate of drug release.

2.1.10. In Vitro Cytotoxicity Assays

Cytotoxicity of free quercetin and curcumin and their loaded micelles of T904 were assessed on the viability of CHO-K1 cells by the 3-(4,5-Dimethylthiazol-2-yl)-2,5-Diphenyltetrazolium Bromide (MTT) assay. These CHO-K1 cells in a logarithmic phase (104 cells/well) were seeded in 96-well plates along with variable concentration of free quercetin and curcumin (0, 12.5, 25, 50 and 100 µg/mL). The cells were also treated with their equivalent doses loaded in T904 micelles. Incubate these cells for 24 h in carbon dioxide incubator (5% CO_2; 37 °C). Afterwards, add the MTT solution (10 µL; 2.5 mg/mL) into a 96-well plate after centrifuging (2000 rpm; 5 min) and dissolved in an equal volume of media. Again, incubate the plate for another 4 h. Finally, the intracellular formazan crystals were settled at the bottom and the supernatant was discarded (again after centrifuging as mentioned above). These crystals were dissolved in dimethyl sulphoxide and its relative growth inhibition was compared with control cells and its optical density was measured at 570 nm. All experiments were set up in triplicates and repeated thrice for statistical analysis. Results were expressed as mean ± S.E.

The percentage growth inhibition was calculated using the following formula:

$$\%\text{Growth Inhibition} = \frac{100 - (\text{Mean absorbance of individual test})}{(\text{Mean absorbance of Control})} \times 100.$$

The half maximal inhibitory concentration (IC_{50}) was estimated using the concentration-response and was expressed in the unit µg/mL. The concentration of the test drug required for the inhibition of cell growth by 50% (CTC50) was generated by the dose-response curves for each cell line.

Statistical Analysis: Data analysis was carried out by two-way ANOVA (SPSS 10.0, SPSS Inc., Chicago, IL, USA). A *p*-value ≤ 0.05 was contemplated as statistically important.

3. Results

3.1. Characterization of Tetronics® Micelles and the Effect of Salt

3.1.1. Cloud Point

The cloud points (CPs) of all the three copolymers were measured as a function of salt concentration and pH and are presented in Figure 2. It has been established from the previous studies in literature that copolymers fabricated with longer PEO blocks usually exhibit more hydrophilic character and undergo micellization at elevated temperatures. The occurrence of longer PEO blocks improves the solubility of the copolymer chains in water. Hence, these exhibit higher phase separation temperatures. Conversely, those with shorter PEO blocks (>30%) usually display poor solubility and undergo micellization at comparatively lower temperatures and experiences phase separation at relatively lower temperatures [16,44]. In the present case, as displayed in Table 1, T908 with 80% PEO in its constitution remains highly hydrophilic and usually displays CP at temperatures >100 °C in water and, as anticipated, T904 with moderate hydrophilicity with 40% EO in its constitution undergoes phase separation at temperature ~73 °C. Surprisingly, in the case of T304, despite having

40% EO in its constitution like T904, due to low molecular weight of the constituting PPO block, does not form micelles but remains as unimers even at higher temperatures (discussed in later sections) and undergoes phase separation at a temperature very close to that of T904.

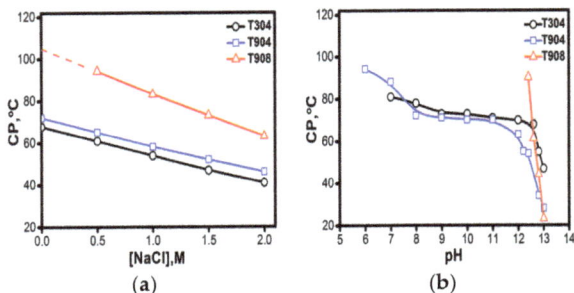

Figure 2. CP of 5% Tetronics® (a) as a function of salt concentration and (b) pH.

For all the copolymers, as shown in Figure 2a, there was almost a linear decrease in CP with an increasing concentration of salt. The presence of salt induces hydrophobicity in the copolymer and makes it prone to form micelles, or micellar growth accounts for the copolymers existing as micelles due to the well-known salting out action of NaCl. A similar CP depressant role of salt has been observed earlier for Pluronics® and other water soluble uncharged polymers that show lower critical solution behavior [52–55]. Bahadur and coworkers [46,48,49,56] have also reported a similar effect of salt on Tetronics® micelles.

The effect of pH was also examined for these copolymers as shown in Figure 2b. The pH values for the solution were adjusted in the range of 6–13 using HCl/NaOH solutions. For T304 and T904, there is only a slight decrease in CP upon changing pH in the range 6-10; however, a steep fall is seen above pH 10. The CPs were higher in acidic pH because acidic pH makes the amino groups protonated and thus induces a more hydrophilic character and increases CP. For T908, CPs were measured only in the alkaline pH (since its CP in the absence of additives remains >100 °C) where low values of CP were observed. The decrease in CP at higher pH is due to the fact that ethylenediamine group gets completely neutralized at this pH and thus induces more hydrophobicity in the copolymer units promoting micelle formation, which consequently lowers the CP values.

3.1.2. Surface Tension

In order to study the surface activity of these copolymers at the water/air interface and to determine the CMC values, surface tension measurements were carried out as a function of the copolymer concentration in aqueous medium and also in salt solutions in the case of T908.

Surface tension → log copolymer concentration plots for the three Tetronics® are shown in Figure 3a. T304 owing to its low molecular weight was least surface active and did not show micelle formation at all the concentrations, as also confirmed through SANS, though a progressive decrease in surface tension (upto 10% w/v) is observed. On the contrary, Gonzalez-Lopez et al. [20] determined the CMC of T304 in acidic media and observed micelle formation at higher concentrations. T908 with more hydrophilic character was less surface active and showed two break points commonly observed for copolymers with more hydrophilic character. Although improved surface activity and a significant decrease in CMC can be made evident in the presence of salt (Figure 3b). The improved hydrophobicity in the presence of salt is further demonstrated by a single break point (Figure 3b) analogous to moderately hydrophilic copolymers. Conversely, T904 with moderately hydrophilic character displayed highest surface activity among the three copolymers and showed a behavior similar to surfactants. Thus, this leads us to a conclusion that, for all of the copolymers, surface activity increased with the increase in molecular weight, decrease in %EO and in the presence of salt.

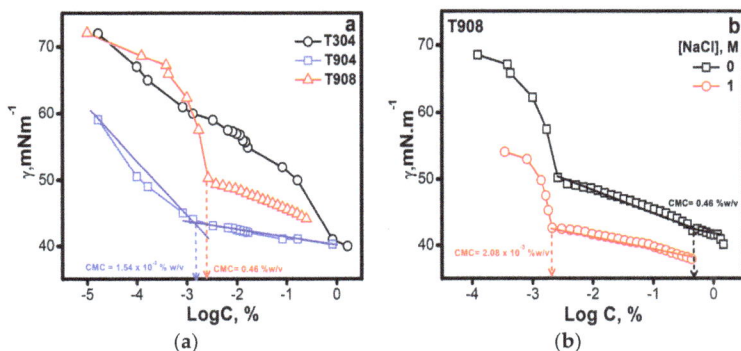

Figure 3. Surface tension plots (**a**) for the aqueous solutions of different Tetronics® and (**b**) T908 solutions in the presence and absence of 1 M salt at 25 °C.

3.1.3. Nuclear Magnetic Resonance (NMR)

^{1}H NMR spectra for the three copolymers were recorded at different temperature intervals and the signals for methylene protons of PEO, methylene and methyl protons of PPO were considered to analyze the spectra (Only spectra for T908 has been included). As evident from Figure 4a, a triplet at 1.16 ppm is assigned to the methyl protons of the PPO, broad peaks around 3.65 to 3.45 ppm correspond to the methylene protons of PPO and intense resonance peak at 3.7 ppm correspond to the methylene protons of PEO. As evident in Figure 4b, an increase in temperature shifts the resonance peaks corresponding to methyl and methylene protons of PPO towards low frequencies with a simultaneous line-width broadening indicating reduced mobility of PPO segments since it forms a hydrophobic core of the micelles.

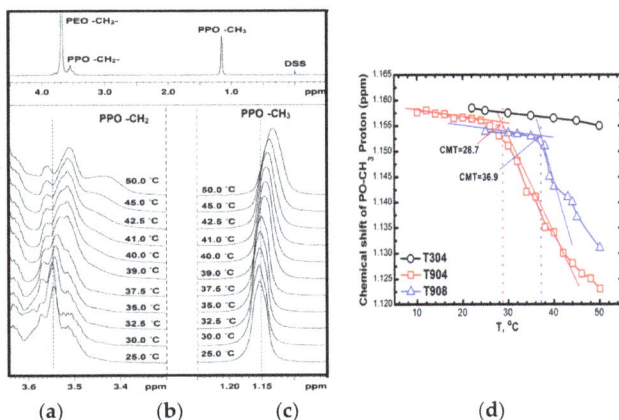

Figure 4. (**a**) ^{1}H-NMR spectra of 10% T908 in D_2O at 25 °C; (**b**) –CH_2– signals of PPO; (**c**) –CH_3 signals of PPO; (**d**) chemical shift vs. temperature.

The chemical shift observed for the PPO –CH_3 peaks were then plotted against temperature as shown in Figure 4c to determine the critical micelle concentration (CMT) of block copolymers. The

CMTs were taken as the inflexion point in the plot. T304 owing to its low molecular weight did not form micelle under the conditions studied despite containing 40% PEO while T904 (with similar %PEO) remains much lower than T908 at the same concentration due to its moderately hydrophobic character. It is interesting to note that T904 and T908 have similar molecular weight of PPO but vary in %PEO, which sufficiently alters their hydrophobicity and micellar and micellization characteristics. A similar trend is also observed for the hydrophilic linear block copolymers, Pluronics® [57,58].

3.1.4. HSDSC

In this study, HSDSC experiments were performed to determine the effect of salt on the CMT of T908. Owing to its hydrophilic character, T908 forms micelles at elevated temperatures as observed in the previous section. Since no micellization was observed in the previous sections for T304, CMT of T304 was not determined. Likewise, measurements for T904, which usually form micelles at lower temperatures, were not carried out since the addition of salt will further decrease the CMT and the practical limitations of the instrument may not allow us to go to such lower temperatures. Hence, the measurements were limited only to T908 to understand the influence of salt on copolymer micelles.

Typical HSDSC thermograms (not shown) with the endothermic peak were obtained signifying the endothermic phase transition from a fully solvated solution of unimers to a solution consisting of solvated micelles with a poly(propylene oxide) microphase inner core. Generally, CMT of the copolymer can be defined by three different methods viz. T_{onset}, T_{inf} and T_m. All three of the methods were explained in detail in our previous reports [44,47]. In the present study, T_m is chosen as the CMT of the block copolymer.

As evident in data presented in Table 2, the CMT of aqueous solutions of 5% T908 decreases in the presence of salt. This is attributed to the fact that addition of salt leads to the dehydration of the EO and PO blocks, which favors micellization and thus reduces CMT significantly. Alexandridis and Holzwarth [52] using DSC scrutinized the effect of different salts on the CMT of Pluronics® solutions and observed similar results. Similarly, Bahadur et al. [56] using fluorescence witnessed a similar decrease in CMT of T1307 solutions in the presence of salt.

Table 2. Thermodynamic parameters for micellization of 5% T908 from high sensitivity differential scanning calorimetry thermograms.

[NaCl], M	T_{onset} (°C)	T_{inf} (°C)	T_m (°C)	ΔH (kJ/mol)	ΔG (kJ/mol)	ΔS (kJ/mol·K)
0	32.00	32.06	37.82	133.38	−26.31	0.51
1	19.03	19.73	24.83	178.45	−25.25	0.68
2	a	a	20.26	a	−24.80	a

[a] The pre-transitional baseline was so short to calculate the T_{onset}, T_{inf}, and ΔH properly.

3.1.5. Viscosity

Viscosities of 10% w/v solutions of Tetronics® in H$_2$O, 1 and 2 M NaCl were measured at different temperatures and the relative viscosities were plotted against temperature as shown in Figure 5.

The viscosity behavior of copolymers was different. For copolymer T304, due to its low molecular weight, there was no effect on viscosity and it remained almost constant at different temperatures. The viscosity did not change even in the presence of salt up to its CP. This is because T304 remains molecularly dissolved and does not show any micelles even at elevated temperatures or in the presence of salt. Conversely, T904 with moderately hydrophilic character occurs as micelles at ambient temperature. Any increase in temperature promotes micellar growth due to increased dehydration of EO and PO chains. Hence, more and more unimers participate in micelle formation, eventually leading to micellar growth. Quite interestingly, T908, due to its characteristic hydrophilic character, forms micelles only at high temperature and in the presence of salt. Accordingly, for various concentrations of salts, different morphologies are assumed by the copolymer molecules. In line with this, the contribution from the unimers and micelles shows different viscosity behavior. As evident in the figure, with the increase in temperature, the relative viscosity increases initially but decreases at higher temperatures. This is attributed to the fact that increase in temperature leads to the formation of spherical micelles from the existing unimolecular form of copolymer, which, as a result, enhances the viscosity significantly until a maxima is reached at ~55 °C. As the temperature is raised further, the dehydration of EO blocks is triggered and more compact micelles are formed, which eventually decreases the viscosity of the solution. The presence of salt promotes micellization as also reported in the earlier sections. Hence, the maxima shift to lower temperatures in the presence of 1 and 2 M NaCl. A similar trend has been reported in literature for hydrophilic copolymers earlier. Thus, the results are in line with literature.

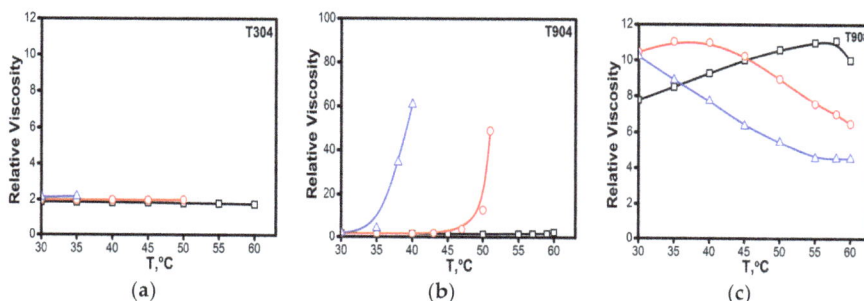

Figure 5. Effect of temperature on different Tetronic® solutions (10% w/v) in the presence of salt (□) 0 M, (○) 1 M and (△) 2 M NaCl. (**a**) relative viscosity of 10% T304 aqueous and salt solutions versus temperature (**b**) relative viscosity of 10% T904 aqueous and salt solutions versus temperature (**c**) relative viscosity of 10% T908 aqueous and salt solutions versus temperature.

3.1.6. DLS

The apparent hydrodynamic diameter of micelles at different temperatures for 10% w/v Tetronic® aqueous and salt solutions was determined using DLS as shown in Figure 6. This figure also displays the effect of temperature (30–50 °C) on the above said copolymer solutions in the presence and absence of salt.

T304 unimer peaks of the size 2–5 nm along with large particles of a few hundred nm are seen. This simply reflects the presence of molecularly dissolved T304 at all temperature and salt concentrations. In the case of T908, the presence of unimers (5–8 nm) along with another peak originating from the presence of micelles and large particles can be clearly seen. With the increase in temperature or in the presence of salt micellar peaks gets sharper while the contribution from large clusters and unimers decreases. Micelles of size ~15–20 nm are the predominant species at higher temperatures and in salt solutions. Single peak arising from the micelles (of ~12–20 nm size) can be seen for moderately hydrophilic T904 at relatively lower temperatures and in the absence of salt. However, micellar growth is observed at elevated temperatures and in the presence of salt. The increased micellar dimensions consequently slow their diffusion in solution, which eventually leads to enhanced solution viscosity.

3.1.7. SANS

SANS curves for all three Tetronics® at different temperature are shown in Figure 7. As evident from Figure 7a, the scattering intensity increases with temperature, though overall remains low for T304, indicating an absence of aggregates and the occurrence of only unimers in the solution. This is attributed to the fact that T304, due to its low molecular weight, is incorporated with shorter PPO blocks and hence cannot form micelles. Likewise, T908 with more hydrophilic character occurs as unimers at 30 °C. However, as the temperature is raised, the overall solubility of the copolymer chain decreases, which, in turn, promotes micellization. Thus, as evident in Figure 7c, the scattering intensity of T908 increases significantly as the temperature is raised from 30 to 60 °C. For T904, core-shell micelles were the species present, which show growth at a higher temperature as well as in the presence of salt. It can be observed from Figure 7b that the scattering intensity increases significantly with the increase in temperature, clearly indicating micellar growth.

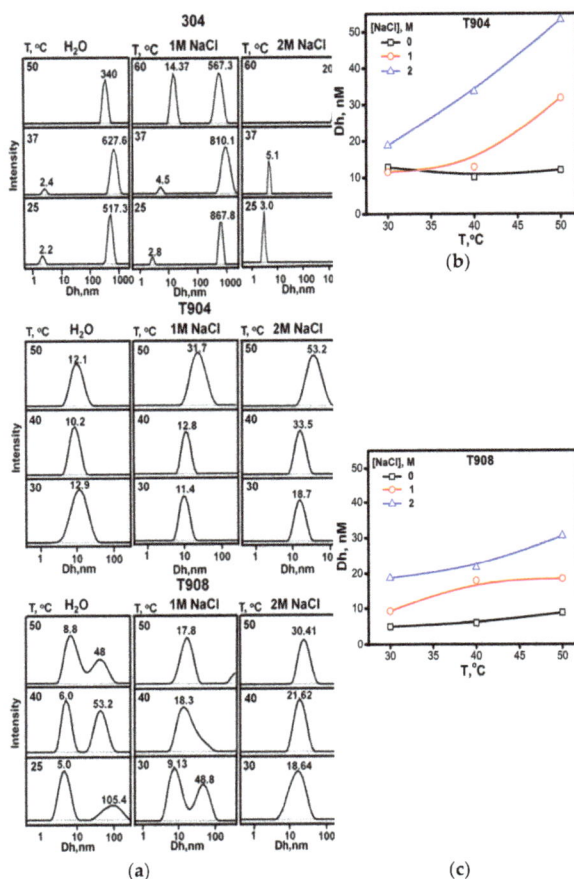

Figure 6. Apparent hydrodynamic diameter of different Tetronics® in aqueous and salt solutions. (**a**) DLS stacks of T304, T904 and T908 in aqueous and salt solutions (**b**) apparent hydrodynamic diameters of aqueous and salt solutions of T904 as a function of temperature (**c**) apparent hydrodynamic diameters of aqueous and salt solutions of T908 as a function of temperature.

A similar trend for all three Tetronics® can also be observed from the data presented in Table. In case of T304, it can be understood from the data that neither increase in temperature (Figure 7a) or concentration (Figure 7d) can induce micelle formation and it remains as unimers for all the concentrations and temperatures measured, while, for T904 with moderate hydrophobicity, core-shell micelles with N_{agg} ~10 occur in solution at ambient temperature and, with an increase in temperature, R_c, R_{hs} and N_{agg} increase significantly, clearly indicating micellar growth. In the case of T908, with a more hydrophilic character, only unimers occur at lower temperatures. However, as the temperature is raised, micelles with N_{agg} ~7 are formed at 40 °C and an increase in micellar parameters is observed at still higher temperatures, clearly indicating micellar growth at higher temperatures.

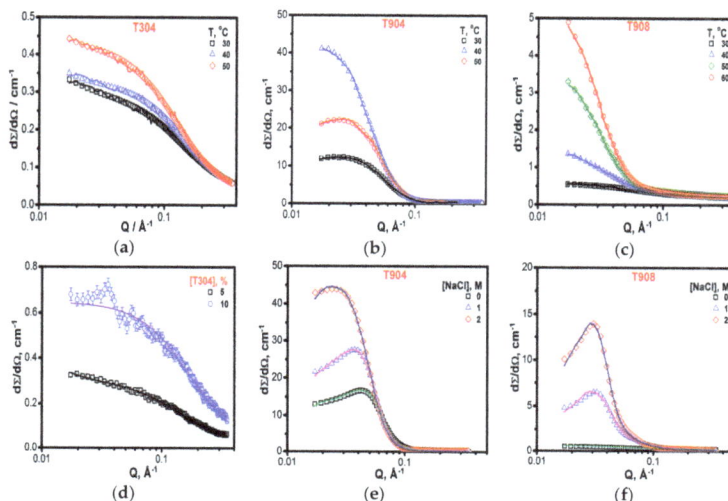

Figure 7. Small angle neutron scattering curves (**a**) for 10% T304 as a function of temperature (**b**) for 5% T904 as a function of temperature (**c**) for 10% T908 as a function of temperature (**d**) for 5% and 10% T304 aqueous solutions (**e**) for 10% T904 aqueous as a function of salt concentration and (**f**) for 10% T908 aqueous solutions as a function of salt concentration.

The presence of salt has a similar effect and leads to micellar growth at low temperatures. As evident in Figure 7e,f, the increase in concentration of salt improves the scattering intensity indicating growth in micelles. This can be further seen from the calculated parameters shown in Table 3. Thus, SANS results indicate micellar growth in copolymer solutions (except T304) with the increase in temperature/salt concentration in line with the results attained by other techniques.

Table 3. Micellar parameters of 10% Tetronics® in D_2O obtained from small angle neutron scattering analysis.

Tetronic®	[NaCl], M	Temperature, °C	R_c, Å	R_{hs}, Å	R_g, Å	N_{agg}
5% T304	0	30	-	-	10.6	-
10% T304	0	30	-	-	11.8	-
10% T304	0	35	-	-	11.1	-
10% T304	0	50	-	-	13.0	-
10% T908	0	30	-	-	25.8	-
10% T908	0	40	39.9	-	-	7
10% T908	0	50	57.2	-	-	10
10% T908	0	60	60.1	-	-	13
5% T904 *	0	30	25.0	52.2	-	10
5% T904 *	0	40	29.9	52.2	-	18
5% T904 *	0	50	33.0	129.0	-	23
10% T904	0	30	34.7	51.6	-	24
10% T904	1 M	30	39.9	59.6	-	36
10% T904	2 M	30	43.61	65.7	-	47
10% T908	1 M	30	25.36	81.6	-	8
10% T908	2 M	30	44.3	84.9	-	43

* From our previous report [29].

3.1.8. Micellar Solubilization

To investigate the solubilization behavior of micelles from different Tetronics® in water and salt solutions, one model of hydrophobic dye Orange OT and two drugs—quercetin and curcumin—were taken. As evident in Figure 8, the solubility of the dye increased in the presence of salt. However, for T304, the solubility of the dye was negligible and did not improve in the presence of salt. In contrast, T908 showed poor solubility of dye in the absence of salt, but the solubility marginally increased with the progressive addition of salt. The maximum solubility of the dye was observed for T904 solutions, which further increased many folds in the presence of salt.

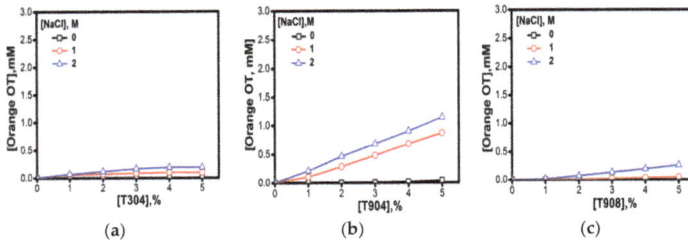

Figure 8. Orange OT Solubility in (**a**) 10% T304 (**b**) 10% T904 and (**c**) 10% T908 aqueous and salt solutions (□) 0 M, (○) 1 M and (△) 2 M NaCl at 30 °C.

A similar solubilizing trend of the copolymers was noted for the two poorly water soluble anti-cancer drugs viz. quercetin and curcumin as presented in Figure 9. Pillai et al. [43] compared the solubilizing behavior of the two copolymers T1304 and T1307 in the presence of glycine and observed many fold increase in the solubility of quercetin. The presence of glycine in the above case has an analogous effect to that of salt in the present study. A similar trend in the presence of salt has also been observed by Parekh et al. [59] for the solubility of nimesulide drug by T904 micelles at different copolymer concentrations.

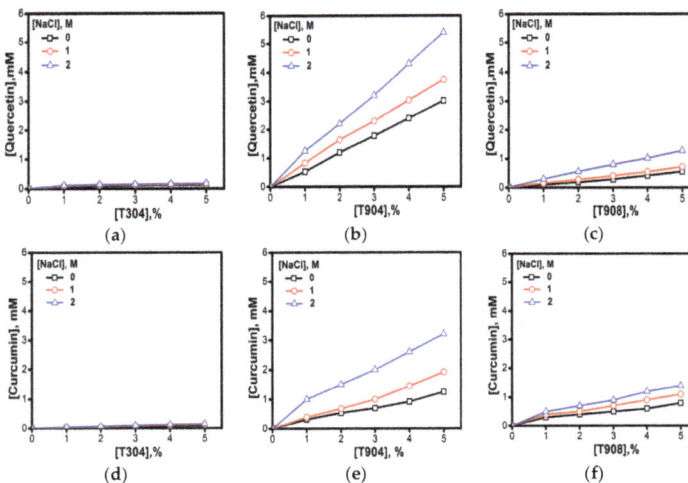

Figure 9. Solubility of quercetin (**a**–**c**) and curcumin (**d**–**f**) in Tetronic® solutions in the presence of (□) 0 M, (○) 1 M and (△) 2 M NaCl at 30 °C.

3.1.9. In Vitro Drug Release

In recent years, nanocarrier delivery systems have gained enormous attention for targeting anti-cancer drugs to tumors [60]. Micelles of natural and synthetic biocompatible polymers have been extensively examined as potential carrier materials for drug delivery [61–63] by virtue of their dimensions in nano range, competence to shield the encapsulated drug, targeting features by means of the enhanced permeability and retention (EPR) effect [64–66], and superior therapeutic capabilities [67]. In addition to their targeting ability, biocompatibility, enhanced circulation time and reduced toxicity are some of the prime factors responsible for the success of these polymers in drug delivery systems. With this approach, in the present work, we have tried to explore the in vitro release behaviors of curcumin and quercetin-loaded T904 micelle formulations under physiological conditions. Since the maximum solubility of the dye and drug in the previous section was observed for the copolymer solutions of T904, the release study was limited only to it.

Figure 10 shows the cumulative release profile of two hydrophobic drugs (quercetin and curcumin) versus time where 2% T904 was taken as the release media for understanding its role in in vitro release of both the drugs in physiological conditions. The time intervals were chosen by trial and error method and the study of release was restricted only for the initial 50%. The release remains slow and sustained for both drugs and only about 7% release is observed for curcumin and 18% of release is observed for quercetin during the early 24 h and approximately 50% of curcumin is released in 11 days while for quercetin in about 14 days. This is due to the fact that both the drugs remain localized within the core of the micelle and results in sustained release from this region.

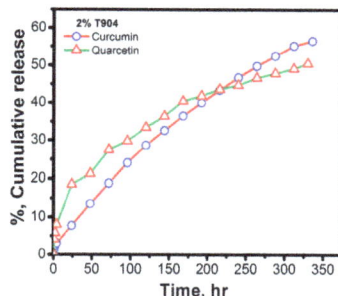

Figure 10. In vitro release profile of ($\triangle$) quercetin and ($\bigcirc$) curcumin from T904 micelles at physiological conditions (pH 7.4, 37 °C).

To understand the drug release kinetics and mechanism, the data was estimated mathematically using zero-order kinetics, first-order kinetics and Higuchi model. The prime condition for choosing the most appropriate model was grounded on best goodness-of-fit (R^2 values). The k_0, k_1, and k_H values were estimated by fitting the data into corresponding equations and are given in Table 4 along with the regression coefficients (R^2).

Table 4. Different kinetic models describing the release pattern of the anticancer drugs, curcumin and quercetin, from T904 micelles.

Drug	Mathematical Models for Drug Release Kinetics					
	Zero Order		First Order		Higuchi	
	k_0 (M. h^{-1})	R^2	k_1 (h^{-1})	R^2	k_H (M. h$^{-1/2}$)	R^2
curcumin	0.169	0.99	0.004	0.76	3.321	0.98
quercetin	0.131	0.90	0.002	0.69	2.69	0.99

From the results shown in Table 4, it is clearly evident that, for T904 micelles, the release of curcumin follows zero order kinetics while that of quercetin follows Higuchi equation. It has been established in literature that Higuchi and zero order kinetics indicate controlled drug diffusion. Hence, we can come to a conclusion that, from T904 micelles, both the drugs were released through diffusion.

3.1.10. In Vitro Cytotoxicity

To investigate the cytotoxicity of curcumin and quercetin from 2% T904 micelles, IC50 was determined on the CHO-K1 cell lines with different concentrations of curcumin and quercetin was evaluated using MTT assay and the data are presented in Table 5. As portrayed in Figure 11, no obvious decline in cell viability was observed after incubation with the blank micelles, indicating that the copolymer was nontoxic for being used as nanocarriers.

Table 5. The Half Maximal Inhibitory Concentration (IC_{50}) of curcumin, quercetin and drug loaded micelles of T904 in the Chinese Hampster Ovarian Cells (CHO-K1 cells).

Solvent	IC_{50}, µg/mL	
	Curcumin	Quercetin
H_2O	18.60	55.2
T904	45.8	48.2

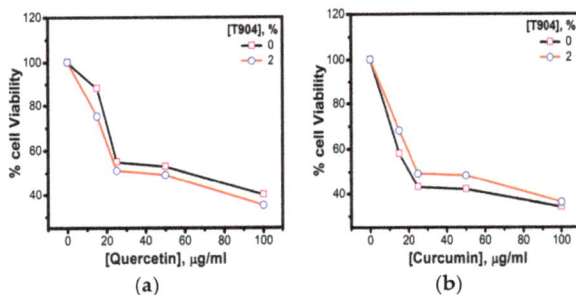

Figure 11. In vitro cell viability of (a) quercetin and (b) curcumin in the presence and absence of 2% T904.

To evaluate the feasibility of using this copolymer for cancer therapy, we compared the anticancer effects of free curcumin and quercetin with the drug-loaded T904 micelles. From Figure 11a, it is evident that the quercetin loaded micelles displayed more inhibition of proliferation of CHO-K1 cells than free drugs, while, in the case of curcumin, the IC_{50} of curcumin loaded micelles was comparatively larger than free form. A similar trend has been observed earlier by Pillai et al. [46] for the release of curcumin from the micelles of T1304. According to them, the possible cause for such deviation is due to the difference of drug release rate from micelles inside the cells.

4. Conclusions

Aqueous solution behaviour of three star shaped EO–PO block copolymers with different molecular characteristics was examined at different temperature and in the presence of salt using a variety of techniques. The self-assembly was found to be markedly dependent on their molecular characteristics and solution conditions. Data on micellar/phase behavior and interfacial characteristics of the copolymers are reported. More hydrophobic T904 formed micelles at low temperature and showed micellar growth with temperature, T908 with more hydrophilic character does not form micelles at ambient temperature but undergoes micellization at elevated temperatures while T304 with moderate hydrophobicity but low molecular weight remains molecularly dissolved in water in all of the conditions, and does not form micelles even at higher temperatures or in the presence of salt. The presence of salt has an analogous effect to that observed with the increase in temperature. Viscosity/DLS/SANS suggests growth in T904 micelles with temperature. NMR yields CMT, which decreases in the presence of salt. The solubility of dye and drugs improved in the presence of salt and with the increase in copolymer concentration. The in vitro release profiles for both of the drugs showed a slow and sustained release pattern from T904 micelles, and IC50 values for quercetin decreased significantly while increasing for curcumin, probably due to the slow release rate from the core of micelles. Thus, fine-tuning in micellar parameters using the copolymer of desired molecular characteristics and salt at desired temperatures and pH may be useful as nanoreservoirs for delivery systems.

Acknowledgments: All of the authors thank V. K. Aswal (BARC, Mumbai, India) for SANS measurements, Junhe Ma (Ashland Inc.) for NMR, Li-Jen Chen (National Taiwan University, Taipei) for HSDSC and Urjita Sheth for release kinetics and toxicity measurements. Pratap Bahadur thanks University Grants Commision, New Delhi (India) for fellowship.

Author Contributions: Pratap Bahadur conceived and designed the experiments; Bijal Vyas performed the experiments; Sadafara A. Pillai analyzed the data and wrote the paper; Anita Bahadur contributed reagents/materials/analysis tools.

Conflicts of Interest: The authors declare no conflict of interest.

References

1. Nakashima, K.; Bahadur, P. Aggregation of water-soluble block copolymers in aqueous solutions: Recent trends. *Adv. Colloid Interface Sci.* **2006**, *123*, 75–96. [CrossRef] [PubMed]
2. Bahadur, P. Block copolymers-their microdomain formation (in solid state) and surfactant behaviour (in solution). *Curr. Sci.* **2001**, *80*, 1002–1007.
3. Riess, G. Micellization of block copolymers. *Prog. Polym. Sci.* **2003**, *28*, 1107–1170. [CrossRef]
4. Sang, L.-C.; Coppens, M.-O. Effects of surface curvature and surface chemistry on the structure and activity of proteins adsorbed in nanopores. *Phys. Chem. Chem. Phys.* **2011**, *13*, 6689–6698. [CrossRef] [PubMed]
5. Chen, J.-K.; Chang, C.-J. Fabrications and applications of stimulus-responsive polymer films and patterns on surfaces: A review. *Materials* **2014**, *7*, 805–875. [PubMed]
6. Singh, V.; Khullar, P.; Dave, P.N.; Kaura, A.; Bakshi, M.S.; Kaur, G. PH and thermo-responsive tetronic micelles for the synthesis of gold nanoparticles: Effect of physiochemical aspects of tetronics. *Phys. Chem. Chem. Phys.* **2014**, *16*, 4728–4739. [CrossRef] [PubMed]
7. Habas, J.-P.; Pavie, E.; Perreur, C.; Lapp, A.; Peyrelasse, J. Nanostructure in block copolymer solutions: Rheology and small-angle neutron scattering. *Phys. Rev. E* **2004**, *70*, 061802. [CrossRef] [PubMed]
8. Fernandez-Tarrio, M.; Yañez, F.; Immesoete, K.; Alvarez-Lorenzo, C.; Concheiro, A. Pluronic and tetronic copolymers with polyglycolyzed oils as self-emulsifying drug delivery systems. *AAPS PharmSciTech* **2008**, *9*, 471–479. [CrossRef] [PubMed]
9. Hedberg, E.L.; Shih, C.K.; Solchaga, L.A.; Caplan, A.I.; Mikos, A.G. Controlled release of hyaluronan oligomers from biodegradable polymeric microparticle carriers. *J. Control. Release* **2004**, *100*, 257–266. [CrossRef] [PubMed]
10. Oh, K.T.; Bronich, T.K.; Kabanov, A.V. Micellar formulations for drug delivery based on mixtures of hydrophobic and hydrophilic pluronic® block copolymers. *J. Control. Release* **2004**, *94*, 411–422. [CrossRef] [PubMed]
11. Sezgin, Z.; Yüksel, N.; Baykara, T. Preparation and characterization of polymeric micelles for solubilization of poorly soluble anticancer drugs. *Eur. J. Pharm. Biopharm.* **2006**, *64*, 261–268. [CrossRef] [PubMed]
12. Csaba, N.; Caamaño, P.; Sánchez, A.; Domínguez, F.; Alonso, M.J. Plga: Poloxamer and plga: Poloxamine blend nanoparticles: New carriers for gene delivery. *Biomacromolecules* **2005**, *6*, 271–278. [CrossRef] [PubMed]
13. Hamley, I.W. *The Physics of Block Copolymers*; Oxford University Press: New York, NY, USA, 1998; Volume 19.
14. Alexandridis, P.; Lindman, B. *Amphiphilic Block Copolymers: Self-Assembly and Applications*; Elsevier: Amsterdam, The Netherlands, 2000.

15. Zhou, Z.; Chu, B. Anomalous association behavior of an ethylene oxide/propylene oxide aba block copolymer in water. *Macromolecules* **1987**, *20*, 3089–3091. [CrossRef]

16. Almgren, M.; Brown, W.; Hvidt, S. Self-aggregation and phase behavior of poly(ethylene oxide)-poly(propylene oxide)-poly(ethylene oxide) block copolymers in aqueous solution. *Colloid Polym. Sci.* **1995**, *273*, 2–15. [CrossRef]

17. Liu, T.; Xu, G.; Gong, H.; Pang, J.; He, F. Effect of alcohols on aggregation behaviors of branched block polyether tetronic 1107 at an air/liquid surface. *Langmuir* **2011**, *27*, 9253–9260. [CrossRef] [PubMed]

18. Nivaggioli, T.; Tsao, B.; Alexandridis, P.; Hatton, T.A. Microviscosity in pluronic and tetronic poly(ethylene oxide) poly(propylene oxide) block-copolymer micelles. *Langmuir* **1995**, *11*, 119–126. [CrossRef]

19. Ganguly, R.; Kadam, Y.; Choudhury, N.; Aswal, V.; Bahadur, P. Growth and interaction of the tetronic 904 micelles in aqueous alkaline solutions. *J. Phys. Chem. B* **2011**, *115*, 3425–3433. [CrossRef] [PubMed]

20. Gonzalez-Lopez, J.; Alvarez-Lorenzo, C.; Taboada, P.; Sosnik, A.; Sandez-Macho, I.; Concheiro, A. Self-associative behavior and drug-solubilizing ability of poloxamine (tetronic) block copolymers. *Langmuir* **2008**, *24*, 10688–10697. [CrossRef] [PubMed]

21. Goy-López, S.; Taboada, P.; Cambón, A.; Juárez, J.; Alvarez-Lorenzo, C.; Concheiro, A.; Mosquera, V.C. Modulation of size and shape of au nanoparticles using amino-x-shaped poly(ethylene oxide)—Poly(propylene oxide) block copolymers. *J. Phys. Chem. B* **2009**, *114*, 66–76. [CrossRef] [PubMed]

22. Dong, J.; Chowdhry, B.Z.; Leharne, S.A. Surface activity of poloxamines at the interfaces between air-water and hexane-water. *Colloids Surf. A* **2003**, *212*, 9–17. [CrossRef]

23. Dong, J.; Armstrong, J.; Chowdhry, B.Z.; Leharne, S.A. Thermodynamic modelling of the effect of pH upon aggregation transitions in aqueous solutions of the poloxamine, T701. *Thermochim. Acta* **2004**, *417*, 201–206. [CrossRef]

24. Dong, J.; Chowdhry, B.Z.; Leharne, S.A. Solubilisation of polyaromatic hydrocarbons in aqueous solutions of poloxamine t803. *Colloids Surf. A* **2004**, *246*, 91–98. [CrossRef]

25. Armstrong, J.K.; Chowdhry, B.Z.; Snowden, M.J.; Dong, J.; Leharne, S.A. The effect of ph and concentration upon aggregation transitions in aqueous solutions of poloxamine T701. *Int. J. Pharm.* **2001**, *229*, 57–66. [CrossRef]

26. Tirnaksiz, F.; Kalsin, O. A topical w/o/w multiple emulsions prepared with tetronic 908 as a hydrophilic surfactant: Formulation, characterization and release study. *J. Pharm. Pharm. Sci.* **2004**, *8*, 299–315.

27. Fernandez-Tarrio, M.; Alvarez-Lorenzo, C.; Concheiro, A. Calorimetric approach to tetronic/water interactions. *J. Therm. Anal. Calorim.* **2007**, *87*, 171–178. [CrossRef]

28. Xin, X.; Xu, G.; Zhang, Z.; Chen, Y.; Wang, F. Aggregation behavior of star-like peo–ppo–peo block copolymer in aqueous solution. *Eur. Polym. J.* **2007**, *43*, 3106–3111. [CrossRef]

29. Kadam, Y.; Singh, K.; Marangoni, D.; Ma, J.; Aswal, V.; Bahadur, P. Induced micellization and micellar transitions in aqueous solutions of non-linear block copolymer Tetronic® T904. *J. Colloid Interface Sci.* **2010**, *351*, 449–456. [CrossRef] [PubMed]

30. Branca, C.; Magazu, S.; Migliardo, F. Star polymer/water solutions: New experimental findings. *Condens. Matter Phys.* **2008**, *5*, 275–284. [CrossRef]

31. Wu, J.; Xu, Y.; Dabros, T.; Hamza, H. Effect of EO and PO positions in nonionic surfactants on surfactant properties and demulsification performance. *Colloids Surf. A* **2005**, *252*, 79–85. [CrossRef]

32. Mansur, C.R.; Barboza, S.P.; González, G.; Lucas, E.F. Pluronic× tetronic polyols: Study of their properties and performance in the destabilization of emulsions formed in the petroleum industry. *J. Colloid Interface Sci.* **2004**, *271*, 232–240. [CrossRef] [PubMed]

33. Tonge, S.; Jones, L.; Goodall, S.; Tighe, B. The ex vivo wettability of soft contact lenses. *Curr. Eye Res.* **2001**, *23*, 51–59. [CrossRef] [PubMed]

34. Subbaraman, L.N.; Bayer, S.; Gepr, S.; Glasier, M.-A.; Lorentz, H.; Senchyna, M.; Jones, L. Rewetting drops containing surface active agents improve the clinical performance of silicone hydrogel contact lenses. *Optom. Vis. Sci.* **2006**, *83*, 143–151. [CrossRef] [PubMed]

35. Cappel, M.J.; Kreuter, J. Effect of nonionic surfactants on transdermal drug delivery: Ii. Poloxamer and poloxamine surfactants. *Int. J. Pharm.* **1991**, *69*, 155–167. [CrossRef]

36. Moghimi, S.M.; Hunter, A.C. Poloxamers and poloxamines in nanoparticle engineering and experimental medicine. *Trends Biotechnol.* **2000**, *18*, 412–420. [CrossRef]

37. Sosnik, A.; Sefton, M.V. Poloxamine hydrogels with a quaternary ammonium modification to improve cell attachment. *J. Biomed. Mater. Res. Part A* **2005**, *75*, 295–307. [CrossRef] [PubMed]

38. Sosnik, A.; Leung, B.; McGuigan, A.P.; Sefton, M.V. Collagen/poloxamine hydrogels: Cytocompatibility of embedded hepg2 cells and surface-attached endothelial cells. *Tissue Eng.* **2005**, *11*, 1807–1816. [CrossRef] [PubMed]

39. De Lisi, R.; Giammona, G.; Lazzara, G.; Milioto, S. Copolymers sensitive to temperature and pH in water and in water+ oil mixtures: A dsc, itc and volumetric study. *J. Colloid Interface Sci.* **2011**, *354*, 749–757. [CrossRef] [PubMed]

40. Larrañeta, E.; Isasi, J.R. Non-covalent hydrogels of cyclodextrins and poloxamines for the controlled release of proteins. *Carbohydr. Polym.* **2014**, *102*, 674–681. [CrossRef] [PubMed]

41. González-Gaitano, G.; Müller, C.L.; Radulescu, A.; Dreiss, C.C.A. Modulating the self-assembly of amphiphilic x-shaped block copolymers with cyclodextrins: Structure and mechanisms. *Langmuir* **2015**, *31*, 4096–4105. [CrossRef] [PubMed]

42. Poellmann, M.J.; Sosnick, T.R.; Meredith, S.C.; Lee, R.C. The pentablock amphiphilic copolymer T1107 prevents aggregation of denatured and reduced lysozyme. *Macromol. Biosci.* **2016**, *17*. [CrossRef] [PubMed]

43. Pillai, S.A.; Bharatiya, B.; Casas, M.; Lage, E.V.; Sandez-Macho, I.; Pal, H.; Bahadur, P. A multitechnique approach on adsorption, self-assembly and quercetin solubilization by Tetronics® micelles in aqueous solutions modulated by glycine. *Colloids Surf. B* **2016**, *148*, 411–421. [CrossRef] [PubMed]

44. Pillai, S.A.; Lee, C.-F.; Chen, L.-J.; Aswal, V.K.; Bahadur, P. Glycine elicited self-assembly of amphiphilic star block copolymers with contradistinct hydrophobicities. *Colloids Surf. A* **2016**, *506*, 234–244. [CrossRef]

45. Pillai, S.A.; Lee, C.-F.; Ray, D.; Aswal, V.K.; Pal, H.; Chen, L.-J.; Bahadur, P. Microstructure of copolymeric micelles modulated by ionic liquids: Investigating the role of the anion and cation. *RSC Adv.* **2016**, *6*, 87299–87313. [CrossRef]

46. Pillai, S.A.; Sheth, U.; Bahadur, A.; Aswal, V.K.; Bahadur, P. Salt induced micellar growth in aqueous solutions of a star block copolymer tetronic® 1304: Investigating the role in solubilizing, release and cytotoxicity of model drugs. *J. Mol. Liq.* **2016**, *224*, 303–310. [CrossRef]

47. Pillai, S.A.; Lee, C.-F.; Chen, L.-J.; Bahadur, P.; Aswal, V.K.; Bahadur, P. Thermal and scattering studies of tetronic® 1304 micelles in the presence of industrially important glycols, their oligomers, cellosolves, carbitols, ethers and esters. *Colloids Surf. A* **2016**, *506*, 576–585. [CrossRef]

48. Patidar, P.; Pillai, S.A.; Sheth, U.; Bahadur, P.; Bahadur, A. Glucose triggered enhanced solubilisation, release and cytotoxicity of poorly water soluble anti-cancer drugs from T1307 micelles. *J. Biotechnol.* **2017**, *254*, 43–50. [CrossRef] [PubMed]

49. Patidar, P.; Pillai, S.A.; Bahadur, P.; Bahadur, A. Tuning the self-assembly of EO-PO block copolymers and quercetin solubilization in the presence of some common pharmacuetical excipients: A comparative study on a linear triblock and a starblock copolymer. *J. Mol. Liq.* **2017**, *241*, 511–519. [CrossRef]

50. Chen, S.-H.; Lin, T.-L.; Price, D. *Methods of Experimental Physics*; Academic Press: New York, NY, USA, 1987; Volume 23, p. 489.

51. Percus, J.K.; Yevick, G.J. Analysis of classical statistical mechanics by means of collective coordinates. *Phys. Rev.* **1958**, *110*. [CrossRef]

52. Alexandridis, P.; Holzwarth, J.F. Differential scanning calorimetry investigation of the effect of salts on aqueous solution properties of an amphiphilic block copolymer (poloxamer). *Langmuir* **1997**, *13*, 6074–6082. [CrossRef]

53. Jain, N.; George, A.; Bahadur, P. Effect of salt on the micellization of pluronic P65 in aqueous solution. *Colloids Surf. A* **1999**, *157*, 275–283. [CrossRef]

54. Pandit, N.; Trygstad, T.; Croy, S.; Bohorquez, M.; Koch, C. Effect of salts on the micellization, clouding, and solubilization behavior of pluronic f127 solutions. *J. Colloid Interface Sci.* **2000**, *222*, 213–220. [CrossRef] [PubMed]

55. Gu, T.; Galera-Gomez, P. Clouding of triton x-114: The effect of added electrolytes on the cloud point of triton x-114 in the presence of ionic surfactants. *Colloids Surf. A* **1995**, *104*, 307–312. [CrossRef]

56. Bahadur, A.; Cabana-Montenegro, S.; Aswal, V.K.; Lage, E.V.; Sandez-Macho, I.; Concheiro, A.; Alvarez-Lorenzo, C.; Bahadur, P. Nacl-triggered self-assembly of hydrophilic poloxamine block copolymers. *Int. J. Pharm.* **2015**, *494*, 453–462. [CrossRef] [PubMed]

57. Alexandridis, P.; Holzwarth, J.F.; Hatton, T.A. Micellization of poly(ethylene oxide)-poly(propylene oxide)-poly(ethylene oxide) triblock copolymers in aqueous solutions: Thermodynamics of copolymer association. *Macromolecules* **1994**, *27*, 2414–2425. [CrossRef]

58. Alexandridis, P.; Nivaggioli, T.; Hatton, T.A. Temperature effects on structural properties of pluronic P104 and f108 peo-ppo-peo block copolymer solutions. *Langmuir* **1995**, *11*, 1468–1476. [CrossRef]

59. Parekh, P.; Singh, K.; Marangoni, D.; Bahadur, P. Micellization and solubilization of a model hydrophobic drug nimesulide in aqueous salt solutions of tetronic® T904. *Colloids Surf. B* **2011**, *83*, 69–77. [CrossRef] [PubMed]

60. Danhier, F.; Feron, O.; Préat, V. To exploit the tumor microenvironment: Passive and active tumor targeting of nanocarriers for anti-cancer drug delivery. *J. Control. Release* **2010**, *148*, 135–146. [CrossRef] [PubMed]

61. Tanner, P.; Baumann, P.; Enea, R.; Onaca, O.; Palivan, C.; Meier, W. Polymeric vesicles: From drug carriers to nanoreactors and artificial organelles. *Acc. Chem. Res.* **2011**, *44*, 1039–1049. [CrossRef] [PubMed]

62. Kedar, U.; Phutane, P.; Shidhaye, S.; Kadam, V. Advances in polymeric micelles for drug delivery and tumor targeting. *Nanomedicine* **2010**, *6*, 714–729. [CrossRef] [PubMed]

63. Elsabahy, M.; Wooley, K.L. Design of polymeric nanoparticles for biomedical delivery applications. *Chem. Soc. Rev.* **2012**, *41*, 2545–2561. [CrossRef] [PubMed]

64. Torchilin, V. Tumor delivery of macromolecular drugs based on the EPR effect. *Adv. Drug Deliv. Rev.* **2011**, *63*, 131–135. [CrossRef] [PubMed]

65. Fang, J.; Nakamura, H.; Maeda, H. The epr effect: Unique features of tumor blood vessels for drug delivery, factors involved, and limitations and augmentation of the effect. *Adv. Drug Deliv. Rev.* **2011**, *63*, 136–151. [CrossRef] [PubMed]

66. Gaucher, G.; Dufresne, M.-H.; Sant, V.P.; Kang, N.; Maysinger, D.; Leroux, J.-C. Block copolymer micelles: Preparation, characterization and application in drug delivery. *J. Control. Release* **2005**, *109*, 169–188. [CrossRef] [PubMed]

67. Tian, H.; Tang, Z.; Zhuang, X.; Chen, X.; Jing, X. Biodegradable synthetic polymers: Preparation, functionalization and biomedical application. *Prog. Polym. Sci.* **2012**, *37*, 237–280. [CrossRef]

polymers

MDPI

Article

Aggregation of Cationic Amphiphilic Block and Random Copoly(vinyl ether)s with Antimicrobial Activity

Yukari Oda [1,2,*], **Kazuma Yasuhara** [3], **Shokyoku Kanaoka** [1,4], **Takahiro Sato** [1], **Sadahito Aoshima** [1,*]and **Kenichi Kuroda** [5,*]

[1] Department of Macromolecular Science, Graduate School of Science, Osaka University, Toyonaka, Osaka 560-0043, Japan; tsato@chem.sci.osaka-u.ac.jp

[2] Current, Department of Applied Chemistry, Kyushu University, Motooka, Nishi-ku, Fukuoka 819-0395, Japan

[3] Graduate School of Materials Science, Nara Institute of Science and Technology, Ikoma, Nara 630-0192, Japan; yasuhara@ms.naist.jp

[4] Current, Department of Materials Science, The University of Shiga Prefecture, Hikone, Shiga 522-8533, Japan; kanaoka.s@mat.usp.ac.jp

[5] Department of Biologic and Materials Science, School of Dentistry, University of Michigan, Ann Arbor, MI 48109, USA

* Correspondence: y-oda@cstf.kyushu-u.ac.jp (Y.O.); aoshima@chem.sci.osaka-u.ac.jp (S.A.); kkuroda@umich.edu (K.K.); Tel.: +81-92-802-2880 (Y.O.); +81-6-6850-5448 (S.A.); +1-734-936-1440 (K.K.)

Received: 29 December 2017; Accepted: 16 January 2018; Published: 19 January 2018

Abstract: In this study, we investigated the aggregation behaviors of amphiphilic poly(vinyl ether)s with antimicrobial activity. We synthesized a di-block poly(vinyl ether), $B38_{26}$, composed of cationic primary amine and hydrophobic isobutyl (*i*Bu) side chains, which previously showed antimicrobial activity against *Escherichia coli*. $B38_{26}$ showed similar uptake behaviors as those for a hydrophobic fluorescent dye, 1,6-diphenyl-1,3,5-hexatriene, to counterpart polymers including homopolymer H44 and random copolymer $R40_{25}$, indicating that the *i*Bu block does not form strong hydrophobic domains. The cryo-TEM observations also indicated that the polymer aggregate of $B38_{26}$ appears to have low-density polymer chains without any defined microscopic structures. We speculate that $B38_{26}$ formed large aggregates by liquid-liquid separation due to the weak association of polymer chains. The fluorescence microscopy images showed that $B38_{26}$ bonds to *E. coli* cell surfaces, and these bacterial cells were stained by propidium iodide, indicating that the cell membranes were significantly damaged. The results suggest that block copolymers may provide a new platform to design and develop antimicrobial materials that can utilize assembled structures and properties.

Keywords: aggregation; amphiphilic block copolymer; poly(vinyl ether); antimicrobial activity

1. Introduction

The emergence of drug-resistant bacteria poses a serious threat to human health [1–3], as the number of treatment options for bacterial infections is significantly reduced. There is urgent need for new antimicrobials effective in controlling drug-resistant bacteria. However, it has been a significant challenge to design and develop such molecules with novel antimicrobial targets in bacteria and mechanisms. To that end, one recent strategy is to design synthetic polymers to mimic the structural features and functions of host-defense antimicrobial peptides (AMPs) found in the innate immune system [4,5], which act directly by disrupting bacterial cell membranes. In general, antimicrobial (co)polymers have cationic and hydrophobic moieties in their side chains to mimic the cationic amphiphilicity of AMPs, which govern the bacterial selectivity and membrane-disrupting mechanism for antimicrobial activity [6,7]. The cationic groups of polymers enhance the binding of polymers to anionic lipids of bacterial membranes by electrostatic interactions. Because the bacterial membranes are more negatively charged than those of human cell membranes, the polymers are expected to selectively bind to bacterial membranes over human cell membranes, imparting the selective activity of polymers to bacteria over human cells. Upon the binding of polymers to membranes, the hydrophobic groups of polymers are inserted into the hydrophobic domain of the membranes, causing membrane disruption and ultimately bacterial cell death. It has been previously demonstrated that the antimicrobial activity of polymers and their toxicity to human cells can be controlled by modulating key structural parameters, including compositions of cationic and hydrophobic monomers [8–11], molecular weight [11,12], the hydrophobicity of side chains [13], and the type of cationic charge [14].

Synthetic polymers with cationic and hydrophobic segments or cationic amphiphilic block copolymers have been utilized as a platform for designing antibacterial polymers [15,16]. Such block copolymers are prepared by living polymerization, their length of polymer chains and block sequences can be precisely designed and controlled, which provides great advantages for the development of materials with target biological functions [17]. We previously synthesized a series of di-block poly(vinyl ether)s composed of cationic and hydrophobic blocks and investigated the relationship of their amphiphilic structures (block vs. random) with their antibacterial activity and lytic activity against human red blood cells (hemolysis) as a measure of undesired toxicity to human cells [15]. We demonstrated that the amphiphilic structures of these copolymers play an important role in their antibacterial and hemolytic activities [15]. The random and di-block copolymers with the same cationic/hydrophobic monomer compositions showed the same level of bactericidal activity against *Escherichia coli*. However, the block copolymers were not hemolytic, while the random copolymers were highly hemolytic. This result suggested that the block copolymers were selective to bacteria over human red blood cells while they remained active against bacteria, which is the desired properties for antimicrobials. A static light scattering (SLS) experiment suggested that the block copolymer formed aggregates with a diameter of ~500 nm in an aqueous media, which may be a vesicle rather than polymer micelles with a single hydrophobic core. Interestingly, the minimum polymer concentration of the block copolymer for bactericidal activity was below its critical (intermolecular) aggregation concentration (CAC), indicating that single-polymer chains were bactericidal. In addition, the copolymer was not hemolytic throughout the polymer concentration range above and below the CAC, suggesting that the selective activity of copolymer to bacteria over human cells was not necessarily the results of polymer aggregation or vesicle formation. We proposed the mechanism that the cationic polymer block wrapped the hydrophobic polymer block to form cationic single chain polymer particles. This particle structure shielded the hydrophobicity of copolymer chains and reduced their non-specific hydrophobic binding to the membranes of human red blood cells, resulting in no significant hemolytic activity [15]. On the other hand, the random copolymers might not be able to effectively shield the hydrophobicity of copolymers, because of the random distribution of cationic and hydrophobic groups in the polymer chains in comparison to block copolymers, and may thus bind to human red blood cells and cause hemolysis. It is generally known that there is an equilibrium between free single-polymer chains and aggregates above the CAC, and the concentration of single-polymer chains remains constant above the CAC. Our results indicate the possibility that single-polymer chains free in solution were responsible for the selective bactericidal activity of copolymer rather than the polymer aggregates.

In this study, we further extend our previous study on antimicrobial copolymers to investigate their aggregation behaviors in an aqueous environment. Amphiphilic copolymers intrinsically form aggregates and/or assemblies in aqueous media [18,19], which may control the interactions with bacterial cell membranes that govern the membrane-disrupting mechanism, thus determining the antimicrobial activity and selectivity. Therefore, it is important to investigate the formation and physicochemical properties of polymer aggregates in order to understand the role of aggregates in their underlying antimicrobial mechanism toward the goal of development of a novel class of antibacterial polymers. Specifically, the objective of this study is to determine the formation of polymer aggregates in water and their structures. In particular, we are interested in the aggregates formed by the block copolymer, because it previously showed potent bactericidal activity with selectivity to bacteria over human cells, which will be a good candidate for a new antimicrobial polymer platform. To that end, we first examined the uptakes of a hydrophobic probe by the copolymers to determine the formation of hydrophobic domains or polymer aggregates. The structure of block copolymer aggregates was further examined by a cryogenic transmission electron microscopy (cryo-TEM) that enables in situ visualization of the polymer assembly in water. The interaction between aggregates and bacterial cells was also examined by using fluorescent microscopy.

2. Materials and Methods

2.1. Materials

All materials for polymerization were prepared and used as described in the previous report [15]. 4-(2-Hydroxyethyl)-1-piperazineethanesulfonic acid (HEPES) and fluorescein isothiocyanate (FITC) were purchased from Fischer Scientific (Waltham, MA, USA) and Sigma-Aldrich (St. Louis, MO, USA), respectively.

2.2. Synthesis of Amphiphilic Copolymers

A series of amphiphilic poly{(isobutyl vinyl ether)-*co*-(2-aminoethyl vinyl ether)}s {poly(IBVE-*co*-AEVE)s} (Figure 1) were prepared by living cationic copolymerization of IBVE and 2-phthalimidoethyl vinyl ether (PIVE), which was a protected monomer for AEVE, and subsequent deprotection as described in the previous report [15,20].

A FITC-labeled block copolymer was prepared by the reaction of the amino-containing block copolymer with FITC in the presence of trimethylamine in *N,N*-dimethylformamide at room temperature for 4 h, as described in the previous report [15]. The obtained FITC-labeled block copolymer was purified by size exclusion chromatography (Sephadex LH-20 gel, Amersham Bioscience, Uppsala, Sweden) using methanol.

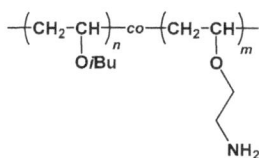

Figure 1. Chemical structure of poly(IBVE-*co*-AEVE)s.

2.3. Dye Uptake Experiment

The dye uptake by the polymer aggregates in the aqueous solution was examined using a fluorescent probe, 1,6-diphenyl-1,3,5-hexatriene (DPH) [21]. Polymer stock solutions were prepared in dimethyl sulfoxide (DMSO) (10 or 20 mg/mL). The stock solution was serially diluted 16 2-fold by 0.01% acetic acid. The polymer stock solutions (20 μL) were mixed with HEPES buffer (10 mM HEPES, 150 mM NaCl, pH 7, 175 μL) on a 96-well black microplate. DPH in tetrahydrofuran (THF) (20 μL, 50 μM) was diluted with HEPES buffer (480 μL). Then this DPH solution (5.0 μL) was added to the polymer solution on the microplate to give a final concentration of 50 nM for DPH, and THF of 0.1 vol %. After a 1 h incubation at 37 °C with orbital shaking (100 rpm), the fluorescence intensity in each well was recorded using a microplate reader (Thermo Scientific Varioskan Flash, Fischer Scientific, Waltham, MA, USA) with excitation and emission wavelengths of 357 and 430 nm, respectively.

2.4. Fluorescence Microscopic Observation

A single colony of *E. coli* was incubated in Mueller-Hinton (MH) broth at 37 °C with gentle shaking overnight. The *E. coli* suspension was diluted by MH broth to OD_{600} = 0.1 (OD_{600}: optical density at 600 nm) and incubated again for 90 min. The bacterial culture in the midlogrithmic phase (OD_{600} ~0.5–0.6) was diluted to OD_{600} = 0.1 with HEPES buffer, corresponding to ~2 × 10^7 cfu/mL (cfu: colony forming unit). This bacterial suspension (40 μL) was mixed with the stock polymer solution containing a small amount of FITC-labeled polymer (200 μg/mL, 50 μL) in a 96-well polypropylene microplate, which was not treated for tissue culture (Corning #3359). After a 45 min incubation at 37 °C, propidium iodide (PI) aqueous solution (16 μM, 10 μL) was added to the mixture and then incubated for additional 15 min. Confocal fluorescence microscopy images of the mixtures were recorded using Eclipse T*i* Confocal Microscope C1 (Nikon, Melville, NY, USA). FITC and PI were excited at 488 and 561 nm, respectively.

2.5. Cryo-TEM Observation

The specimen for cryo-TEM was prepared by rapid freezing of a polymer solution at a concentration of 10 mg/mL. A 200 mesh copper microgrid was used and pretreated with a glow-discharger (HDT-400, JEOL, Tokyo, Japan) to make the microgrid surface hydrophilic. An aliquot (3.0 μL) of a polymer sample was placed on the mesh and immediately plunged into liquid propane using a specimen preparation machine (EM CPC, Leica, Wetzlar, Germany). The temperature of the specimen was maintained below −140 °C during the observation using a cryo-transfer holder (Model 626.DH, Gatan, Pleasanton, CA, USA). Microscopic observations were carried out using a transmission electron microscope (JEM-3100FEF, JEOL, Tokyo, Japan) at an acceleration voltage of 300 kV in zero-loss imaging mode. The microscopic image was recorded using a CCD camera (Model 794, Gatan, Pleasanton, CA, USA) installed in the microscope.

3. Results and Discussion

3.1. Polymer Design, Synthesis, and Antimicrobial Activity

In this study, amphiphilic block ($B38_{26}$) and random ($R40_{25}$) poly(IBVE-*co*-AEVE)s with almost the same degree of polymerization (DP ~40) and compositions of hydrophobic IBVE (~25 mol %) were used. The synthesis and antimicrobial activities of these copolymers have been reported previously [15]. Briefly, the copolymers were synthesized by living cationic polymerization using protected monomer, PIVE, followed by removing the phthalate groups to give primary amine groups. The deprotected copolymers were denoted as R/BX_y (R: random, B: block, X: total DP, y: mol % of IBVE) using the values of protected polymers (Table 1). We also prepared a cationic homopolymer H44 for comparison.

Table 1. Characterization, bactericidal activity and hydrophobic dye uptake behaviors for poly(IBVE-*co*-AEVE)s.

Polymer	Copolymer Structure	DP[1]	MP_{IBVE}[1] (mol %)	$BC_{99.9}$[2] (μg/mL)	HC_{50}[3] (μg/mL)	C_{DPH}[4] (μg/mL)	CAC[5] (μg/mL)	R_H[6], R_g[7] (nm)
H44	Homopolymer	44	0	1.6 ± 0.0	>1000 (42.5 ± 6.3%)[3]	90	N.D.	N.D.
$B38_{26}$	Block copolymer	38	26	2.4 ± 0.91	>1000 (37.7 ± 2.8%)[3]	124	36	250[6]
$R40_{25}$	Random copolymer	40	25	1.6 ± 0.0	0.49 ± 0.17	125	380	27[7]

[1] See [15]; [2] Determined in HEPES buffer against *E. coli*; [3] Local minimum values of hemolysis induced by each polymer; [4] Determined by dye uptake experiment in HEPES buffer; [5] Critical (intermolecular) aggregation concentration, determined by SLS; [6] Hydrodynamic radius, determined by DLS; [7] Radius of gyration, determined by SLS.

These copolymers showed a bactericidal activity against *E. coli* [15]. The lowest polymer concentration to kill *E. coli* at least 99.9% of initial seeding concentration after 4-h incubation in

HEPES buffer at 37 °C (BC$_{99.9}$) was determined as a measure of the bactericidal activity of copolymers. We used a non-growth defined medium of HEPES buffer for our antimicrobial assay, as well as for characterization of the polymer aggregation. The BC$_{99.9}$ values of B38$_{26}$ and R40$_{25}$ were very similar, indicating that the copolymer structures (random vs. block) do not determine the antimicrobial activity against *E. coli*. On the other hand, R40$_{25}$ was highly hemolytic, showing a small HC$_{50}$ value, while B38$_{26}$ did not cause significant hemolysis (Table 1) [15]. Here, the HC$_{50}$ values were defined as the polymer concentration required to cause 50% hemolysis relative to the positive control.

3.2. Dye Uptakes by Copolymers

In the previous study, we determined the formation of aggregates of B38$_{26}$ and R40$_{25}$ by static and dynamic light scattering (SLS and DLS) [15]. We found that B38$_{26}$ formed large spherical aggregates with a diameter of 400–500 nm above CAC of 36 μg/mL, whereas R40$_{25}$ formed smaller aggregates with a diameter of 54 nm above CAC of 380 μg/mL (Table 1).

To further examine the role of hydrophobic side chains in copolymer aggregation, we first determined the critical aggregation concentration of polymers (C$_{DPH}$) by monitoring uptake of a hydrophobic dye, DPH into the hydrophobic domains of formed polymer aggregates. The DPH probe has been widely used in the field to determine the critical aggregation concentrations of polymers, because its fluorescence property is sensitive to the polarity of the surrounding environment; the fluorescence of DPH increases upon partitioning into a non-polar or hydrophobic environment, while DPH in aqueous media is only slightly or not at all fluorescent [21]. The fluorescence intensity would increase when the polymer chains associate to form hydrophobic domains, and then take up the dye. Therefore, the DPH uptake would reflect the formation of microscopic hydrophobic domains due to association of hydrophobic side chains or block segments of polymers studied here.

All the polymers showed similar DPH uptake behaviors, resulting in the similar C$_{DPH}$ values of 90–125 μg/mL (Table 1, Figure 2). This result indicates that the formation of aggregates of these polymers is not dependent on (1) the hydrophobicity of polymers (homopolymer vs. amphiphilic copolymers) and (2) copolymer amphiphilic structures (random vs. block copolymers). Other block and random copolymers with larger MP$_{IBVE}$ values also showed similar DPH uptake behaviors (Table S1 and Figure S1), supporting the conclusion.

Interestingly, the homopolymer H44 exhibited DPH uptake, although this polymer has no hydrophobic *i*Bu side chains. This result suggests that the cationic homopolymer can form hydrophobic domains and bind DPH molecules, likely as a result of their hydrophobic polymer backbones. Such hydrophobic domains can be formed by single polymer chains intramolecularly, or association of multiple polymer chains (intermolecular aggregation). Therefore, the C$_{DPH}$ value may reflect either the onset of DPH binding curves by single polymer chains or the formation of intramolecular aggregates, but not necessarily formation of large polymer aggregates such as micelles.

On the other hand, B38$_{26}$ and R40$_{25}$ also showed similar DPH uptake behaviors to H44, indicating that the hydrophobic *i*Bu side chains or blocks are not involved in the DHP binding. Therefore, the DHP uptake was likely a result of the intrinsic hydrophobicity of polymer backbones as postulated for H44 above. In the literature, amphiphilic polymers are reported to show the DHP uptake by the formation of aggregates due to the association of hydrophobic side chains [22,23]. However, the reported polymers generally have strong hydrophobic moieties such as long alkyl chains and/or higher molecular weights, which are likely to readily form hydrophobic domains in water. However, our copolymers used in this study are relatively short (DP ~40), and the *i*Bu group is relatively small, so that these copolymers may not be able to form strong hydrophobic domains. Instead, the intrinsic hydrophobicity of the polymer backbone is likely to play a more dominant role in the DHP uptake. Therefore, the observed C$_{DHP}$ values may not present the critical concentration for the formation of polymer aggregates. Taken together, the results of the DPH uptake experiments suggest that the *i*Bu side chains or blocks do not form strong microscopic hydrophobic domains. In addition, the results also indicate that the polymer aggregates previously observed by SLS and DLS are not conventional aggregates formed by strong microscopic hydrophobic domains.

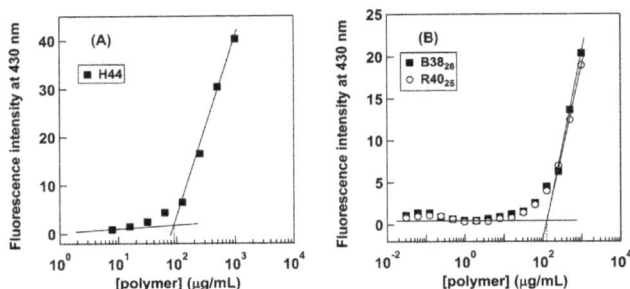

Figure 2. Fluorescence intensity of DPH (50 nM) versus polymer concentrations of (**A**) homopolymer, H44 and (**B**) poly(IBVE-*co*-AEVE)s with MP_{IBVE} ~25 mol % in HEPES buffer (pH 7). The data points represent the average from duplicate measurements.

3.3. Cryo-TEM Observations of the Block Copolymer Aggregates

The results of the DHP uptake experiments indicated that the hydrophobicity of the PIBVE blocks of $B38_{26}$ is not sufficient for the DHP uptakes. However, our previous study demonstrated that the $B38_{26}$ polymer chains were able to form large aggregates with diameters of 400–500 nm. To investigate the aggregation mechanism of $B38_{26}$, we examined the structure of the aggregates at 10 mg/mL, which is substantially higher than the critical concentration observed in the DHP uptake experiments using cryo-TEM (Figure 3). The aggregate particle in the cryo-TEM image presented as a spherical blur shadow with no clear boundaries. The diameter of the particle was found to be around 500 nm, which is consistent with the results of the SLS and DLS measurements (Table 1). In our previous study, the SLS data suggested that the density of the polymer chains in the $B38_{26}$ aggregates was relatively low, and the aggregates were relatively large, such that we speculated that $B38_{26}$ formed a vesicle (polymer bilayers). However, the aggregate structure presented in the cryo-TEM image does not appear to have any polymer bilayers, but seems rather to consist of low-density polymer aggregates without any defined structures.

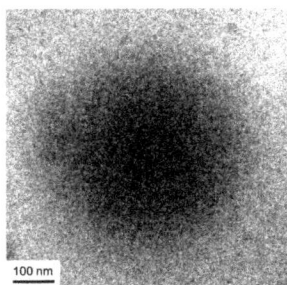

Figure 3. The cryo-TEM image of $B38_{26}$ rapidly freeze-dried from 10 mg/mL solution in HEPES buffer.

Recently, Takahashi et al. demonstrated both experimentally [24–26] and theoretically [27] that if the amphiphilicity of a block copolymer is not strong enough, the copolymer does not form micelles; rather, a liquid-liquid phase separation takes place in the solution. The amphiphilicity of $B38_{26}$ may be too weak to form micelles, and the large aggregate of a 500-nm diameter may be colloidal droplets of the phase-separated concentrated phase. If the concentration of the concentrated phase is not high, the droplet will contain a considerable amount of water, which prevents DPH uptake, and thus the contrast between the concentrated and dilute phases may be so weak that the cryo-TEM image may be blurred.

3.4. Fluorescnt Study of Block Copolymer Aggregates

We further investigated the formation of B38$_{26}$ aggregates and interaction with bacteria using fluorescence spectroscopy. Here, the block copolymer B38$_{26}$ was labeled with FITC (FITC-labeled B38$_{26}$: F-B38$_{26}$) [15]. The molar absorbance coefficient of F-B38$_{26}$ was 37,000 M^{-1} cm^{-1} in HEPES buffer. Based on the molar absorbance coefficient of F-B38$_{26}$ and the free fluorescein (83,000 M^{-1} cm^{-1}), the average number of FITC molecule per B38$_{26}$ chain was estimated to be 0.45, assuming no significant difference in the absorbance of fluorescein before and after FITC conjugation.

First, we investigated the concentration dependence of fluorescence emission from F-B38$_{26}$. A small amount of F-B38$_{26}$ was added to non-labeled B38$_{26}$ with in HEPES buffer. Based on the absorbance of 20 µg/mL polymer solution and free fluorescein absorbance, the FITC content in this mixture was estimated to be 5.3 mol % or 5.3 FITC in 100 polymer chains. The fluorescence intensity increased proportionally as a function of polymer concentration, and it exhibited a flexion point at 83 µg/mL, which may indicate that the surrounding environment of FITC in polymer chains might be changed, whereas the maximum absorbance was almost insensitive to changes in polymer concentration (Figure 4). This might reflect the onset of the formation of polymer aggregates, which change the polymer conformation and density as compared to the polymer chains free in solution.

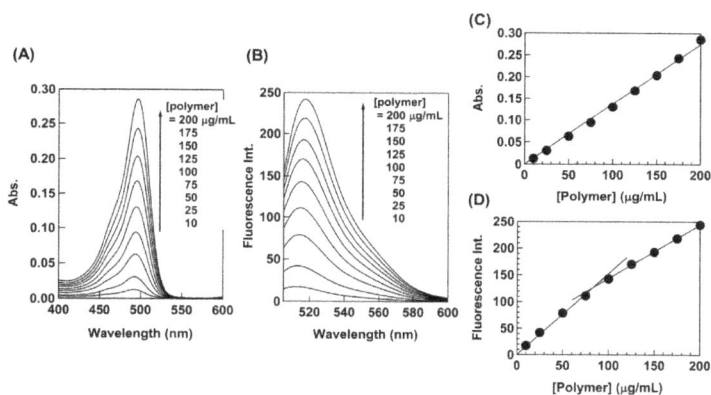

Figure 4. (**A**) Absorption and (**B**) emission spectra of B38$_{26}$ containing F-B38$_{26}$ in HEPES (1% DMSO), and (**C**) the maximum absorbance and (**D**) maximum fluorescent intensity versus polymer concentration.

Finally, we examined the interaction between the polymer aggregates and bacterial cells. The *E. coli* cells were incubated with B38$_{26}$ containing a small amount of F-B38$_{26}$ at 100 µg/mL, which was a higher concentration than CAC. We previously demonstrated that the polymer aggregates can be seen in fluorescence images as fluorescent particles with ~500 nm in diameter (Figure 5A) [15], which is close to the aggregate size estimated by DLS (R_H = 250 nm). The perimeters of *E. coli* cells treated with F-B38$_{26}$ were fluorescent green, indicating the binding of the polymer on the cell surfaces (Figure 5B). However, the resolution of the images was not sufficient to identify the structure of the polymer aggregates bound on the bacterial cell surfaces. The *E. coli* cells were also stained by PI, which can only penetrate cells with damaged membranes, and shows red fluorescence [28]. The *E. coli* cells bound with F-B38$_{26}$ showed red fluorescence, indicating that the cell membranes were damaged (Figure 5C). Liu et al. speculated that cationic polymer nanoparticles with a diameter of 177 nm caused steric hindrance and crosslinking of peptideglycans in the cell wall, disrupting cell membranes and cell death [29]. The cationic particles reported here are relatively large (400–500 nm of diameter), so the aggregates may not be able to penetrate into the cell wall structure. However, we have previously demonstrated that the $BC_{99.9}$ values are smaller than the CAC values, suggesting that the free single polymer chain could be responsible for the bactericidal activity. Therefore, although the polymer aggregates may not be directly active against bacterial cell membranes, the polymer chains may dissociate from the polymer aggregates, and the free polymer chains may penetrate the cell wall and disrupt bacterial cell membranes to kill bacteria. The polymer aggregates are likely to have a high net-positive charge, which would facilitate the binding of aggregates onto anionic bacterial cell surfaces. The results of DPH uptake experiments and cryo-TEM observations indicate that the polymers may weakly associate to form aggregates or colloidal droplets. Therefore, the polymer chains may be able to readily dissociate to attack bacterial cell membranes after the aggregates bind to bacterial cell surfaces. The polymer aggregates may serve as a reservoir that can deliver active polymer chains to the bacterial cell surface and release them for antimicrobial actions. Our previous computational model of cationic amphiphilic methacrylate copolymers also demonstrated that the copolymer formed aggregates in an aqueous environment, but the aggregate dissociated to individual polymer chains upon binding to bacterial cell membranes [30]. Then, the free polymer chains bound to the bacterial cell membrane for antimicrobial action. These previous data also support the new perception of polymer aggregates as a delivery reservoir proposed in this study.

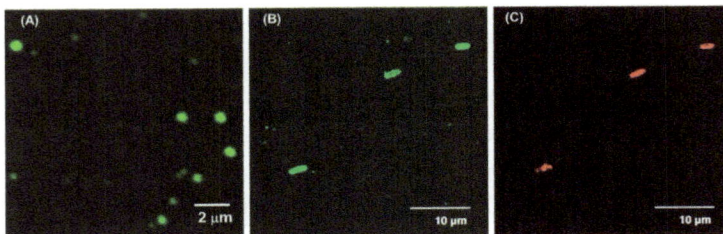

Figure 5. Confocal fluorescent microscopic images of (**A**) 50 μg/mL solution of $B38_{26}$ containing F-$B38_{26}$ (FITC: 5.3 mol %) in HEPES buffer, (**B**) *E. coli* (OD_{600} ~0.05) incubated with 100 μg/mL solutions of $B38_{26}$ containing F-$B38_{26}$ and (**C**) PI (1.6 μM) in HEPES buffer (0.5% DMSO). The images are projected images of 42 image stacks acquired with a z-step of 0.1 μm (total height: 4.2 μm).

4. Conclusions

In this study, we studied the aggregation behaviors of amphiphilic poly(vinyl ether)s with antimicrobial activity using fluorescent dye, DPH uptake assay, and fluorescent microscopy. The results of the DPH uptake experiments indicated that the hydrophobic side chains of our polymers may not form microscopic strong hydrophobic domains. The cryo-TEM images also indicated that the polymer aggregate of $B38_{26}$ appears to have a low density of polymer chains without any defined microscopic structures. We speculate that the block copolymer, $B38_{26}$, formed large aggregates by liquid-liquid separation due to the weak association of polymer chains, rather than the conventional core-shell type micelles or vesicles. The fluorescence microcopy images showed that $B38_{26}$ bounds to *E. coli* cell surfaces although it was not clear that the structure of aggregates remained when bound on the cell surface. The *E. coli* cells with $B38_{26}$ were stained by PI, indicating that the cell membranes were significantly damaged. These results suggest that the polymer aggregates may act directly by disrupting bacterial cell membranes. It is also possible that the polymer aggregates may not act directly, but that free polymer chains released from the aggregates may attack the bacterial cell membranes.

This study showed the discrepancy between methods for determining the CACs of copolymers. The CAC values determined by different methods are likely to reflect different dimensions and molecular processes (microscopic hydrophobic domains, polymer chain association, and particle formation) in the formation of polymer aggregates. It would be a subject for a future study to link the CAC values to the aggregation mechanism using different methods and determine the cause of the discrepancies. The expected results would also shed light into the polymer aggregate structures and dynamics, which would be useful for designing new antimicrobial polymer aggregates.

Many polymer platforms have been studied, including random and block copolymers, star-shaped polymers, and graft copolymers [15,31,32]. However, the role of aggregates in their antimicrobial mechanisms is not clear yet. The physicochemical properties (charge density, size, etc.) and dynamics (exchange between polymer chains and aggregates) of polymer aggregates are likely to control the interactions with bacterial cell membranes, thus determining their antimicrobial activity. In particular, this study proposes the role of polymer aggregates as a delivery reservoir for antimicrobial action. Such properties and dynamics of polymer aggregates can be tuned by chemical compositions and structures of polymer chains. Therefore, block copolymers may provide a new programmable platform to design and develop antimicrobial materials that can utilize assembled structures and properties.

Supplementary Materials: The following are available online at www.mdpi.com/2073-4360/10/1/93/s1, Table S1: Characterization, bactericidal activity and hydrophobic dye uptake behaviors for poly(IBVE-*co*-AEVE)s with different MP$_{IBVE}$s, Figure S1: Fluorescence intensity of DPH (50 nM) versus polymer concentrations for poly(IBVE-*co*-AEVE)s with different MP$_{IBVE}$s.

Acknowledgments: This research was partly supported by the Global Center of Excellence (G-COE) Program "Global Education and Research Center for Bio-Environmental Chemistry" at Osaka University (to Yukari Oda), NSF CAREER Award (DMR-0845592) (to Kenichi Kuroda), and Department of Biologic and Materials Sciences, University of Michigan School of Dentistry. We thank Takeshi Suwabe at School of Dentistry, University of Michigan, for his help on the fluorescence microscopic observations.

Author Contributions: Kenichi Kuroda designed all the experiments and directed the work. Sadahito Aoshima and Shokyoku Kanaoka contributed to the polymer design and synthesis, Takahiro Sato contributed to the characterization of polymer aggregates, Kazuma Yasuhara performed and helped the cryo-TEM observation, and Yukari Oda performed all the experiments. All the authors contributed to the writing of the manuscript.

Conflicts of Interest: The authors declare no conflict of interest.

References

1. Fischbach, M.A.; Walsh, C.T. Antibiotics for emerging pathogens. *Science* **2009**, *325*, 1089–1093. [CrossRef] [PubMed]

2. Fernandes, P. Antibacterial discovery and development—The failure of success? *Nat. Biotechnol.* **2006**, *24*, 1497–1503. [CrossRef] [PubMed]

3. Levy, S.B. *The Antibiotic Paradox. How Miracle Drugs Are Destroying the Miracle*; Springer: New York, NY, USA, 1992; ISBN 978-1-4899-6042-9. [CrossRef]

4. Hancock, R.E.W.; Lehrer, R. Cationic peptides: A new source of antibiotics. *Trends Biotechnol.* **1998**, *16*, 82–88. [CrossRef]

5. Zasloff, M. Antimicrobial peptides of multicellular organisms. *Nature* **2002**, *415*, 389–395. [CrossRef] [PubMed]

6. Takahashi, H.; Caputo, G.A.; Vemparala, S.; Kuroda, K. Synthetic random copolymers as a molecular platform to mimic host-defense antimicrobial peptides. *Bioconj. Chem.* **2017**, *28*, 1340–1350. [CrossRef] [PubMed]

7. Tew, G.N.; Scott, R.W.; Klein, M.L.; DeGrado, W.F. De novo design of antimicrobial polymers, foldamers, and small molecules: From discovery to practical applications. *Acc. Chem. Res.* **2010**, *43*, 30–39. [CrossRef] [PubMed]

8. Ilker, M.F.; Nüesslein, K.; Tew, G.N.; Coughlin, E.B. Tuning the hemolytic and antibacterial activities of amphiphilic polynorbornene derivatives. *J. Am. Chem. Soc.* **2004**, *126*, 15870–15875. [CrossRef] [PubMed]

9. Mowery, B.P.; Lee, S.E.; Kissounko, D.A.; Epand, R.F.; Epand, R.M.; Weisblum, B.; Stahl, S.S.; Gellman, S.H. Mimicry of antimicrobial host-defense peptides by random copolymers. *J. Am. Chem. Soc.* **2007**, *129*, 15474–15476. [CrossRef] [PubMed]

10. Kuroda, K.; Caputo, G.A.; Degradol, W.F. The role of hydrophobicity in the antimicrobial and hemolytic activities of polymethacrylate derivatives. *Chem. Eur. J.* **2009**, *15*, 1123–1133. [CrossRef] [PubMed]

11. Mowery, B.P.; Lindner, A.H.; Weisblum, B.; Stahl, S.S.; Gellman, S.H. Structure-activity relationships among random nylon-3 copolymers that mimic antibacterial host-defense peptides. *J. Am. Chem. Soc.* **2009**, *131*, 9735–9745. [CrossRef] [PubMed]

12. Kuroda, K.; DeGrado, W.F. Amphiphilic polymethacrylate derivatives as antimicrobial agents. *J. Am. Chem. Soc.* **2005**, *127*, 4128–4129. [CrossRef] [PubMed]

13. Sovadinova, I.P.; Palermo, E.F.; Urban, M.; Mpiga, P.; Caputo, G.A.; Kuroda, K. Activity and mechanism of antimicrobial peptide-mimetic amphiphilic polymethacrylate derivatives. *Polymers* **2011**, *3*, 1512–1532. [CrossRef]

14. Palermo, E.F.; Lee, D.K.; Ramamoorthy, A.; Kuroda, K. Role of cationic group structure in membrane binding and disruption by amphiphilic copolymers. *J. Phys. Chem. B* **2011**, *115*, 366–375. [CrossRef] [PubMed]

15. Oda, Y.; Kanaoka, S.; Sato, T.; Aoshima, S.; Kuroda, K. Block versus random amphiphilic copolymers as antibacterials agents. *Biomacromolecules* **2011**, *12*, 3581–3591. [CrossRef] [PubMed]

16. Su, X.; Zhou, X.; Tan, Z.; Zhou, C. Highly efficient antibacterial diblock copolypeptides based on lysine and phenylalanine. *Biopolymers* **2017**, *107*, e23041. [CrossRef] [PubMed]

17. Venkataraman, S.; Tan, J.P.K.; Ng, V.W.L.; Tan, E.W.P.; Hedrick, J.L.; Yang, Y.Y. Amphiphilic and hydrophilic block copolymers from aliphatic N-substituted 8-membered cyclic carbonates: A versatile macromolecular platform for biomedical applications. *Biomacromolecules* **2017**, *18*, 178–188. [CrossRef] [PubMed]

18. Sato, T.; Matsuda, Y. Macromolecular assemblies in solution: Characterization by light scattering. *Polym. J.* **2009**, *41*, 241–251. [CrossRef]

19. Nakashima, K.; Bahadur, P. Aggregation of water-soluble block copolymers in aqueous solutions: Recent trends. *Adv. Colloid Interface Sci.* **2006**, *123–126*, 75–96. [CrossRef] [PubMed]

20. Oda, Y.; Kanaoka, S.; Aoshima, S. Synthesis of dual pH/temperature-responsive polymers with amino groups by living cationic polymerization. *J. Polym. Sci. Part A Polym. Sci.* **2010**, *48*, 1207–1213. [CrossRef]

21. Chattopadhyay, A.; London, E. Fluorimetric determination of critical micelle concentration avoiding interference from detergent charge. *Anal. Biochem.* **1984**, *139*, 408–412. [CrossRef]

22. Szczubia-lka, K.; Ishikawa, K.; Morishima, Y. Associating behavior of sulfonated polyisoprene block copolymers with short polystyrene blocks at both chain ends. *Langmuir* **2000**, *16*, 2083–2092. [CrossRef]

23. Sugihara, S.; Hashimoto, K.; Okabe, S.; Shibayama, M.; Kanaoka, S.; Aoshima, S. Stimuli-responsive diblock copolymers by living cationic polymerization: Precision synthesis and highly sensitive physical gelation. *Macromolecules* **2004**, *37*, 336–343. [CrossRef]

24. Takahashi, R.; Sato, T.; Terao, K.; Qiu, X.-P.; Winnik, F.M. Self-association of a thermosensitive poly(alkyl-2-oxazoline) block copolymer in aqueous solution. *Macromolecules* **2012**, *45*, 6111–6119. [CrossRef]

25. Sato, T.; Tanaka, K.; Toyokura, A.; Mori, R.; Takahashi, R.; Terao, K.; Yusa, S. Self-association of a thermosensitive amphiphilic block copolymer poly(N-isopropylacrylamide)-b-poly(N-vinyl-2-pyrrolidone) in aqueous solution upon heating. *Macromolecules* **2013**, *46*, 226–235. [CrossRef]

26. Takahashi, R.; Qiu, X.-P.; Xue, N.; Sato, T.; Terao, K.; Winnik, F.M. Self-association of the thermosensitive block copolymer poly(2-isopropyl-2-oxazoline)-b-poly(N-isopropylacrylamide) in water-methanol mixtures. *Macromolecules* **2014**, *47*, 6900–6910. [CrossRef]

27. Sato, T.; Takahashi, R. Competition between the micellization and the liquid-liquid phase separation in amphiphilic block copolymer solutions. *Polym. J.* **2017**, *49*, 273–277. [CrossRef]

28. Krishan, A. Rapid flow cytofluorometric analysis of mammalian cell cycle by propidium iodide staining. *J. Cell Biol.* **1975**, *66*, 188–193. [CrossRef] [PubMed]

29. Liu, L.; Xu, K.; Wang, H.; Tan, P.K.J.; Fan, W.; Venkatraman, S.S.; Li, L.; Yang, Y.Y. Self-assembled cationic peptide nanoparticles as an efficient antimicrobial agent. *Nat. Nanotechnol.* **2009**, *4*, 457–463. [CrossRef] [PubMed]

30. Ivanov, I.; Vemparala, S.; Pophristic, V.; Kuroda, K.; DeGrado, W.F.; McCammon, J.A.; Klein, M.L. Characterization of nonbiological antimicrobial polymers in aqueous solution and at water-lipid interfaces from all-atom molecular dynamics. *J. Am. Chem. Soc.* **2006**, *128*, 1778–1779. [CrossRef] [PubMed]

31. Song, A.; Walker, S.G.; Parker, K.A.; Sampson, N.S. Antibacterial studies of cationic polymers with alternating, random, and uniform backbones. *ACS Chem. Biol.* **2011**, *6*, 590–599. [CrossRef] [PubMed]
32. Totani, M.; Ando, T.; Terada, K.; Terashima, T.; Kim, I.Y.; Ohtsuki, C.; Xi, C.; Kuroda, K.; Tanihara, M. Utilization of star-shaped polymer architecture in the creation of high-density polymer brush coatings for the prevention of platelet and bacteria adhesion. *Biomater. Sci.* **2014**, *2*, 1172–1185. [CrossRef] [PubMed]

polymers

Article

Synergistic Effect of Binary Mixed-Pluronic Systems on Temperature Dependent Self-assembly Process and Drug Solubility

Chin-Fen Lee [1], Hsueh-Wen Tseng [1], Pratap Bahadur [2] and Li-Jen Chen [1,*]

[1] Department of Chemical Engineering, National Taiwan University, Taipei 10617, Taiwan; nattianan24@gmail.com (C.-F.L.); r00524034@ntu.edu.tw (H.-W.T.)
[2] Department of Chemistry, Veer Narmad South Gujarat University, Surat 395007, India; pbahadur2002@yahoo.com
* Correspondence: ljchen@ntu.edu.tw; Tel.: +886-2-3366-3049

Received: 29 December 2017; Accepted: 19 January 2018; Published: 22 January 2018

Abstract: Mixed Pluronic micelles from very hydrophobic and very hydrophilic copolymers were selected to scrutinize the synergistic effect on the self-assembly process as well as the solubilization capacity of ibuprofen. The tendency of mixing behavior between parent copolymers was systematically examined from two perspectives: different block chain lengths at same hydrophilicity (L92 + F108, +F98, +F88, and +F68), as well as various hydrophobicities at the same PPO moiety (L92 + F88, +F87, and +P84). Temperature-dependent micellization in these binary systems was clearly inspected by the combined use of high sensitivity differential scanning calorimeter (HSDSC) and dynamic light scattering (DLS). Changes in heat capacity and size of aggregates at different temperatures during the whole micellization process were simultaneously observed and examined. While distinction of block chain length between parent copolymers increases, the monodispersity of the binary Pluronic systems decreases. However, parent copolymers with distinct PPO moieties do not affirmatively lead to non-cooperative binding, such as the L92 + P84 system. The addition of ibuprofen promotes micellization as well as stabilizes aggregates in the solution. The partial replacement of the hydrophilic Pluronic by a more hydrophobic Pluronic L92 would increase the total hydrophobicity of mixed Pluronics used in the system to substantially enhance the solubility of ibuprofen. The solubility of ibuprofen in the 0.5 wt % L92 + 0.368 wt % P84 system is as high as 4.29 mg/mL, which is 1.4 times more than that of the 0.868 wt % P84 system and 147 times more than that in pure water at 37 °C.

Keywords: polymer micelles; Pluronic; mixed micelle; micellization; synergistic effect; differential scanning calorimetry

1. Introduction

Block copolymer molecules undergo self-assembly in solvent, which has preferential solubility for one of the blocks. With a progressive increase in concentration, polymer micelles transform to hexagonal, lamellar structures and further spontaneously pack into crystal lattices. In some amphiphilic block copolymer aqueous solutions, temperature plays a crucial role in solvent selectivity of certain polymer blocks. Micelles exhibit the property of thermosreversibility between unimers and organized aggregates. For example, Pluronics (or Poloxamers), a kind of amphiphilic block copolymers of poly (ethylene oxide)$_n$-poly(propylene oxide)$_m$-poly(ethylene oxide)$_n$, are well known to form various kinds of aggregates and biocompatible properties.

Micellization arises from the entropy-driven process where the middle poly (propylene oxide) block gets dehydrated and shrinks to form core-shell aggregates while system temperature increases.

Micelles formed by amphiphilic block copolymers would increase hydrophobic drug solubility, metabolic stability, and circulation time [1]. As a promising nanomedicine carrier for anti-cancer drugs, polymer micelles have been evaluated in several clinical trials [2,3]. Due to poor aqueous solubility and systematic toxicity, anticancer drugs are extremely limited in terms of clinical application. Most chemotherapeutic drugs have a narrow therapeutic window and short elimination half-life so that higher doses are needed. However, the toxicity in the formulation would limit the maximum intravenous dose that can be used safely. Hence, Pluronics have been more and more attractive as drug carriers due to their non-toxicity.

A wide array of Pluronics is accessible depending on their molecular characteristics through varying the propylene oxide (PO)/ethylene oxide (EO) composition ratio and/or its molecular weight. Certainly, thermophysical properties of Pluronics micelles and their applications to controlled drug release have been extensively examined [4,5]. It has been found that Pluronic block copolymers are able to interact with multi–drug resistance (MDR) cancer cells leading to drastic sensitization of these tumors with respect to doxorubicin and other anticancer agents [6]. Individual Pluronic micelles have been studied for the solubilization capacity of drugs in pharmaceutical applications [7–10].

Mixed Pluronic systems have also been investigated as well to search for better formulations. For example, Gaisford et al. [11] suggested that similar PPO moiety would perform cooperative binding between two copolymers, while a distinct length of PPO does not. Oh et al. [12] conducted a series of tests of system stability on several mixtures consisting of extremely hydrophobic and hydrophilic Pluronics. It was found that the combination of L121 + F127 in a 50/50 mixture of a similar PPO moiety is the most stable one and shows an outstanding capacity to solubilize hydrophobic dyes. Many studies were then conducted to study the practical use of binary Pluronic systems based on the perspective of two copolymers with a similar PPO moiety [1,8,13,14]. From those reports, it was evidenced that the combination of a hydrophobic copolymer and a hydrophilic copolymer with a similar PPO chain length could be an ideal candidate for drug delivery and pharmaceutical use. In order to overcome low drug loading efficiency and the MDR-caused failure of chemotherapy in the treatment of cancer, Chen et al. [8] chose the formulation of P105 + F127 for the high sensitizing MDR cancer efficacy of P105, as well as the relatively similar hydrophobic moieties of F127.

On the other hand, the distinct PPO chain length between parent copolymers performing unimodal behavior to form mixed micelles was also observed. The F127 + L61 mixed system was used as one of the formulations for doxorubicin and is under clinical trial phase escalation [2]. Note that the F127 and L61 did not have a similar PPO moiety. Two combinations, F127 + P105 and F127 + L64, were applied to study the influence of the non-identical PPO block chain on the mixing behavior [15]. It was pointed out that not only Pluronics with a similar PPO moiety show cooperative binding, but distinct PPO chain lengths could also exhibit unimodal behavior via cooperative binding.

Micelles with spherical conformation are usually stable in aqueous solutions because of their core-shell structure. On the other hand, the lamellar and cylindrical morphology provides stable nano-environments from the interstitials between copolymers on the continuous architectures. It was calculated by the molecular theory of solubilization [16] and then experimentally evidenced for highly capable drug loading, but it was usually large and unstable in an aqueous solution [12]. Herein, use of mixtures of Pluronic block copolymers is an alternative to compensate for drawbacks of a neat system. The stability of the more hydrophobic copolymer in an aqueous solution could be substantially improved by adding another hydrophilic one. In other words, the solubilization capacity of the hydrophilic copolymer for drugs could be enhanced by the presence of the more hydrophobic one. Hence, the main advantage of the binary mixed Pluronic systems is to overcome the limitations of each neat system, allowing copolymer rearrangement to form stable nano-environments to enhance drug solubility.

Intrigued by those interesting studies, we conducted a series of experiments to study mixing behavior from two different perspectives: the different block chain length at the same hydrophilicity (L92 + F108, +F98, +F88, and +F68), as well as various hydrophobicities at same PPO moiety (L92 +

F88, +F87, and +P84). Pluronic L92 was chosen as the hydrophobic parent copolymer and the other constituent was selected based on the two perspectives to systematically vary the resemblance between parent copolymers. Temperature-dependent micellization in binary Pluronic systems was inspected by the combined use of the high-sensitivity, differential scanning calorimeter (HSDSC) and dynamic light scattering (DLS). Heat capacity changes and aggregate sizes at different temperatures during the whole micellization process were scrutinized and correlated. Moreover, we used a nonsteroidal anti- inflammatory drug ibuprofen into the binary systems to examine the size distribution and solubilization capacity. This study endeavored to clarify the tendencies between mixing behaviors and tried to find out some principles for establishing binary systems.

2. Materials and Methods

2.1. Materials

Pluronics L92 and F108 were purchased from Sigma-Aldrich, and Pluronics F98, F88, F68, F87, and P84 were purchased from BASF Corporation. All these Pluronics were used as received without further purification. Their physicochemical and molecular characteristics are described in Table 1 and the molecular structure of Pluronic is schematically illustrated in Figure 1a. The drug α-methyl-4-(isobutyl) phenylacetic acid, known as ibuprofen, was bought from Alfa-Aesar and used as received. The molecular structure of ibuprofen is illustrated in Figure 1b. NaOH was bought from SHOWA, and its purity is 96%. Water was purified by double distillation followed by a PURELAB Maxima Series (ELGA Lab Water) purification system with a resistivity better than 18.2 MΩ cm.

Table 1. Physicochemical characteristics of Pluronics used.

Pluronic	Average (Mw [a])	Structure [11,17–21]	HLB [a]	CP at 1% [a] (°C)
L92	3650	$(EO)_8(PO)_{50}(EO)_8$	1–7	26
F108	14,600	$(EO)_{127}(PO)_{48}(EO)_{127}$	>24	>100
F98	13,000	$(EO)_{118}(PO)_{45}(EO)_{118}$	28	>100
F88	11,400	$(EO)_{97}(PO)_{39}(EO)_{97}$	28	>100
F68	8400	$(EO)_{80}(PO)_{30}(EO)_{80}$	>24	>100
F87	7700	$(EO)_{61}(PO)_{40}(EO)_{61}$	>24	>100
P84	4200	$(EO)_{19}(PO)_{39}(EO)_{19}$	12–18	74

[a] Information from BASF.

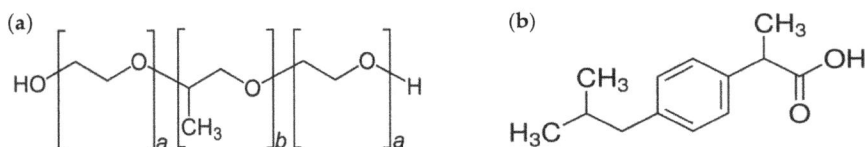

Figure 1. Molecular structures of (**a**) Pluronic and (**b**) Ibuprofen. Molecular weight of ibuprofen: 206.28 g/mol; solubility in water at 20 °C: 0.021 mg/mL [22]; partition coefficient (logP): 2.48 [23]; pKa: 5.38 at 25 °C [24].

2.2. High Sensitivity Differential Scanning Calorimetry (HSDSC)

HSDSC (VP-DSC, MicroCal) was used to determine the critical micelle temperature (CMT). After loading the reference and sample cells, the cell port air space was compressed to reach positive pressures up to approximately 0.2 MPa by using the pressurizing cap equipped with an o-ring-sealed piston. The detailed experimental procedure can be found in our previous studies [25–27]. All experiments were conducted at a scanning rate of 30 or 60 °C/h from 5 to 120 °C for 7 scans and reproduced in duplicate. The standard deviation of the CMT is always within $\pm 0.15\ ^\circ$C for replicate measurements.

2.3. Dynamic Light Scattering (DLS)

Zetasizer Nano system equipped with a He-Ne laser operating at a wavelength 633 nm (Nano-ZS, Malvern, UK) was used to determine particle size distributions of Pluronic aggregates in solution. The system temperature can be controlled in a range of 0–80 $^\circ$C. The intensity basis of the DLS data was adopted and presented in the study.

2.4. UV/Vis Spectroscopy and Drug Solubility

To prepare saturated drug-loaded solutions, an excess of drug (ibuprofen) powder was mixed with Pluronic aqueous solutions through stirring at 140 rpm at 37 $^\circ$C for one day. Solutions were filtered by a syringe fitted with a 0.22 µm PTFE filter (Millipore) to remove unsolubilized drug before UV detection. The dissolved ibuprofen concentration in the solution was determined by measuring absorbance at 264.8 nm by using UV-Vis double beam spectrophotometer (CARY 100nc, Agilent Technologies, Santa Clara, CA, USA). Blank experiments without copolymer were conducted for the determination of drug solubility. Dilute solutions of ibuprofen dissolved in 0.1 N NaOH solution were used for the calibration according to the Beer-Lambert law, as illustrated in Figure S1 in the Supplementary Materials. The absorbance peak maximum of ibuprofen in neat and binary Pluronic solution was consistent with that in 0.1 N NaOH solution [28], as illustrated in Figure S2 in the Supplementary materials, which validates the applicability of this calibration curve. Some other details about this calibration curve can be found in the Supplementary Materials. For each neat and binary Pluronic system, at least 3 samples were prepared to perform the solubility measurements. Standard deviation (sd) of these multiple measurements for each system was determined by sd $= \sqrt{\frac{\sum_{i=1}^{N}(x_i - \bar{x})^2}{N-1}}$, where $\bar{x}$ stands for the average value of solubility.

3. Results and Discussion

In this study, six Pluronics, F108, F98, F88, F68, F87, and P84, in water, were chosen to examine their self-assembly process and drug (ibuprofen) solubility by using HSDSC, DLS, and UV-Vis spectroscopy. In addition, the synergistic effect of binary mixed Pluronic systems by adding L92 to each Pluronic mentioned above on the self-assembly process and drug (ibuprofen) solubility were also explored. Four Pluronics, F108, F98, F88, and F68, in water, were applied to systematically examine the effect of molecular weight of Pluronics with same hydrophobicity on the aggregation behavior of pure and mixed Pluronic L92 + Fx8 in water. On the other hand, three Pluronics F88, F87 and P84, were used to examine effect of hydrophobicity of Pluronics at a fixed PPO block length on the aggregation behavior of pure and mixed Pluronic L92 + F8x in water. In addition, the molecular weight and hydrophobicity of Pluronic copolymers were also manipulated to explore how to enhance the solubility of drug (ibuprofen).

3.1. Effect of Molecular Weight of Pluronic Fx8 (x = 6, 8, 9 and 10) at a Fixed PEO/PPO (80/20) Mass Ratio on the Thermophysical Properties of Pure and Mixed Pluronic L92 + Fx8 in Water

First of all, the influence of molecular weight of Pluronic Fx8 on the self-assembly process of mixed Pluronic L92 + Fx8 mixtures was systematically examined. Four very hydrophilic Pluronics: F108, F98, F88, and F68, with a fixed PEO/PPO mass ratio (80/20), were chosen. The HSDSC thermograms of pure 1.0 wt % solutions of these Pluronics along with 0.5 wt % L92 in water are shown in Figure 2. Molar ratios of Fx8/L92 are 0.50, 0.56, 0.64, and 0.87 for x = 10, 9, 8 and 6, respectively. Usually, there are three methods applied to determine the CMT, i.e., the onset temperature T_{onset}, the inflection point temperature T_{inf}, and the peak maximum temperature T_m from the thermograms [29]. For example, the CMT (T_{onset}) of F108, F98, F88, and F68 are 29.4, 30.7, 35.9, and 47.4 °C respectively. It is obvious that the CMT increases with decrease in the molecular weight for the Pluronic Fx8.

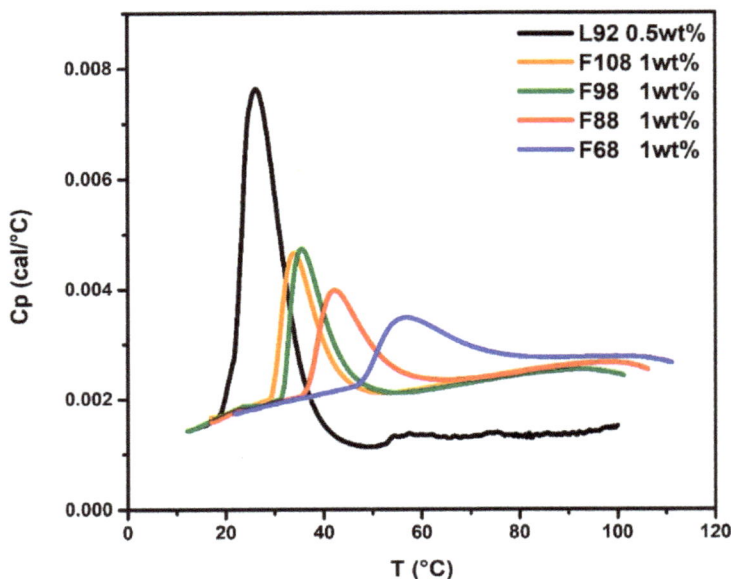

Figure 2. The HSDSC thermograms of pure Pluronic solutions.

For 0.5 wt % solution of L92 system, phase separation occurred around 25 °C to form large aggregates of size of 450 nm, due to its hydrophobic nature as illustrated by DLS results in Figure 3c. The variation of aggregate size as a function of temperature for all systems is listed and reported in Tables S1–S6 in the Supplementary Materials. Nagarajan [16] pointed out that Pluronic L92 aggregates are lamellar structures because of their short PEO chain. The apparent hydrodynamic diameter of micelles formed by pure 1.0 wt % F108 is approximately 23.8 nm. The mixed Pluronic 0.5 wt % L92 + 1.0 wt % F108 system still exhibits large aggregates of ~190 nm in size below 25 °C, comparable to the size of aggregates formed by neat L92. Further increase in temperature up to 30 °C would dramatically decrease the micelle size to around 30.3 nm. It is interesting to note that the HSDSC thermogram illustrated in Figure 3a demonstrates a second peak with T_m = 27.2 °C appearing right after the 1st endothermic peak with T_{onset} = 19.0 °C (initiation of phase separation due to L92). This second peak is located between the endothermic peaks of neat L92 (T_{onset} = 18.9 °C) and F108 (T_{onset} = 29.4 °C). The appearance of this second peak suggests that F108 actively participates in the micellization process at T_m = 27.2 °C by integrating the L92 into mixed F108/L92 micelles via breaking down the large aggregates of L92, as revealed by sudden drop in micelle size around 30 °C illustrated in Figure 3c. The size of mixed F108/L92 micelles remains almost constant ~30 nm within the temperature range from 30 to 60 °C. The DLS results of the L92 + F108 system show a single sharp peak with a narrow distribution (polydispersity index (PDI) < 0.2) [30]: PDI = 0.17 at 25 °C and PDI = 0.10 at 35 °C. Note that the PPO block length of F108 (with 48 PO units) is almost equal to that of L92 (with 50 PO units). The similar PPO moiety between F108 and L92 indeed enhances the cooperative binding between these two copolymers to form mixed micelles, consistent with the findings of Oh et al. [12].

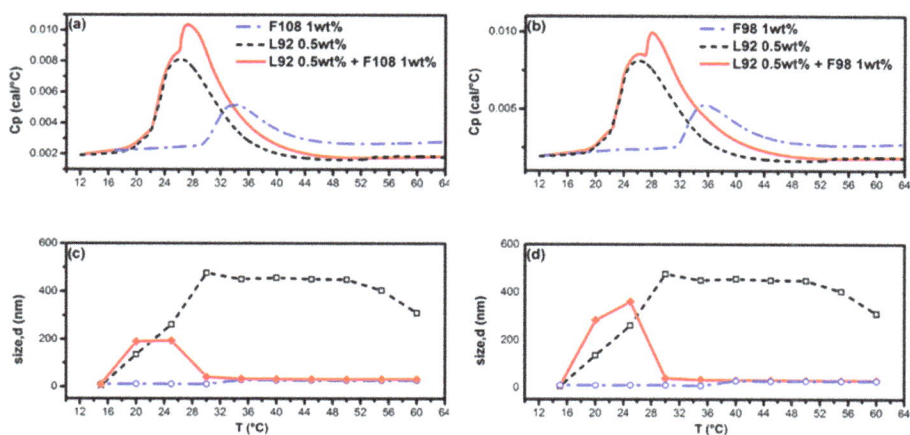

Figure 3. HSDSC thermograms (a,b) and evolution of size of aggregates (c,d) as a function of temperature from DLS; (a,c) Mixed Pluronic 0.5 wt % L92 + 1 wt % F108 system; (b,d) Mixed Pluronic 0.5 wt % L92 + 1 wt % F98 system. Red solid line and diamond (◆) for mixed Pluronic system. Black dashed line and open square (□) for L92. Blue dashed-dotted line and open circle (○) for F108/F98.

Instead of F108, a slightly shorter PPO block length of F98 (with 45 PO units) was applied to examine the self-assembly process of the mixed Pluronic 0.5 wt % L92 + 1.0 wt % F98 system. Figure 3b illustrates the HSDSC thermogram of the mixed Pluronic L92 + F98 system. The micellization process of the mixed Pluronic L92 + F98 system is rather similar to that of the L92 + F108 system. A second peak appears at 28.0 °C (T_m), which is also between the endothermic peaks of neat L92 (T_{onset} = 19.0 °C) and F98 (T_{onset} = 30.7 °C). The diameter of micelle formed by pure F98 is approximately 25.0 nm, as the DLS data shown in Figure 3d. Addition of F98 to the L92 system breaks down the size of L92 aggregates from 358.4 nm to 27.9 nm after the occurrence of the second peak (i.e., $T > 30$ °C). The size of mixed F98/L92 micelles remains almost constant ~28 nm within the temperature ranging from 30 to 60 °C. However, PDI value of the solution is 0.70 at 27 °C and remains around 0.55 at high temperatures. This implies that the F98 is not as capable as F108 to stabilize the mixed micelles with a narrow size distribution due to a shorter PPO block length of F98.

In these two binary mixed Pluronic L92 + F108 and L92 + F98 systems, the size of mixed micelles is slightly larger than the one formed by neat F108 and F98 systems, respectively, and pronouncedly much smaller than L92 aggregates. Furthermore, the size of mixed micelles is stable and remains almost constant even at high temperatures without further agglomeration.

Pluronic F88 with a shorter PPO block length (39 PO units) than that of F98 was then used to examine its capability of breaking down the L92 aggregates to form mixed micelles. It is interesting to find out that there appear two additional kinks at 31.2 °C (the second peak maximum temperature T_{m2}) and 40.1 °C (the third peak maximum temperature T_{m3}) in the HSDSC thermogram of the mixed Pluronic 0.5 wt % L92 + 1.0 wt % F88 system, as illustrated in Figure 4a. The DSC thermogram of the L92 + F88 system almost coincides with that of neat L92 system from 10 to 30 °C, as well as the temperature-dependent size of aggregates, as the DLS results shown in Figure 4b. This implies the presence of aggregates in the L92 + F88 system at temperatures below 30 °C are mainly formed by pure L92. Note that the second peak with T_m = 31.2 °C is located in-between the endothermic peaks of neat L92 (T_{onset} = 18.9 °C) and neat F88 (T_{onset} = 35.9 °C). Around this first kink at 31.2 °C in the DSC thermogram, the L92 + F88 system starts to deviate from that of neat L92 system. It is plausible to conjecture that F88 also participates in the aggregates to increase the size of aggregates up to 632 nm, which is larger than the size of aggregates formed by neat L92 system.

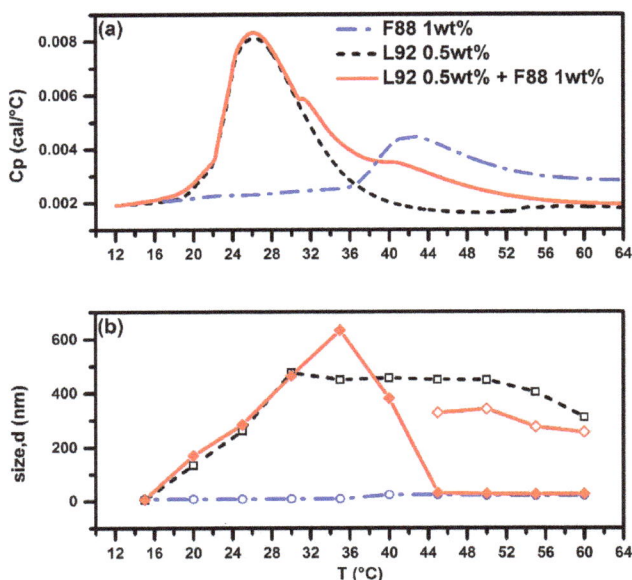

Figure 4. HSDSC thermograms (**a**) and evolution of size of aggregates (**b**) as a function of temperature from DLS for pure 0.5 wt % L92, pure 1 wt % F88, and mixed Pluronic 0.5 wt % L92 + 1 wt % F88 systems. Red solid line and diamond (◆) for L92 + F88. Red solid line and open diamond (◇) for L92 + F88 (the other coexisting aggregates). Black dashed line and open square ((□) for pure L92. Blue dashed-dotted line and open circle (○) for pure F88.

While the third peak (the second kink) with $T_{m3} = 40.1\,°C$ is located in-between the $T_{onset} = 35.9\,°C$ and $T_m = 42.2\,°C$ of the neat F88 system. When the temperature is increased up to 40.1 °C and beyond, F88 is able to break down the L92 aggregates of into smaller ones. With a further increase in temperature up to 44 °C and beyond, the L92 + F88 system exhibits bimodal behavior: small mixed F88/L92 micelles (~26.6 nm in diameter) and large aggregates (~300 nm in size) coexist. Note that the size of mixed L88/L92 micelles (~26.6 nm) is obviously larger than the size of micelles (~20.4 nm) formed by neat 1.0 wt % F88. The capability of F88 to break down the aggregates formed by L92 is not as efficient as for F108 and F98 as described before. Most of F88 molecules are incorporated into the mixed micelles (~26.6 nm) and only relatively small amount of F88 molecules join the aggregates formed by L92 (~300 nm).

There are only two peaks observed in the DSC thermogram of the 0.5 wt % L92 + 1.0 wt % F68 system, as shown in Figure 5a, compared to three peaks for the L92 + F88 system. Large aggregates (>400 nm) were observed at temperatures above 30 °C. Micelle size of neat F68 system is around 16.7 nm. It is obvious that the addition of F68 to L92 could not break down the size of large aggregates according to the DLS results shown in Figure 5b. It is interesting to note that the size of aggregates decreases to 350 nm at 45 °C while approaching the onset temperature of neat F68 system, but increases back to 467 nm at 55 °C. It may be attributed to the micellization of F68 by partially integrating in L92 aggregates. Very few small aggregates (~55 nm), less than 0.1% of the total number of aggregates, were observed in the temperature ranging from 40 to 60 °C. In other word, L92 + F68 system also exhibits bimodal behavior: small aggregates (~55 nm in size) and large aggregates (~400 nm in size) coexist within the temperature ranging from 40 to 60 °C.

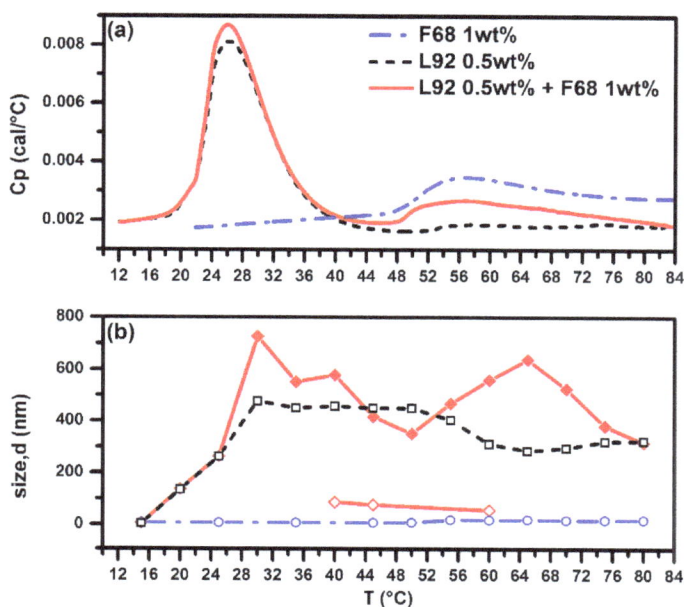

Figure 5. HSDSC thermograms (**a**) and evolution of size of aggregates (**b**) as a function of temperature from DLS for pure 0.5 wt % L92, pure 1 wt % F68, and mixed Pluronic 0.5 wt % L92 + 1 wt % F68 systems. Red solid line and diamond (◆) for L92 + F68. Red solid line and open diamond (◇) for L92 + F68 (the other coexisting aggregates). Black dashed line and open square (□) for pure L92. Blue dashed-dotted line and open circle (○) for pure F68.

It is pronounced that the evolution of self-assembly process of mixed Pluronic L92 + Fx8 systems as a function of temperature depends on the molecular weight of copolymers. On initial heating of the mixed copolymer solutions, more hydrophobic L92 aggregates first form in the solution, and agglomeration may bring some hydrophilic Fx8 together into the aggregates. This could be evidenced by the larger aggregate sizes of mixed systems compared to the size of neat L92 system below 25 °C. With a further increase in temperature close to critical micelle temperature (CMT) of hydrophilic copolymer, dehydration of hydrophobic PPO block triggers the copolymers starting to aggregate. The performance of this second aggregation is then dominated by the block chain length of hydrophilic copolymer relative to the hydrophobic one. Large aggregates formed by neat L92 system could be suppressed by the introduction of hydrophilic Pluronic F108 with a large molecular weight. At a fixed PEO/PPO mass ratio of 80/20, Pluronic Fx8 with long PPO block length would disintegrate large aggregates of L92 dramatically to be integrated into the mixed Fx8/L92 micelles. In addition, the longer the PPO block length of hydrophilic copolymer Fx8, the more powerful the ability of Fx8 of breaking down the size of large aggregates to form stable mixed micelles in aqueous solution. For hydrophilic copolymers with a shorter PPO block, mixed Pluronic systems exhibit the coexistence of large aggregates and mixed micelles. Figure 6 summarizes the evolution of self-assembly process of pure and mixed Pluronic L92 + Fx8 systems as a function of temperature.

Number of PO Units	Systems			
48	F108	Unimers		Micelles
45	F98	Unimers		Micelles
39	F88	Unimers		Micelles
30	F68	Unimers		Micelles
40	F87	Unimers		Micelles
39	P84	Unimers		Micelles
50	L92	Unimers	Aggregates	
48 + 50	F108+L92	Unimers	Aggregates	Micelles
45 + 50	F98+L92	Unimers	Aggregates	Micelles
39 + 50	F88+L92	Unimers	Aggregates	Bimodal
30 + 50	F68+L92	Unimers	Aggregates	Bimodal
40 + 50	F87+L92	Unimers		Aggregates
39 + 50	P84+L92	Unimers	Aggregates	Micelles	Bimodal

Temperature (°C): 10 20 30 40 50 60 70

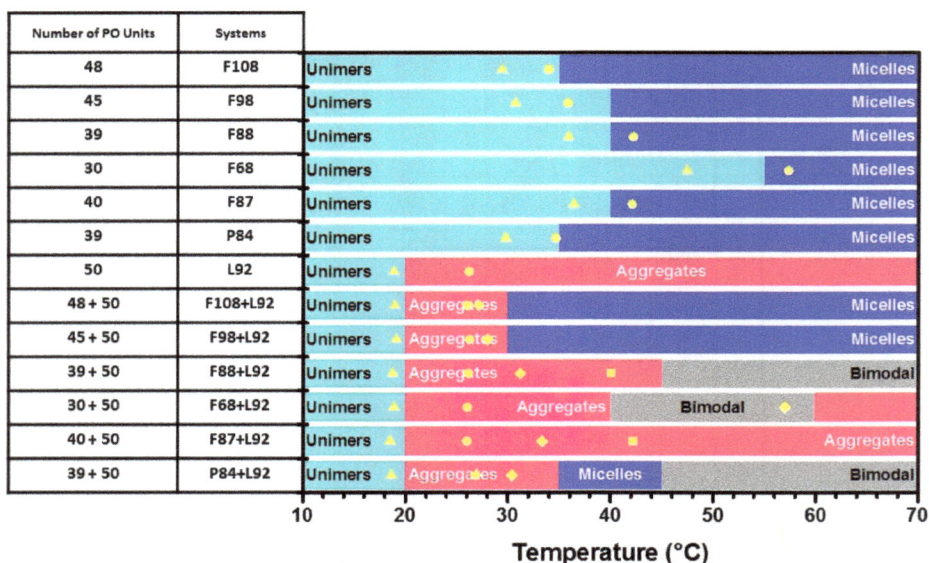

Figure 6. The evolution of self-assembly behavior of pure and mixed Pluronic L92 + Fx8 and L92 + F8x systems as a function of temperature. Symbols on the stacking bars represent characteristic temperatures measured by HSDSC. For pure Pluronic systems, T_{onset} (yellow triangle) and T_m (yellow circle) of the endothermic peak in the HSDSC thermograms. For the mixed Pluronic systems, T_{onset} (yellow triangle) of the first endothermic peak, T_m (yellow circle) of the 1st peak, T_{m2} (yellow diamond) of the 2nd peak, and T_{m3} (yellow square) of the 3rd peak in the HSDSC thermograms.

3.2. Effect of PEO Block Length of F8x (x = 8, 7 and 4) at a Fixed PPO Block Length on the Thermophysical Properties of Pure and Mixed Pluronic L92 + F8x in Water

The effect of PEO block length of Pluronic F8x (x = 8, 7, and 4) on the self-assembly process of mixed Pluronic L92 + F8x systems was systematically examined. Three Pluronics F88, F87, and P84 with a fixed PPO block length (~39 PO units) but varying PEO block length (respectively, 97, 61, and 19 EO units) were chosen in this study. In order to fix a constant amount of PO units of F8x added into the system, 1.0 wt % F88 was converted into 8.77×10^{-7} m F88 in water. Therefore, 8.77×10^{-7} m F87 and P84 aqueous solutions were equivalent to 0.675 wt % F87 and 0.368 wt % P84 in water. The HSDSC thermograms of pure 1.0 wt % F88, 0.675 wt % F87, 0.368 wt % P84, and 0.5 wt % L92 are compared and shown in Figure 7. Molar ratios of F8x/L92 are fixed at 0.64 for x = 8, 7 and 4. Note that the HSDSC thermogram of 0.675 wt % F87 system almost coincides with that of 1.0 wt % F88 system. It is obvious that CMT of 0.368 wt % P84 system is lower than that of the other two F8x systems. The CMTs (T_{onset}) of the 1.0 wt % F88, 0.675 wt % F87, and 0.368 wt % P84 are 35.9, 36.4, and 29.8 °C, respectively.

There exist two kinks after the first endothermic peak in the HSDSC thermogram of the mixed Pluronic 0.5 wt % L92 + 0.675 wt % F87, as shown in Figure 8a, similar to that of 0.5 wt % L92 + 1 wt % F88 system (Figure 4a).

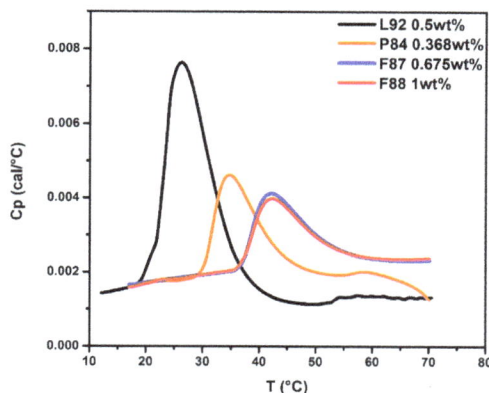

Figure 7. The HSDSC thermograms of pure Pluronic F88, F87, P84, and L92 system. 1.0 wt % F88, red line; 0.675 wt % F87, blue line; 0.368 wt % P84, orange line; and 0.5 wt % L92, black line.

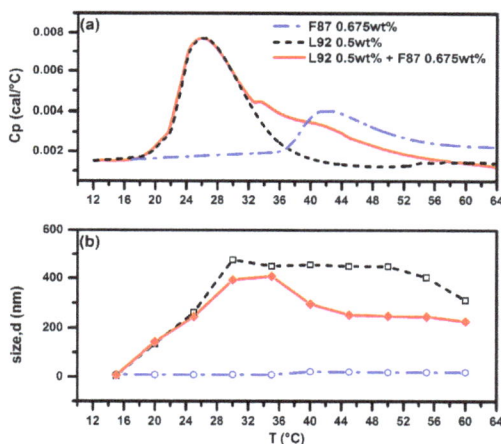

Figure 8. HSDSC thermograms (**a**) and evolution of size of aggregates (**b**) as a function of temperature from DLS for pure 0.5 wt % L92, pure 0.675 wt % F87, and mixed Pluronic 0.5 wt % L92 + 0.675 wt % F87 systems. Red solid line and diamond (◆) for L92 + F87. Black dashed line and open square (□) for pure L92. Blue dashed-dotted line and open circle (○) for pure F87.

For the 0.5 wt % L92 + 0.675 wt % F87 system, the second and third peaks (two kinks) appear at $T_{m2} = 33.3\ °C$ and $T_{m3} = 42.2\ °C$, respectively. However, evolution of size of aggregates as a function of temperature from the DLS results for the L92 + F87 system (Figure 8b) is rather different from that of the L92 + F88 system (Figure 4b). For the 0.5 wt % L92 + 1.0 wt % F88 system, major aggregates (in terms of number of aggregates) detected in the solution are mixed micelles of ~26 nm in size coexisting with very few large aggregates of ~300 nm in size when system temperature is higher than 45 °C. In contrast, for 0.5 wt % L92 + 0.675 wt % F87 system, only large aggregates around 245 nm are detected without any small aggregates when system temperature is higher than 40 °C. The PDI value after the second peak was around 0.37 and aggregate size distribution could be found in Figures S3 and S4 in the Supplementary Materials. Obviously, adding F87 to the L92 system does not break down the size of large aggregates like F88 does, since F87 is slightly more hydrophobic than F88 (the PEO block length of F87 is shorter than that of F88).

Instead of F87, an even shorter PEO block length P84 (19 EO units) was used to prepare the mixed Pluronic 0.5 wt % L92 + 0.368 wt % P84 aqueous solution for comparison. The HSDSC thermogram of this L92 + P84 system, as illustrated in Figure 9a, shows that as the temperature is increased from 10 °C to 18.6 °C (=T_{onset}), the first endothermic peak occurs to trigger phase separation due to the Pluronic L92 to form large aggregates. When the temperature is further increased up to ~30 °C, another endothermic peak appears for P84, precipitating the self-assembly process to form mixed P84/L92 micelles. Simultaneously, the size of aggregates dramatically drops from 492 nm down to about 18.3 nm, which is slightly larger than the size of micelles formed by neat P84 (16.5 nm), as revealed by DLS results shown in Figure 9b. PDI value is around 0.55, implying the polydispersity of size of aggregates in the solution, but the system is dominated by the mixed micelles (18.3 nm). When the temperature is further increased up to 45 °C and beyond, large aggregates about 450 nm are detected, but only in the proportion of less than 0.1% (on number basis), and coexisting with the mixed micelles. The observed large aggregates above 45 °C may arise from the intrinsic intention of phase separation for neat 1% P84 in water (cloud point = 74 °C). The hydrophobicity of these Pluronics increases along with temperature.

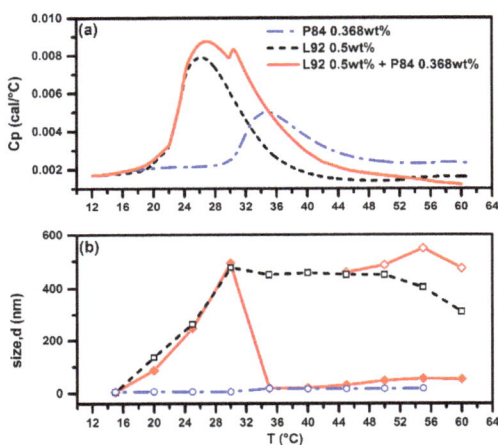

Figure 9. HSDSC thermograms (**a**) and evolution of size of aggregates (**b**) as a function of temperature from DLS for pure 0.5 wt % L92, pure 0.368 wt % P84, and mixed Pluronic 0.5 wt % L92 + 0.368 wt % P84 systems. Red solid line and diamond (◆) for L92 + P84. Red solid line and open diamond (◇) for L92 + P84 (the other coexisting aggregates). Black dashed line and open square (□) for pure L92. Blue dashed-dotted line and open circle (○) for pure P84.

It is surprising that the addition of P84 would break down large aggregates and develop small aggregates (micelles) that stably exist in the solution. P84, more hydrophobic than F87 and F88, was predicted not to be as stable in the solution as F88 for the shorter PEO chain length, nor as being powerful enough to entangle with L92 molecules that break down the size of large aggregates as F108. At lower temperatures, Pluronic L92 stacks into a lamellar form. Increasing temperature to the CMT of 0.368 wt % P84 promotes molecular interaction between P84 and L92, owing to the driving force of aggregation for P84. This rather hydrophilic and slightly larger copolymer compared to L92 may be incorporated into lamellar structure mainly formed by L92 and gradually breaks down size of aggregate. Due to the dehydration of P84 molecules, micellization process of P84 would probably be accompanied by L92 molecules to form mixed micelles. However, owing to the shorter PPO block length of P84, the ability of entangling with L92 is limited and large aggregates still exist in the system. Raising temperature would enlarge the micelles and make them more hydrophobic.

For a binary-mixed Pluronic system including an extremely hydrophobic Pluronic copolymer (L92 used in this study) and a hydrophilic one, Pluronics with large molecular weight would disintegrate large aggregates pronouncedly, as well as diminish the cloud point at room temperature efficiently. Under the condition of a fixed PPO block length, Pluronic copolymers with a shorter PEO block length would, in contrast, disintegrate large aggregates more efficiently. The evolution of aggregation behavior of pure and mixed Pluronic L92 + F8x systems as a function of temperature is also summarized and illustrated in Figure 6.

3.3. Enhancement of Solubility of Ibuprofen in Neat and Mixed Pluronic Systems

The drug (ibuprofen) solubility in neat and mixed Pluronic systems was then carefully measured to examine the effect of molecular weight and hydrophilicity of copolymers. Firstly, solubility of ibuprofen in 1.5 wt % neat Pluronic F108, F98, F88, and F68 system at 37 °C was examined. It is obvious that solubilization capacity of F68 for ibuprofen is inferior to the other three Pluronics, and no significant differences exist among neat F108, F98, and F88 systems. However, based on the experimental results listed in Table 2, the capability of drug incorporation for the four copolymers could still be distinguishable. It is interesting to find out that the solubilization capacity of neat Pluronics with different block chain lengths at the same hydrophilicity is in the order of F98 > F108 > F88 > F68. It seems that the micelles formed by longer block chain solubilize more ibuprofen, except F108. This may be attributed to the long PEO block length of F108 inducing a steric hindrance for the entrapment efficiency. Liveri et al. [31] performed a systematic spectrophotometric study of the kinetic of solubilization process of the poorly water soluble drug tamoxifen. They pointed out that PEO corona may act as the steric barrier that hampers the transfer of tamoxifen into the micelle core. Pluronic F108 has 127 EO units on each side of the copolymer, which may contribute a stable environment for the hydrophobic core inside the micelle with PEO block chains stretching toward the solution, but simultaneously may probably hinder the transfer of ibuprofen leading to limited incorporating amount of drugs.

Table 2. Hydrodynamic Diameter of aggregates (D_h, nm) and solubility of ibuprofen in neat Pluronic Fx8 (1.5 wt %) and binary mixed Pluronic 0.5 wt % L92 + 1.0 wt % Fx8 systems at 37 °C.

Neat Pluronic	F108 (1.5 wt %)	F98 (1.5 wt %)	F88 (1.5 wt %)	F68 (1.5 wt %)
D_h (nm) without ibuprofen	27.5 ± 0.4	25.5 ± 0.4	6.8 ± 0.2	5.6 ± 0.2
D_h (nm) Saturated ibuprofen	26.6 ± 0.2	24.6 ± 0.2	23.9 ± 0.1	25.2 ± 1.8
Solubility of ibuprofen (mg/mL)	1.61 ± 0.03	1.75 ± 0.03	1.53 ± 0.03	0.67 ± 0.02
Mixed Pluronic	L92 (0.5 wt %) + F108 (1 wt %)	L92 (0.5 wt %) + F98 (1 wt %)	L92 (0.5 wt %) + F88 (1 wt %)	L92 (0.5 wt %) + F68 (1 wt %)
D_h (nm) * without ibuprofen	30.3 ± 0.1	30.3 ± 1.5	632 ± 25	550 ± 85
D_h (nm) Saturated ibuprofen	327 ± 20	314 ± 11	254 ± 8	590 ± 27
Solubility of ibuprofen (mg/mL)	3.33 ± 0.11	3.37 ± 0.05	3.34 ± 0.12	2.58 ± 0.18

* The data measured at 35 °C.

Particle size distribution and PDI values for the systems with/without ibuprofen at 37 °C were also measured and used as an index to evaluate system stability. The micelle hydrodynamic diameters of neat Pluronic F108 and F98 system are 27.5 nm and 25.5 nm, respectively, whereas for ibuprofen-loaded micelles they are 26.6 nm and 24.6 nm, respectively. There is no obvious change in particle size after drug incorporation. It should be noted that the CMTs of neat (1.5 wt %) F88 and F68 system are higher than 37 °C. Therefore, the F88 and F68 molecules exist as unimers at 37 °C in neat F88 and F68 system without ibuprofen, as the particle size reported in Table 2. The addition of ibuprofen would trigger the micellization process at lower temperature 37 °C to form uniform micelles with around 25 nm in diameter, larger than the micelle size of the neat Pluronic system without ibuprofen (say at 60 °C, see Tables S3 and S4 in the Supplementary Materials). All of the PDI values drop down below 0.15 for these systems after adding ibuprofen.

In addition to the 1.5 wt % F88 system, solubility of ibuprofen in 1.175 wt % F87 and 0.868 wt % P84 system at 37 °C was carefully measured. Solubilization capacity of neat Pluronic F8x at a fixed PPO block length with various hydrophilicity is in the order of F88 ≅ F87 < P84, as reported in Table 3. It is pronounced that the solubility of ibuprofen increases along with an increase in the hydrophobicity of Pluronic used. Similar to the 1.5 wt % F88 system, the CMT of neat 1.175 wt % F87 system is also higher than 37 °C. Pluronic F87 molecules exist as unimers at 37 °C in neat F87 system without ibuprofen. Addition of ibuprofen to the Pluronic F87 system promotes the micellization at 37 °C to form uniform micelles of ~20 nm in diameter, larger than the micelle size of pure F87 system without ibuprofen, e.g., the micelle size of 0.675 wt % F87 system at 45 °C is ~18 nm. On the other hand, neat Pluronic F84 system already forms micelles of ~18 nm in diameter at 37 °C, and addition of ibuprofen to neat P84 system would increase the size of aggregates up to 88 nm to enhance the solubility of ibuprofen as high as 3.02 mg/mL, larger than that in the 1.5 wt % F108 system (1.61 mg/mL). This finding is consistent with that of Singla et al. [32]. They explored the solubilization of hydrophobic drug (oxcarbazepine) in different Pluronics F108, F127, and P84, revealing that the solubilization capacity of P84 is the highest among these three copolymers.

Table 3. Hydrodynamic Diameter of aggregates (D_h, nm) and solubility of ibuprofen in neat Pluronic F8x and binary mixed Pluronic L92 + F8x systems at the same total mass concentration at 37 °C.

Neat Pluronic	F88 (1.5 wt %)	F87 (1.175 wt %)	P84 (0.868 wt %)
D_h (nm) without ibuprofen	6.8 ± 0.2	5.7 ± 0.1	17.9 ± 0.8
D_h (nm) Saturated ibuprofen	23.9 ± 0.1	20.1 ± 0.1	88.2 ± 0.04
Solubility of ibuprofen (mg/mL)	1.53 ± 0.03	1.80 ± 0.07	3.02 ± 0.12
Mixed Pluronic	L92 (0.5 wt %) + F88 (1 wt %)	L92 (0.5 wt %) + F87 (0.675 wt %)	L92 (0.5 wt %) + P84 (0.368 wt %)
D_h (nm) * without ibuprofen	632 ± 25	406 ± 18	18.3 ± 1.0
D_h (nm) Saturated ibuprofen	254 ± 8	144 ± 4	130 ± 7
Solubility of ibuprofen (mg/mL)	3.34 ± 0.12	3.26 ± 0.12	4.29 ± 0.18

* The data measured at 35 °C.

Note that the solubility of ibuprofen in pure water is as low as 0.0206 mg/mL at 35 °C and 0.0264 mg/mL at 40 °C [33], which can be linearly interpolated to estimate the solubility of ibuprofen at 37 °C around 0.0229 mg/mL. The solubility of ibuprofen in 1.5 wt % F98 system (1.75 ± 0.03 mg/mL) is dramatically increased by 75 times more than that in pure water. Furthermore, the solubility of ibuprofen in 0.886 wt % P84 is even increased up to 130 times more than that in pure water. All these Pluronics demonstrate the outstanding solubilization capacity for incorporating hydrophobic drug (ibuprofen).

Based on our experimental results, Pluronics with larger molecular weight (F98 in Fx8 series) and *more hydrophobic* characteristics (F84 in F8x series) exhibit better solubilization capacity for ibuprofen. To follow along this line, it is plausible to conjecture that the solubility of ibuprofen can be enhanced by simply increasing the hydrophobicity of Pluronic through replacing partially the Pluronic (0.5 wt % Fx8 or F8x) by a more hydrophobic Pluronic (0.5 wt % F92). Thus, all the 1.5 wt % Fx8 systems are replaced by the 1.0 wt % Fx8 + 0.5 wt % F92 systems. On the other hand, the 1.175 wt % F87 (and 0.868 wt % P84) system is replaced by the 0.675 wt % F87 + 0.5 wt % F92 (and 0.368 wt % P84 + 0.5 wt % F92) system for further exploration of enhancement of ibuprofen solubility.

For the binary mixed Pluronic L92 + Fx8 systems (see Table 2), L92 can be stabilized by F108 (and F98) to form mixed F108/L92 (and F98/L92) micelles of ~30 nm in diameter via cooperative binding between parent copolymers. Addition of ibuprofen dramatically increases the size of aggregates from 30 to 320 nm, forming a stable and rather monodisperse system where the PDI value is around 0.20 for these two L92 + F108 and L92 + F98 systems. For the L92 + F88 system, the size of aggregates drops down from 632 to 254 nm at 37 °C after ibuprofen is loaded. The size of aggregates was stabilized to around 30 nm after 45 °C for the L92 + F88 system without ibuprofen, as mentioned above. Below 45 °C, large aggregates (above 600 nm) with broad distribution were still observed in the solution. Inclusion of ibuprofen makes the size of aggregates decrease and become more uniform, and the PDI value dropped from 1.00 to 0.19. That is, the large aggregates (632 nm) in the solution could be stabilized by ibuprofen developing monodisperse distribution (254 nm). Meznarich and Love [34] revealed that methylparaben, a common food and drugs additive, enhances monodisperse micelles formed by Pluronic F127 and promotes its gelation behaviors on the kinetic studies. Similar trend was observed from the L92 + F68 system for the decrease in PDI value from 0.34 to 0.12. However, the size of aggregates after drug loaded does not change (from 550 to 590 nm), indicating that the block chain length of Fx8 is strongly associated with the capability to stabilize the aggregates. This observation again supports our conclusions: the longer the block chain length of the hydrophilic parent copolymer, the more capable this copolymer is to break down large aggregates. The addition of ibuprofen to the L92 + Fx8 system induces large aggregates in the solutions due to the insufficient amount of hydrophilic PEO chains that is necessary to assemble into micelles.

For all the mixed Pluronic L92 + Fx8 (x = 10, 9, 8, and 6) systems, the solubility of ibuprofen indeed is dramatically enhanced at least 2 times higher than that of neat Pluronic systems, as can be seen in the solubility data listed in Table 2. It is interesting to find out that the solubilization capacity of the mixed Pluronic L92 + Fx8 systems with different block chain lengths at the fixed PEO/PPO mass ratio (80/20) is in the order of F98 ≅ F108 ≅ F88 > F68. The solubility of ibuprofen in the mixed L92 + F98 system is 3.37 mg/mL, around 147 times larger than that in pure water.

For the mixed Pluronic L92 + F8x (x = 8, 7 and 4) systems, the addition of ibuprofen to the L92 + F87 and L92 + F88 systems would decrease the size of aggregates (see Table 3), as well as the PDI value compared to that without ibuprofen. Replacement of 0.5 wt % F87 by L92 in the system would enhance the solubility of ibuprofen by 1.8 times than that of neat F87 system. Self-assembly behavior of the mixed L92 + P84 system is similar to that of the mixed L92 + F108 system. The size of aggregates is increased from 18 to 130 nm before and after ibuprofen loaded, but the aggregates still remain rather monodisperse (PDI = 0.24) after ibuprofen loaded. Replacement of 0.5 wt % P84 by L92 would enhance the solubility of ibuprofen up to 4.29 mg/mL, around 147 times higher than that in pure water.

Lee et al. also demonstrated that solubilization capacity of pure P123 system was lower than that of the mixed Pluronic L121 + P123 system by using a water-insoluble dye Sudan III [1]. The lamella-forming L121 provides a hydrophobic pool to increase in the solubilization capacity compared to pure P123 system. Dutra et al. reported that solubilization capacity of P123 + F127 system for griseofulvin is higher than that of pure F127 and increases along with the proportion of P123 [35]. These observations [1,35] are in good agreement with our experimental results that the replacement of hydrophilic Pluronic by L92 in the systems increases solubility of ibuprofen compared to that in neat Pluronic systems. Indeed, either the mixed Pluronic L121 + P123 system or the P123 + F127 system is composed of parent copolymers with similar PPO moiety. It should be noted that the Pluronic L92 has 50 PO units and F68 has 30 PO units, as can be seen in the molecular structures of neat Pluronics illustrated in Table 1. That is, the difference of the numbers of PO units between L92 and F68 is as high as 20. The solubility of ibuprofen in the mixed 0.5 wt % L92 + 1.0 wt % F68 system is around 4 times larger than that in the 1.5 wt % F68 system. In other words, introducing more hydrophobic Pluronic L92 into the neat F68 system would obviously enhance the solubility of ibuprofen, even having quite different PPO block lengths between L92 and F68.

For all the six mixed Pluronic systems substituted by 0.5 wt % L92, the solubility of ibuprofen indeed is dramatically enhanced compared to that of neat Pluronic systems, as shown in Tables 2 and 3. The partial replacement of the Pluronic (Fx8 or F8x) by a more hydrophobic Pluronic (F92) would increase the total hydrophobicity of Pluronics used in the system to enhance the solubility of ibuprofen.

4. Conclusions

Through the investigations of a series of binary mixtures concerning various block chain lengths and hydrophobicity, observations from the temperature-dependent self-assembly process using DLS and HSDSC demonstrated the tendencies between mixing behaviors. By simultaneously detecting heat capacity change and particle sizes, the evolution of PDI values reflect the mixing process along with temperature. The PDI value was around 0.50 at a lower temperature for the system L92 + F108, while it gradually decreased to 0.08, reflecting uniform mixed-micelles formation. On the other hand, the PDI value remained around 0.20 at a higher temperature for the system L92 + F68. For the binary Pluronic systems, while the distinction of block chain length between the parent copolymers increases, the monodispersity of the binary Pluronic systems decreases. However, the parent copolymers with distinct PPO moieties do not affirmatively lead to non-cooperative binding, such as L92 + P84. Small particle sizes around 30 nm were still detected after the CMT of P84 in the mixed system.

Ibuprofen was added into the mixed systems to observe solubilization capacity and stability. The addition of ibuprofen promotes micellization and stabilizes aggregates in the solution. Also, participation of the drug unifies the distribution of particle size developing a nearly monodisperse system. The solubility of ibuprofen in the mixed Pluronic systems is dramatically enhanced compared to that of the neat Pluronic systems for all the six mixed systems substituted by 0.5 wt % L92. The capability to incorporate ibuprofen into the system L92 + P84 is the most outstanding one: 4.29 mg/mL, which is 147 times more than that in pure water at 37 °C. In addition, the smallest particle size (130 nm) was measured from the L92 + P84 system.

From our systematic studies, the system of parent copolymers with a similar PPO moiety certainly leads to synergistic mixing (L92 + F108 or + F98), while the parent copolymer with intermediate hydrophobicity (P84) also has the ability to cooperatively bind with L92, forming well dispersing systems. With the participation of the model drug—ibuprofen—the solubilization capacity of the mixed Pluronic systems increases as the hydrophobicity of the system increases (L92 + F87 or + P84). It is interesting to point out that the particle size of the mixed systems decreases as the hydrophobicity of Pluronic increases. This may imply that the parent copolymers with a similar PPO moiety are suitable for drug encapsulation. Systems with distinct block chain lengths of parent copolymers, such as L92 + P84, demonstrate high solubilization capacity, as well as stability after drug incorporation. Additionally, the long-term stability of these binary Pluronic systems is crucial with regard to the practical applications of such systems. We are still in the process of examining the long-term stability of these binary Pluronic systems for further applications. Up until now, these binary Pluronic systems remain homogeneous at room temperature for at least one month. Our result also indicates that synergistic mixing of the systems with incorporated drugs could not only depend on the combinations of copolymers, but also the characteristics of drug-forming stable systems. Systematic studies from the series of experiments prove the influence of block chain length, as well as hydrophobicity, on the two copolymers.

Supplementary Materials: The following are available online at http://www.mdpi.com/2073-4360/10/1/105/s1. Figure S1. Calibration line of Ibuprofen in 0.1 N NaOH. Figure S2. UV absorbance of ibuprofen vs. wavelength in different solvents. Figure S3. Volume based Size Distribution of L92 + F87 system at 50 °C. Figure S4. Number based Size Distribution of L92 + F87 system at 50 °C. Table S1. Hydrodynamic Diameters of neat F108, L92 and binary mixed 0.5 wt % L92 + 1 wt % F108 system. Table S2. Hydrodynamic Diameters of neat F98, L92 and binary mixed 0.5 wt % L92 + 1 wt % F98 system. Table S3. Hydrodynamic Diameters of neat F88, L92 and binary mixed 0.5 wt % L92 + 1 wt % F88 system. Table S4. Hydrodynamic Diameters of neat F68, L92 and binary mixed 0.5 wt % L92 + 1 wt % F68 system. Table S5. Hydrodynamic Diameters of neat F87, L92 and binary mixed 0.5 wt % L92 + 1 wt % F87 system. Table S6. Hydrodynamic Diameters of neat P84, L92 and binary mixed 0.5 wt % L92 + 1 wt % F84 system.

Polymers **2018**, *10*, 105

Acknowledgments: This work was supported by the Ministry of Science and Technology of Taiwan (NSC102-2923-E-002-005-MY3 and MOST105-2221-E-002-209-MY3).

Author Contributions: Hsueh-Wen Tseng, Pratap Bahadur, and Li-Jen Chen conceived and designed the experiments. Hsueh-Wen Tseng and Chin-Fen Lee conducted the experiments. Chin-Fen Lee, Li-Jen Chen, and Pratap Bahadur wrote the paper.

Conflicts of Interest: The authors declare no conflict of interest.

References

1. Lee, E.S.; Oh, Y.T.; Youn, Y.S.; Nam, M.; Park, B.; Yun, J.; Kim, J.H.; Song, H.-T.; Oh, K.T. Binary mixing of micelles using Pluronics for a nano-sized drug delivery system. *Colloids Surf. B Biointerfaces* **2011**, *82*, 190–195. [CrossRef] [PubMed]
2. Danson, S.; Ferry, D.; Alakhov, V.; Margison, J.; Kerr, D.; Jowle, D.; Brampton, M.; Halbert, G.; Ranson, M. Phase I dose escalation and pharmacokinetic study of Pluronic polymer-bound doxorubicin (SP1049C) in patients with advanced cancer. *Br. J. Cancer* **2004**, *90*, 2085–2091. [CrossRef] [PubMed]
3. Matsumura, Y. Preclinical and clinical studies of anticancer drug-incorporated polymeric micelles. *J. Drug Target.* **2007**, *15*, 507–517. [CrossRef] [PubMed]
4. Alexandridis, P.; Holzwarth, J.F.; Hatton, T.A. Micellization of poly (ethylene oxide)-poly (propylene oxide)-poly (ethylene oxide) triblock copolymers in aqueous solutions: Thermodynamics of copolymer association. *Macromolecules* **1994**, *27*, 2414–2425. [CrossRef]
5. Cagel, M.; Tesan, F.C.; Bernabeu, E.; Salgueiro, M.J.; Zubillaga, M.B.; Moretton, M.A.; Chiappetta, D.A. Polymeric mixed micelles as nanomedicines: Achievements and perspectives. *Eur. J. Pharm. Biopharm.* **2017**, *113*, 211–228. [CrossRef] [PubMed]
6. Kabanov, A.V.; Batrakova, E.V.; Alakhov, V.Y. Pluronic® block copolymers as novel polymer therapeutics for drug and gene delivery. *J. Controll. Release* **2002**, *82*, 189–212. [CrossRef]
7. Basak, R.; Bandyopadhyay, R. Encapsulation of hydrophobic drugs in Pluronic F127 micelles: Effects of drug hydrophobicity, solution temperature, and PH. *Langmuir* **2013**, *29*, 4350–4356. [CrossRef] [PubMed]
8. Chen, Y.; Zhang, W.; Gu, J.; Ren, Q.; Fan, Z.; Zhong, W.; Fang, X.; Sha, X. Enhanced antitumor efficacy by methotrexate conjugated Pluronic mixed micelles against KBv multidrug resistant cancer. *Int. J. Pharm.* **2013**, *452*, 421–433. [CrossRef] [PubMed]
9. Kadam, Y.; Yerramilli, U.; Bahadur, A. Solubilization of poorly water-soluble drug carbamezapine in Pluronic® micelles: Effect of molecular characteristics, temperature and added salt on the solubilizing capacity. *Colloids Surf. B Biointerfaces* **2009**, *72*, 141–147. [CrossRef] [PubMed]
10. Kadam, Y.; Yerramilli, U.; Bahadur, A.; Bahadur, P. Micelles from PEO–PPO–PEO block copolymers as nanocontainers for solubilization of a poorly water soluble drug hydrochlorothiazide. *Colloids Surf. B Biointerfaces* **2011**, *83*, 49–57. [CrossRef] [PubMed]
11. Gaisford, S.; Beezer, A.E.; Mitchell, J.C. Diode-array UV spectrometric evidence for cooperative interactions in binary mixtures of Pluronics F77, F87, and F127. *Langmuir* **1997**, *13*, 2606–2607. [CrossRef]
12. Oh, K.T.; Bronich, T.K.; Kabanov, A.V. Micellar formulations for drug delivery based on mixtures of hydrophobic and hydrophilic Pluronic® block copolymers. *J. Controll. Release* **2004**, *94*, 411–422. [CrossRef]
13. Wang, Y.; Yu, L.; Han, L.; Sha, X.; Fang, X. Difunctional Pluronic copolymer micelles for paclitaxel delivery: Synergistic effect of folate-mediated targeting and Pluronic-mediated overcoming multidrug resistance in tumor cell lines. *Int. J. Pharm.* **2007**, *337*, 63–73. [CrossRef] [PubMed]
14. Wei, Z.; Hao, J.; Yuan, S.; Li, Y.; Juan, W.; Sha, X.; Fang, X. Paclitaxel-loaded Pluronic P123/F127 mixed polymeric micelles: Formulation, optimization and in vitro characterization. *Int. J. Pharm.* **2009**, *376*, 176–185. [CrossRef] [PubMed]
15. Pragatheeswaran, A.M.; Chen, S.B.; Chen, C.-F.; Chen, B.-H. Micellization and gelation of PEO-PPO-PEO binary mixture with non-identical PPO block lengths in aqueous solution. *Polymer* **2014**, *55*, 5284–5291. [CrossRef]
16. Nagarajan, R. Solubilization of hydrocarbons and resulting aggregate shape transitions in aqueous solutions of Pluronic®(PEO–PPO–PEO) block copolymers. *Colloids Surf. B Biointerfaces* **1999**, *16*, 55–72. [CrossRef]
17. Croy, S.R.; Kwon, G.S. The effects of Pluronic block copolymers on the aggregation state of nystatin. *J. Controll. Release* **2004**, *95*, 161–171. [CrossRef] [PubMed]

18. Guo, C.; Liu, H.-Z.; Chen, J.-Y. A Fourier transform infrared study on water-induced reverse micelle formation of block copoly (oxyethylene–oxypropylene–oxyethylene) in organic solvent. *Colloids Surf. A Physicochem. Eng. Asp.* **2000**, *175*, 193–202. [CrossRef]

19. Kabanov, A.V.; Batrakova, E.V.; Melik-Nubarov, N.S.; Fedoseev, N.A.; Dorodnich, T.Y.; Alakhov, V.Y.; Chekhonin, V.P.; Nazarova, I.R.; Kabanov, V.A. A new class of drug carriers: Micelles of poly (oxyethylene)-poly (oxypropylene) block copolymers as microcontainers for drug targeting from blood in brain. *J. Controll. Release* **1992**, *22*, 141–157. [CrossRef]

20. Sakai, T.; Alexandridis, P. Single-step synthesis and stabilization of metal nanoparticles in aqueous Pluronic block copolymer solutions at ambient temperature. *Langmuir* **2004**, *20*, 8426–8430. [CrossRef] [PubMed]

21. Tanner, S.A.; Amin, S.; Kloxin, C.J.; van Zanten, J.H. Microviscoelasticity of soft repulsive sphere dispersions: Tracer particle microrheology of triblock copolymer micellar liquids and soft crystals. *J. Chem. Phys.* **2011**, *134*, 174903. [CrossRef] [PubMed]

22. Yalkowsky, S.H.; Dannenfelser, R.M. *Aquasol Database of Aqueous Solubility*; College of Pharmacy, University of Arizona: Tucson, AZ, USA, 1992.

23. Scheytt, T.; Mersmann, P.; Lindstädt, R.; Heberer, T. 1-Octanol/water partition coefficients of 5 pharmaceuticals from human medical care: Carbamazepine, clofibric acid, diclofenac, ibuprofen, and propyphenazone. *Water Air Soil Pollut.* **2005**, *165*, 3–11. [CrossRef]

24. Domanska, U.; Pobudkowska, A.; Pelczarska, A.; Gierycz, P. PKa and Solubility of Drugs in Water, Ethanol, and 1-Octanol. *J. Phys. Chem. B* **2009**, *113*, 8941–8947. [CrossRef] [PubMed]

25. Tsui, H.-W.; Wang, J.-H.; Hsu, Y.-H.; Chen, L.-J. Study of heat of micellization and phase separation for Pluronic aqueous solutions by using a high sensitivity differential scanning calorimetry. *Colloid Polym. Sci.* **2010**, *288*, 1687–1696. [CrossRef]

26. Hsu, Y.-H.; Tsui, H.-W.; Lee, C.-F.; Chen, S.-H.; Chen, L.-J. Effect of alcohols on the heat of micellization of Pluronic F88 aqueous solutions. *Colloid Polym. Sci.* **2015**, *293*, 3403–3415. [CrossRef]

27. Hsu, Y.-H.; Tsui, H.-W.; Lin, S.-Y.; Chen, L.-J. The origin of anomalous positive heat capacity change upon micellization of Pluronic triblock copolymer F108 in aqueous solutions: Effect of PEO-PPO diblock impurities. *Colloids Surf. A Physicochem. Eng. Asp.* **2016**, *509*, 109–115. [CrossRef]

28. Kesur, B.R.; Salunkhe, V.; Magdum, C. Development and validation of UV spectrophotometric method for simultaneous estimation of ibuprofen and famotidine in bulk and formulated tablet dosage form. *Int. J. Pharm. Pharm. Sci.* **2012**, *4*, 271–274.

29. Tsui, H.-W.; Hsu, Y.-H.; Wang, J.-H.; Chen, L.-J. Novel behavior of heat of micellization of Pluronics F68 and F88 in aqueous solutions. *Langmuir* **2008**, *24*, 13858–13862. [CrossRef] [PubMed]

30. Zhang, Y.; Mintzer, E.; Uhrich, K.E. Synthesis and characterization of PEGylated bolaamphiphiles with enhanced retention in liposomes. *J. Colloid Interface Sci.* **2016**, *482*, 19–26. [CrossRef] [PubMed]

31. Liveri, M.T.; Licciardi, M.; Sciascia, L.; Giammona, G.; Cavallaro, G. Peculiar mechanism of solubilization of a sparingly water soluble drug into polymeric micelles. Kinetic and equilibrium studies. *J. Phys. Chem. B* **2012**, *116*, 5037–5046. [CrossRef] [PubMed]

32. Singla, P.; Chabba, S.; Mahajan, R.K. A systematic physicochemical investigation on solubilization and in vitro release of poorly water soluble oxcarbazepine drug in pluronic micelles. *Colloids Surf. A Physicochem. Eng. Asp.* **2016**, *504*, 479–488. [CrossRef]

33. Garzón, L.C.; Martínez, F. Temperature dependence of solubility for ibuprofen in some organic and aqueous solvents. *J. Solut. Chem.* **2004**, *33*, 1379–1395. [CrossRef]

34. Meznarich, N.A.; Love, B.J. The kinetics of gel formation for PEO—PPO—PEO triblock copolymer solutions and the effects of added methylparaben. *Macromolecules* **2011**, *44*, 3548–3555. [CrossRef]

35. Dutra, L.M.U.; Ribeiro, M.E.N.P.; Cavalcante, I.M.; Brito, D.H.A.D.; Semião, L.D.M.; Silva, R.F.D.; Fechine, P.B.A.; Yeates, S.G.; Ricardo, N.M.P.S. Binary mixture micellar systems of F127 and P123 for griseofulvin solubilisation. *Polímeros* **2015**, *25*, 433–439. [CrossRef]

![polymers logo] *polymers*

MDPI

Article

Elucidation of Spatial Distribution of Hydrophobic Aromatic Compounds Encapsulated in Polymer Micelles by Anomalous Small-Angle X-ray Scattering

Shota Sasaki, Ginpei Machida, Ryosuke Nakanishi, Masaki Kinoshita and Isamu Akiba *

Department of Chemistry and Biochemistry, The University of Kitakyushu, 1-1 Hibikino, Wakamatsu, Kitakyushu 8080135, Japan; x6maa006@eng.kitakyu-u.ac.jp (S.S.); w5maa014@eng.kitakyu-u.ac.jp (G.M.); w5maa003@eng.kitakyu-u.ac.jp (R.N.); w5maa008@eng.kitakyu-u.ac.jp (M.K.)
* Correspondence: akiba@kitakyu-u.ac.jp; Tel.: +81-93-695-3295

Received: 22 December 2017; Accepted: 11 February 2018; Published: 12 February 2018

Abstract: Spatial distribution of bromobenzene (BrBz) and 4-bromophenol (BrPh) as hydrophobic aromatic compounds incorporated in polymer micelles with vesicular structure consisting of poly(ethylene glycol)-*b*-poly(*tert*-butyl methacrylate) (PEG-*b*-PtBMA) in aqueous solution is investigated by anomalous small-angle X-ray scattering (ASAXS) analyses near Br K edge. Small-angle X-ray scattering (SAXS) intensities from PEG-*b*-PtBMA micelles containing BrBz and BrPh were decreased as the energy of incident X-ray approached to Br K edge corresponding to the energy dependence of anomalous scattering factor of Br. The analysis for the energy dependence of SAXS profiles from the PEG-*b*-PtBMA micelles containing BrBz revealed that BrBz molecules were located in hydrophobic layer of PEG-*b*-PtBMA micelles. On the contrary, it was found by ASAXS that BrPh existed not only in the hydrophobic layer but also in the shell layer. Since ASAXS analysis successfully accomplished to visualize the spatial distribution of hydrophobic molecules in polymer micelles, it should be expected to be a powerful tool for characterization of drug delivery vehicles.

Keywords: anomalous small-angle X-ray scattering; polymer micelle; drug delivery system

1. Introduction

When amphiphilic block copolymers are dissolved in aqueous solution, they undergo self-assembly into polymer micelles consisting of hydrophobic core and hydrated corona [1]. The polymer micelles have been expected to be drug carriers in drug delivery system (DDS) because they can uptake hydrophobic drug compounds in their hydrophobic cores in aqueous solution [2–4]. For the DDS particles, stable retention of drug compounds is the critical issue to reduce side effects [5]. The stability of retention of hydrophobic compounds should be related to their spatial distributions in polymer micelles. Therefore, it should be important to reveal the spatial distribution of hydrophobic compounds in polymer micelles.

In this study, we focus on anomalous small-angle X-ray scattering (ASAXS), which is small-angle X-ray scattering (SAXS) depending on the energy of incident X-ray near absorption edge of a targeted element, to reveal the spatial distribution of hydrophobic compounds incorporated in polymer micelles [6–9]. The ASAXS method has been applied for hard materials containing high atomic number elements because the K edges of high atomic number elements are generally located within the energy range of available synchrotron X-ray. Recently, the ASAXS method has also been applied for structural analyses of soft materials, such as micelles and polymer nanoparticles [9–14]. ASAXS corresponds to the variation of the atomic scattering factor f of a targeted element near its X-ray absorption edge. The energy dependence of f is described by the following equation.

$$f(E) = f_0 + f'(E) + f''(E),\qquad(1)$$

where E is the energy of incident X-ray, f_0 is the normal atomic form factor and f' and f'' are the real and imaginary parts of anomalous scattering factor, respectively. Owing to the energy dependence of f' and f'', X-ray scattering intensity shows energy dependence, which corresponds to scattering contribution of the targeted element, near the absorption edge. Therefore, analysis of the energy dependence of X-ray scattering intensity in small-angle region leads to the spatial distribution of targeted element. Thus, in this study, we apply the ASAXS method to reveal the spatial distribution of Br-labeled aromatic compounds incorporated in polymer micelles. Further, we discuss relation between molecular characteristics of aromatic compounds and their spatial distribution in polymer micelles.

2. Materials and Methods

Reagents. Poly(ethylene glycol) methyl ether (4-cyano-4-pentanoate dodecyl trithiocarbonate) ($M_n = 5.4 \times 10^3$, $M_w/M_n = 1.1$) as a chain transfer agent (CTA) was purchased from Sigma-Aldrich Co. Ltd. (St. Louis, MO, USA). *tert*-Butyl methacrylate (tBMA) was purchased from Tokyo Chemical Industry Inc. (Tokyo, Japan). 2,2′-Azobisisoputyronitrile (AIBN) was purchased from Kanto Chemicals Co. Ltd. (Tokyo, Japan). 4-Bromophenol (BrPh), bromobenzene (BrBz) and other solvents for syntheses were purchased from Wako Pure Chemicals Co. Ltd. (Tokyo, Japan). tBMA was purified by distillation under reduced pressure before polymerization. AIBN was recrystallized before it was used. The other reagents were used as obtained.

Synthesis of Poly(ethylene glycol)-*block*-poly(*tert*-butyl methacrylate) (PEG-*b*-PtBMA). Scheme 1 indicates the synthesis of PEG-*b*-PtBMA as an amphiphilic block copolymer. PEG-*b*-PtBMA was synthesized by reversible addition-fragmentation transfer (RAFT) radical polymerization. CTA (300 mg, 6.00×10^{-2} mmol) and AIBN (1.41 mg, 6.00×10^{-3} mmol) and *N,N*-dimethylformamide (DMF) (20 mL) were added into a flame-dried Schrenk Flask and purged with dry N_2. Then, tBMA (853 mg, 6.00 mmol) was added to the flask with stirring under dry N_2 atmosphere. After all reagents were dissolved in DMF, the flask was subjected to three freeze-pump-thaw cycles. RAFT polymerization was allowed to proceed for 20 h at 80 °C under dry N_2 atmosphere and then quenched by immersion in liquid N_2. PEG-*b*-PtBMA was isolated by precipitation from DMF solution to *n*-hexane. The number- and weight averaged molecular weight (M_n and M_w) determined by ^{1}H-NMR and size-exclusion chromatography (SEC) were 1.4×10^4 and 1.7×10^4, respectively. SEC elugrams and ^{1}H-NMR spectrum of the resulting PEG-*b*-PtBMA were shown in Figures S1 and S2, respectively, in supplementary materials. ^{1}H-NMR measurements for PEG-*b*-PtBMA in chloroform-d$_1$ were performed by using a JEOL JNM ECP500 spectrometer. SEC measurements were performed by using a Shodex GPC K-804 column (eluent: chloroform, range of molar mass: 7000~300,000) combined with a JASCO RI-4030 differential refractive index detector and a JASCO PU-2087 HPLC pump at a flow rate of 1 mL·min^{-1}.

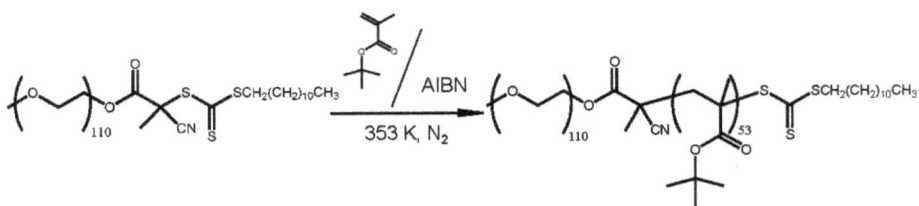

Scheme 1. Synthesis of PEG-*b*-PtBMA by RAFT polymerization.

Preparation of PEG-*b*-PtBMA micelles incorporating aromatic compounds. BrBz and BrPh were used as Br-labeled hydrophobic compounds. The PEG-*b*-PtBMA and BrPh or BrBz were mixed together at 10 wt % of aromatic compound against PEG-*b*-PtBMA. Micellization of the mixtures was induced by dissolving in tetrahydrofran (THF), stirring at an ambient temperature to ensure a homogeneous solution, followed by the gradual addition of 5 times the amount of ion exchanged water at a rate of 0.1 mL·min^{-1} via a syringe pump. The resulting solution was transferred to a dialysis tube (10 kDa MWCO) and dialyzed in a large amount of ion exchanged water for 3 days to remove THF. Finally, ion exchanged water was added to the micelle solution to a micelle concentration of 1.5 mg·mL^{-1}. This concentration of hydrophobic compounds is the lower limit to obtain satisfactory data of anomalous small-angle X-ray scattering.

Small-angle X-ray scattering (SAXS) and anomalous SAXS (ASAXS) measurements. SAXS and ASAXS measurements were performed at the BL-40B2 beamline of SPring-8, Japan. A 30 cm × 30 cm imaging plate (R-AXIS VII, Rigaku, Japan) was placed at a distance of 2 m from the sample position to cover a q range from 0.06 to 2.0 nm^{-1} at λ = 1.0 nm. Here, $q = (4\pi/\lambda)\sin(\theta/2)$, where θ is the scattering angle and λ is the wavelength of the incident X-ray. A sample solution was packed in a quartz capillary with a light path length of 2.0 mm (Hilgenberg GmbH, Malsfeld, Germany). The X-ray transmittance of the samples was measured with ion chambers located in front of and behind the sample. The two-dimensional SAXS images obtained with an imaging plate were converted into one dimensional scattering intensity versus q profiles by circular averaging. To obtain excess scattering intensity $I(q)$ at each q, scattering from the background were subtracted from the raw scattering data after an appropriate correction of transmittance. $I(q)$ was corrected to the absolute scale using the absolute scattering intensity of water (1.632 × 10^{-2}·cm^{-1}) [9]. The exposure time of each SAXS measurement was kept at 1 min to avoid radiation damage and data were accumulated five times for each sample to obtain high signal-to-noise ratio. In addition, we have confirmed that SAXS profiles for just prepared samples agree with those for the sample aged for several weeks. Therefore, leakage of aromatic compounds and degradation have not occurred. Conventional SAXS measurements were carried out at an energy (wavelength) of incident X-ray of 12.40 keV (0.1 nm). For ASAXS measurements, three different energies 13.383, 13.463 and 13.473 keV were used as incident X-rays. Figure 1 shows the energy-dependence of the anomalous scattering factor of Br [9]. The K-edge of Br was determined as 13.483 keV. Because the three energies used here gave large differences of f' near the Br K-edge, they were selected as incident X-ray energies to obtain sufficient energy-dependence of scattering intensities for accurate ASAXS analyses. Table 1 summarizes the f' and f'' values of Br at 13.383, 13.463 and 13.473 keV. The numerical analyses for SAXS and ASAXS data were carried out self-made programs on Igor 7 software.

Table 1. Values of real and imaginary parts of anomalous scattering factors (f' and f'') at 13.383, 13.463 and 13.473 keV.

Energy/keV	f'	f''
13.383	−4.56	0.508
13.463	−6.76	0.503
13.473	−9.33	0.502

Transmission electron microscope (TEM) observation. TEM was performed by using JEM-3010 (JEOL, Akishima shi, Japan). An accelerated voltage of 200 kV was used. The freeze-dried samples were placed on a carbon-spattered Cu grid and stained with Ti blue and lead acetate (II) to enhance TEM contrast.

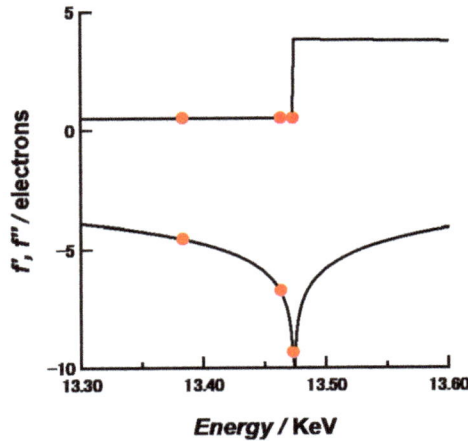

Figure 1. Energy dependences of f' and f'' of Br (solid lines) obtained by convolution of theoretical anomalous dispersions [15] and energy distribution in BL-40B2 station. Red circles indicate the energy of incident X-ray used in this study and corresponding f' and f''.

3. Results and Discussion

Figure 2 shows SAXS profiles of PEG-*b*-PtBMA micelles with and without aromatic compounds. Because of dilute concentration of PEG-*b*-PtBMA micelles, the SAXS profiles are regarded as form factors of micelles. The SAXS intensity of PEG-*b*-PtBMA micelle in low q region shows q^{-2} dependence. The q^{-2} dependence of SAXS intensity indicates that PEG-*b*-PtBMA micelles take disk-like or vesicular forms. TEM photo for freeze-dried PEG-*b*-PtBMA micelles shown in Figure 3 reveals that the PEG-*b*-PtBMA forms vesicular micelles. The SAXS profiles of PEG-*b*-PtBMA micelles containing BrBz and NrPh also show the q^{-2} dependence in low q regions. Therefore, the vesicular structure of PEG-*b*-PtBMA micelles is maintained by incorporating aromatic compounds in their inside. Although the radius of PEG-b-PtBMA micelles cannot be estimated by SAXS because of their extremely large size as shown in TEM photo, the thickness of the membrane of PEG-*b*-PtBMA micelles can be estimated. The solid lines in Figure 2 show the same theoretical SAXS curves calculated for a core-shell platelet model as described by the following equation [16]:

$$I(q) \propto \left[\frac{4\pi}{q^4} \left\{ (\rho_C - \rho_S) \sin\left(q\frac{t_C}{2}\right) + (\rho_S - \rho_0) \sin\left[q\left(\frac{t_S}{2} + t_S\right)\right] \right\}^2 \right], \tag{2}$$

where t_C and t_S are the half thicknesses of a core and overall of a plate, respectively and ρ_C, ρ_S and ρ_0 are the electron densities of core, shell and solvent, respectively. By assuming the cross-sectional electron density profile as shown in Figure 4, the calculated SAXS curve can be fitted to experimental SAXS profiles of PEG-*b*-PtBMA micelles with or without aromatic compounds. The half-thickness of hydrophobic core layer and thickness of shell layer are estimated to 4.7 and 7.8 nm, respectively. As shown in Figure 2, the scattering curve calculated by the cross-sectional electron density profile shown in Figure 4 can describe all SAXS data of PEG-*b*-PtBMA micelles and the micelles with BrBz and BrPh. These thicknesses well agree with the radii of gyration estimated by molecular weights of PtBMA and PEG chains. Hence, PEG-*b*-PtBMA micelles are consisting of single core layer sandwiched by hydrophilic shell layers. Therefore, effect of addition of small amounts of hydrophobic compounds on structure of polymer micelles can be ignored. For the PEG-*b*-PtBMA micelles incorporating BrBz and BrPh, ASAXS measurements near Br K edge are applied to figure out the spatial distribution of BrBz and BrPh in the micelles.

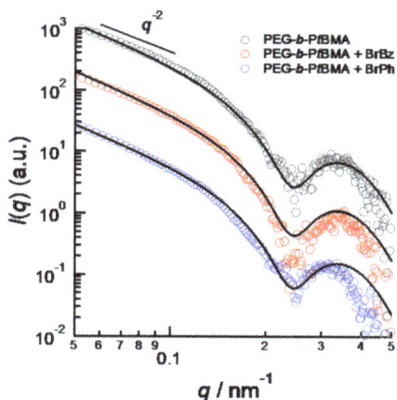

Figure 2. SAXS profiles from PEG-*b*-PtBMA micelles (gray), BrBz-containing PEG-*b*-PtBMA micelles (red) and BrPh-containing PEG-*b*-PtBMA micelles (blue).

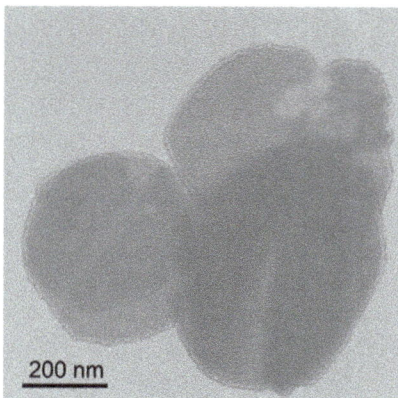

Figure 3. TEM micrograph of freeze-dried PEG-*b*-PtBMA micelles.

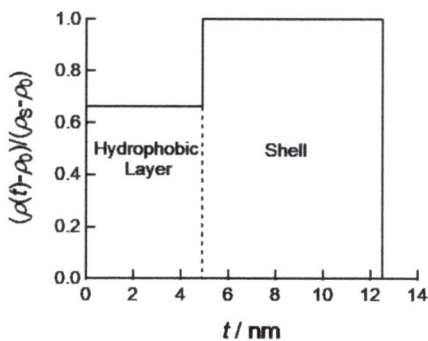

Figure 4. Cross-sectional electron density profile giving the best result in fitting analyses for SAXS profiles of PEG-*b*-PtBMA with and without Br-labeled compounds.

Figures 5 and 6 show the energy dependences of SAXS profile from BrBz- and BrPh-containing PEG-*b*-PtBMA micelles, respectively, measured at 13.383 keV (0.927 nm), 13.463 keV (0.921 nm) and 13.473 keV (0.920 nm). In both systems, SAXS intensities are decreasing as the energy of incident X-ray approaches from lower energy to Br K edge, although the features of energy resonance are different. The energy dependence of SAXS intensity originated from that of the f' of Br as shown in Figure 1 is described by the following equation.

$$I(q, E) = N\left\{ P_0(q) + 2f'(E)A(q)V(q) + \left(f'^2(E) + f''^2(E) \right)V^2(q) \right\},$$ (3)

where $P_0(q)$ is the non-resonant (normal) form factor of micelle and $V^2(q)$ is the form factor of resonant part. In this case, the $V^2(q)$ corresponds to the form factor of the area where Br atoms are existing. Consequently, difference of $V^2(q)$ profiles causes the difference of features of energy resonance in SAXS profiles. Therefore, it is strongly suggested that the spatial distribution of BrBz in PEG-b-PtBMA micelles is much different from that of BrPh. By dissolving simultaneous equation of the SAXS profiles measured at three different energies of incident X-ray, we can obtain $V^2(q)$ profile as described by the following equation.

$$V^2(q) = \frac{1}{K}\left[\frac{\Delta I(q, E_1, E_2)}{f'_{Br}(q, E_1) - f'_{Br}(q, E_2)} - \frac{\Delta I(q, E_1, E_3)}{f'_{Br}(q, E_1) - f'_{Br}(q, E_3)} \right],$$ (4)

where

$$K = f'_{Br}(E_2) - f'_{Br}(E_3) + \frac{f''_{Br}^2(E_1) - f''_{Br}^2(E_2)}{f'_{Br}(E_1) - f'_{Br}(E_2)} - \frac{f''_{Br}^2(E_1) - f''_{Br}^2(E_3)}{f'_{Br}(E_1) - f'_{Br}(E_3)},$$ (5)

Figure 7 shows $V^2(q)$ profile of BrBz-containing and BrPh-containing PEG-*b*-PtBMA micelles. Because of low resolution of ASAXS treatment in high q region, detail analysis for $V^2(q)$ by using scattering function of model particle is difficult. However, both $V^2(q)$ profiles obviously show q^{-2} dependences in low q region. This means the area where Br-labeled aromatic compounds are distributed takes a platelet form. Therefore, the vesicular structure of PEG-*b*-PtBMA micelle is reflected to the spatial distribution of BrBz and BrPh, like templates. For platelet particles like a membrane of vesicle, relation between scattering intensity $I(q)$ (or resonant scattering term $V^2(q)$) and q in low q region is given by the following equation.

$$q^2 I(q) \propto \exp\left(-t^2 q^2 \right),$$ (6)

where t is the half-thickness of a platelet particle. Therefore, the slope of plots of $q^2 V^2(q)$ against q^2—so-called thickness Guinier plots—half-thicknesses of platelet particles can be estimated. From the thickness Guinier plots as shown in the insert of Figure 7, the half-thicknesses of the platelet areas are estimated to 4.7 and 7.2 nm for BrBz and BrPh, respectively. As shown in the insert of Figure 7, the thicknesses estimated in low q region are obtained with sufficient accuracy, although the SAXS data in high q region include large uncertainty. By using these results, scattering curves for $V^2(q)$ profiles for BrBz- and NrPh-containing micelles are calculated by using disk-like particles. The solid lines in Figure 7 are calculated curves. Although the $V^2(q)$ profiles in high q region contain large noise, the calculated curves agree with $V^2(q)$ profiles in low q region. The thickness of BrBz area is close to that of hydrophobic layer of PEG-*b*-PtBMA micelles shown in Figure 4. Therefore, it should be considered that BrBz molecules are homogeneously dispersed in hydrophobic layer of PEG-*b*-PtBMA micelles. On the contrary, the half-thickness of BrPh area is much larger than that of hydrophobic layer of PEG-*b*-PtBMA micelle. This result means that a part of BrPh filtrates to hydrated shell layer of PEG-*b*-PtBMA micelles. This result is consistent with the spatial distribution of a phenol derivative incorporated in spherical polymer micelles reported by Sanada et al. [14] and some of our previous studies [13,17]. BrPh can form hydrogen bonds with both PtBMA and PEG. In addition, the number density of PEG chains near core-shell interface is much higher than in the outer region of shell. Consequently, BrPh can exist in shell layer composed of PEG near the core-shell interface. According to the result in this study, it is strongly suggested that spatial distribution of hydrophobic compounds incorporated in polymer micelles can be controlled by tuning interactions not only between hydrophobic compounds and not only hydrophobic polymers but also water-soluble polymers of amphiphilic block copolymers.

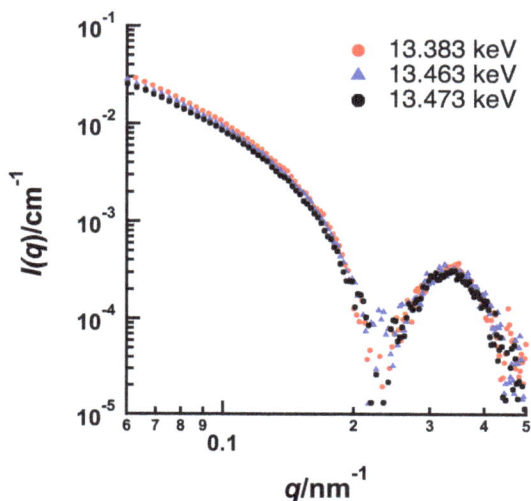

Figure 5. Energy dependence of SAXS profiles of PEG-*b*-PtBMA containing BrBz near Br K edge (13.483 keV).

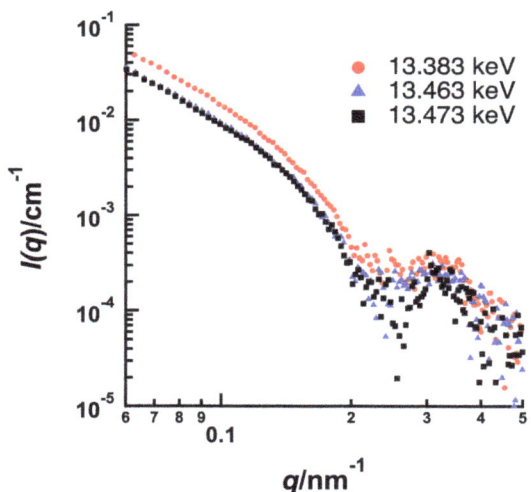

Figure 6. Energy dependence of SAXS profiles of PEG-*b*-PtBMA containing BrPh near Br K edge (13.483 keV).

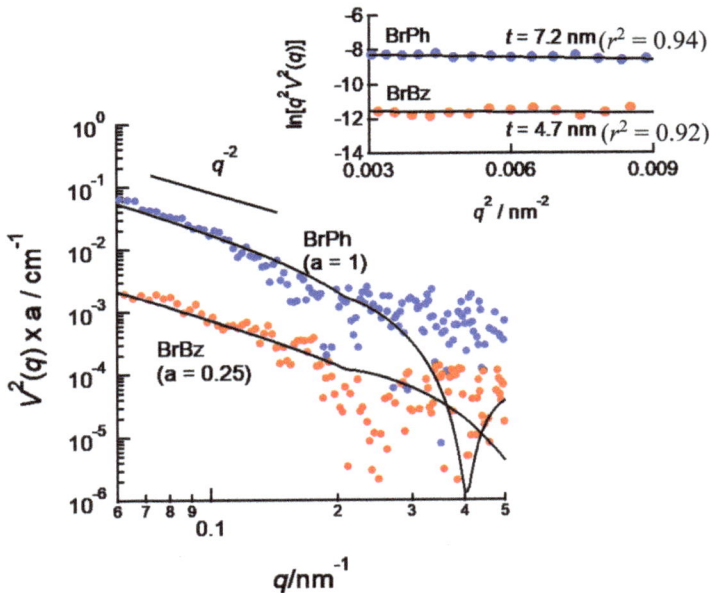

Figure 7. $V^2(q)$ profiles of BrBz- (red) and BrPh-containing (blue) PEG-*b*-PtBMA micelles. Solid lines are calculated by theoretical scattering functions for disk-like particles with 4.7 and 7.2 nm of half-thickness estimated from thickness Guinier plots. Insert shows thickness Guinier plot of $V^2(q)$ of BrBz- and BrPh-containing PEG-*b*-PtBMA micelles. The half-thickness of BrBz and BrPh areas are estimated to 4.7 and 7.2 nm, respectively.

4. Conclusions

ASAXS near Br K edge was applied for analyses of the spatial distribution of hydrophobic aromatic compounds (BrBz and BrPh) encapsulated in polymer micelles consisting of PEG-*b*-PtBMA. It was found that BrBz was dispersed in hydrophobic layer of PEG-*b*-PtBMA micelles, while BrPh was infiltrated to hydrophilic shell layer. This result inevitably leads the conclusion that the spatial distribution of hydrophobic compounds encapsulated in polymer micelles can be controlled by tuning intermolecular interactions, such as hydrogen bonds, between hydrophobic compounds and each chain of amphiphilic block copolymers.

Supplementary Materials: The following are available online at http://www.mdpi.com/2073-4360/10/2/180/s1, Figure S1: GPC elugrams of PEG (blue) and PEG-*b*-PtBMA (Red), Figure S2: ^{1}H-NMR spectrum of PEG-*b*-PtBMA in CDCl$_3$.

Acknowledgments: This study was financially supported by JST-CREST and JSPS Grant-in-Aid for Scientific Research. SAXS and ASAXS experiments at SPring-8 were performed under the approval of the SPring-8 Advisory Committee (approved numbers: 2013B1683, 2014A1676, 2015B1724, 2015A1562, 2015B1464, 2016B1300 and 2017A7231).

Author Contributions: Isamu Akiba, Shota Sasaki, Ryosuke Nakanishi and Ginpei Machida conceived and designed the experiments; Shota Sasaki, Ginpei Machida Ryosuke Nakanishi and Masaki Kinoshita performed the experiments; Shota Sasaki, Ginpei Machida and Isamu Akiba analyzed the data; Shota Sasaki, Ginpei Machida and Masaki Kinoshita contributed synthesis of block copolymer; Shota Sasaki and Isamu Akiba wrote the paper.

Conflicts of Interest: The authors declare no conflict of interest.

References

1. Zhang, L.F.; Eisenberg, A. Multiple Morphologies of "Crew-Cut" Aggregates of Polysyrene-*b*-poly(acrylic acid) Block Copolymers. *Science* **1995**, *268*, 1728–1731. [CrossRef] [PubMed]
2. Yokoyama, M. Polymeric Micelles for the Targeting of Hydrophibic Drugs. In *Polymeric Drug Delivery Systems*; Kwon, G.S., Ed.; Drugs and the Pharmaceutical Sciences 148; Taylor & Francis: Boca Raton, FL, USA, 2005; pp. 533–575.
3. Monfardini, C.; Veronese, F.M. Stabilization of Substances in Circulation. *Bioconjugate Chem.* **1998**, *9*, 418–450. [CrossRef] [PubMed]
4. Matsumura, Y.; Kimura, M.; Yamamoto, T.; Maeda, H. Involvement of the Kinin-generating Cascade in Enhanced Vascular Permeability in Tumor Tissue. *Jpn. J. Cancer Res.* **1988**, *79*, 1327–1334. [CrossRef] [PubMed]
5. Yamamoto, T.; Yokoyama, M.; Opanasopit, P.; Hayama, A.; Kawano, K.; Maitani, Y. What Are Determining Factors for Stable Drug Incorporation into Polymeric Micelle Carriers? Consideration on Physical and Chemical Characters of the Micelle Inner Core. *J. Control. Release* **2007**, *123*, 11–18. [CrossRef] [PubMed]
6. *Resonant Anomalous X-ray Scattering: Theory and Applications*; Materlik, G.; Sparks, C.J.; Fischer, K. (Eds.) Elsevier Science: Amsterdam, The Netherlands, 1994.
7. Waseda, Y. *Anomalous X-ray Scattering for Materials Characterization: Atomic-Scale Structure Determination*; Springer: Berlin, Germany, 2002.
8. Stuhrmann, H.B. Resonance scattering in macromolecular structure research. *Adv. Polym. Sci.* **1985**, *67*, 123–163.
9. Akiba, I.; Takechi, A.; Sakou, M.; Handa, M.; Shinohara, Y.; Amemiya, Y.; Yagi, N.; Sakurai, K. Anomalous Small-Angle X-Ray Scattering Study of Structure of Polymer Micelles Having Bromines in Hydrophobic Core. *Macromolecules* **2012**, *45*, 6150–6157. [CrossRef]
10. Patel, M.; Rosenfeldt, S.; Ballauff, M.; Dingenouts, N.; Pontoni, D.; Narayanan, T. Analysis of the Correlation of Counterions to Rod-like Macroions by Anomalous Small-angle X-ray Scattering. *Phys. Chem. Chem. Phys.* **2004**, *6*, 2962–2967. [CrossRef]
11. Sztucki, M.; Cola, E.D.; Narayanan, T. New Opportunities for Anomalous Small-Angle X-Ray Scattering to Characterize Charged Soft Matters Systems. *J. Phys. Conf. Ser.* **2011**, *272*, 012004. [CrossRef]
12. Sakou, M.; Takechi, A.; Murakami, S.; Sakurai, K.; Akiba, I. Study on the Internal Structure of Polymer Micelles by Anomalous Small-angle X-ray Scattering at Two Edges. *J. Appl. Cryst.* **2013**, *46*, 1407–1413. [CrossRef]
13. Sanada, Y.; Akiba, I.; Sakurai, K.; Shiraishi, K.; Yokoyama, M.; Mylonas, E.; Ohta, N.; Yagi, N.; Shinohara, Y.; Amemiya, Y. Hydrophobic Molecules Infiltrating into the Poly(ethylene glycol) Domain of the Core/Shell Interface of a Polymeric Micelle: Evidence Obtained with Anomalous Small-Angle X-ray Scattering. *J. Am. Chem. Soc.* **2013**, *135*, 2574–2582. [CrossRef] [PubMed]
14. Nakanishi, R.; Machida, G.; Kinoshita, M.; Sakurai, K.; Akiba, I. Anomalous Small-angle X-ray Scattering Study on the Spatial Distribution of Hydrophobic Molecules in Polymer Micelles. *Polym. J.* **2016**, *48*, 801–806. [CrossRef]
15. Sasaki, S. *Anomalous Scattering Factors for Synchrotron Radiation Users, Calculated Using Cromer and Liberman's Method*; KEK Report 88-14; Technical Information & Library; National Laboratory for High Energy Physics: Tsukuba, Japan, 1989.
16. Pons, R.; Valiente, M.; Montalvo, G. Structure of Aggregates in Diluted Aqueous Octyl Glucoside/Tetraethylene Glycol Monododecyl Ether Mixtures with Different Alkanols. *Langmuir* **2010**, *26*, 2256–2262. [CrossRef] [PubMed]
17. Nakanishi, R.; Kinoshita, M.; Sasaki, S.; Akiba, I. Spatial Distribution of Hydrophobic Compounds in Polymer Micelles as Explored by Anomalous Small-angle X-ray Scattering near Br K-edge. *Eur. Polym. J.* **2016**, *81*, 634–640. [CrossRef]

MDPI AG

St. Alban-Anlage 66

4052 Basel, Switzerland

Tel. +41 61 683 77 34

Fax +41 61 302 89 18

http://www.mdpi.com

Polymers Editorial Office

E-mail: polymers@mdpi.com

http://www.mdpi.com/journal/polymers